Recent Progress in Biomass I and High Value Utilization of Pyrolytic Carbon

Recent Progress in Biomass Pyrolysis and High Value Utilization of Pyrolytic Carbon

Guest Editors

Marcin Sajdak
Roksana Muzyka
Grzegorz Gałko

MDPI

Basel • Beijing • Wuhan • Barcelona • Belgrade • Novi Sad • Cluj • Manchester

Guest Editors
Marcin Sajdak
Department of Air Protection
Silesian University of Technology
Gliwice
Poland

Roksana Muzyka
Department of Air Protection
Silesian University of Technology
Gliwice
Poland

Grzegorz Gałko
The Institute for Ecology of Industrial Areas (IETU)
Katowice
Poland

Editorial Office
MDPI AG
Grosspeteranlage 5
4052 Basel, Switzerland

This is a reprint of the Special Issue, published open access by the journal *Energies* (ISSN 1996-1073), freely accessible at: https://www.mdpi.com/journal/energies/special_issues/EAN08J1870.

For citation purposes, cite each article independently as indicated on the article page online and as indicated below:

Lastname, A.A.; Lastname, B.B. Article Title. *Journal Name* **Year**, *Volume Number*, Page Range.

ISBN 978-3-7258-4925-3 (Hbk)
ISBN 978-3-7258-4926-0 (PDF)
https://doi.org/10.3390/books978-3-7258-4926-0

Cover image courtesy of Marcin Sajdak

Contents

energies

MDPI

Editorial

Recent Progress in Biomass Pyrolysis and High Value Utilisation of Pyrolytic Carbon

Marcin Sajdak [1,*], Roksana Muzyka [1] and Grzegorz Gałko [2]

1 Faculty of Energy and Environmental Engineering, Silesian University of Technology, Akademicka 2A, 44-100 Gliwice, Poland; roksana.muzyka@polsl.pl
2 Institute for Ecology of Industrial Areas (IETU), Stanisława Kossutha 6, 40-844 Katowice, Poland; ggalko@itpe.pl
* Correspondence: marcin.sajdak@polsl.pl; Tel.: +48-322371540

1. Introduction

The modern world is facing an unprecedented challenge of energy transition, which requires a radical reorientation of how energy is produced and used. In this context, biomass pyrolysis technologies are gaining importance as a key element of decarbonisation strategies, offering the possibility of converting organic waste into high-value energy products. The increasing interest in these technologies arises not only from the necessity to discover sustainable alternatives to fossil fuels but also from the imperative to efficiently manage the escalating quantities of biomass waste. Pyrolysis, as a process of thermochemical conversion of biomass under limited oxygen conditions, enables the production of pyrolytic carbon with high energy value and a wide range of applications. This process occurs at temperatures between several hundred and over a thousand degrees Celsius, resulting in a substantial alteration of the chemical structure of biomass, producing a material with a markedly higher energy density than the feedstock. In line with the European Union's ambitious climate policy, which aims to reduce greenhouse gas emissions by at least 55% by 2030, biomass technologies are playing an increasingly key role in the energy transition. Biochar serves as a particularly important substitute for conventional carbon fuels in heavy industry, which accounts for about 7% of global CO_2 emissions and is one of the most difficult sectors to decarbonise [1]. This article provides a summary of research conducted by scientists who have contributed new and complementary knowledge in the field of biomass pyrolysis and biochar applications, combining the technological perspective with the environmental and economic aspects of this promising technology.

2. A Short Review of the Contributions in This Special Issue

The utilisation of biochar has a history spanning millennia; however, modern investigations into the pyrolytic processing of biomass signify a fundamentally novel phase in the evolution of this technology. Rejdak et al. (1), in their research on bio-coke production, undertook a systematic analysis of the effect of biocomponent additives on the coking properties of coal mixtures. The work of these scientists is an example of the evolution of thinking from simple biomass combustion to its use as a component in advanced metallurgical processes. Researchers looked at different types of biomass, such as wood materials, including beech and alder chips, and sawmill waste, including pine sawdust, because the way these materials shrink during the plastic phase can be quite different. Different types of biomass undergo different shrinkage during pyrolysis [2], which causes the formation of empty spaces; therefore, adequate expansion potential is necessary [3,4]. A part of that is that they also consider food industry by-products such as hazelnut shells and olive pits.

 https://doi.org/10.3390/en18153919

The diversity of raw materials exemplifies a pragmatic approach to utilising mainly locally available biomass resources, which is crucial for the sustainable development of pyrolysis technologies. The Rejdak's team showed that the addition of up to 20% of biocomponents causes a systematic decrease in the Gieseler fluidity, dilatation, and Roga index of coal mixtures. However, the researchers did not stop at observing this phenomenon but developed a mathematical model describing the effect of the additive's content and its properties on the decrease in fluidity. This modelling approach opens up the prospect of the predictive design of bio-coke mixtures with specified properties.

Building upon these foundational studies of biomass properties, researchers have simultaneously focused their efforts on developing more sophisticated processing technologies [5–8]. In parallel with research into product properties, scientists are focusing on optimising the pyrolysis processes themselves. Illankoon et al. have developed an innovative two-chamber pyrolysis reactor that represents a significant step forward in the democratisation of pyrolysis technology (2). Their design, aimed at producing biochar from rice husks in domestic and small agricultural contexts, fulfils the requirement for technology accessible to local communities in developing nations. The two-chamber reactor design, consisting of an outer cylinder for energy generation and an inner cylinder for pyrolysis, together with a synthesis gas circulation system, represents an elegant solution to the problem of energy self-sufficiency of the process. The parameters achieved by the team, e.g., a biochar yield of 49% with an average production rate of 1.8 kg/h, prove the practical feasibility of this concept. Particularly impressive is the achievement of a biochar specific surface area of 182 m^2/g, which indicates a high product quality comparable to commercial activated carbons. The energy value of the pyrolysis gases, amounting to 23.3 MJ/m^3, further confirms the energy efficiency of the entire process.

The transition from laboratory-scale innovations to real-world industrial implementation represents a critical phase in the development of pyrolysis technologies. This evolution is particularly evident in the steel industry, where revolutionary changes are taking place (3) [9–11]. Kieush and Schenk's research on the use of biochar in electric arc furnace processes is an example of translating basic research into heavy industry (3,4). Their systematic approach to analysing the phenomenon of slag foaming using various carbon sources—from traditional anthracite and calcined coke, through high-temperature coke, to biochar as a single component and mixtures of biochar with coke—represents a comprehensive approach to the issue of replacing conventional fossil fuels. The team's significant finding was that biochar can facilitate stable slag foaming at alkalinity levels between 1.2 and 3.4, which enables its potential application in the steel industry (4). Moreover, it was equally significant to discover that a 1:1 blend of biochar and coke demonstrates superior foaming properties compared to biochar alone, suggesting synergistic interactions among various carbon forms. The XRD analysis conducted by researchers showed that there are differences in the crystalline phases between the upper and lower sections of the foamed slag (3). The presence of maghemite in the upper section, indicative of oxidation processes, alongside metallic iron in the lower section, offers critical insights into the mechanisms underlying the foaming of slag with biochar.

Beyond metallurgical applications, pyrolysis technologies are finding significant applications in waste management and energy production sectors [12–15]. Carotenuto et al. undertook the ambitious task of optimising the sewage sludge gasification process for combined heat and power production [15]. The methodology employed by these scientists, utilising sophisticated modelling inside the Aspen Plus framework, exemplifies a contemporary approach to the design of biomass-derived energy processes. The researchers identified the optimal process conditions—an air heating temperature of 150 °C and an equivalence factor of 0.2—which allow an energy production potential of 2.54 kW per kilogram of dry sludge to be achieved. Equally important is the achievement of a CO_2 emission reduction of 0.59 kg per kilogram of dry matter compared to natural gas, confirming the environmental benefits of this technology. As pyrolysis technologies mature, researchers are exploring increasingly diverse and innovative applications that extend far beyond traditional energy production. The wine industry has emerged as an unexpected source of valuable biomass materials [16]. D'Eusanio et al. have initiated a novel phase in biochar research, concentrating on waste from the wine industry as a precursor for the synthesis of materials with electrical properties (5). Their research on grape seeds, defatted grape seeds, wooden stems, and whole grape seeds under pyrolysis conditions at 300 °C for 3 to 24 h represents a systematic approach to characterising the electrical properties of biochar. The results indicating the best thermal stability and higher electrical conductivity open up prospects for the use of this material in electronics and composite materials. The relationship between the development of carbonaceous structures and electrical conductivity offers significant insights into the mechanisms underlying electrical property formation during pyrolysis. Complementing these developments in electrical applications, alternative processing methods are being explored to maximise the potential of various biomass waste streams [17]. Wądrzyk et al. introduced an innovative approach to the conversion of coffee grounds through a hydrothermal process, which is an alternative to traditional pyrolysis (6). Their research on the influence of process temperature, retention time, and the use of a binary solvent on the yield and quality of products represents a holistic approach to process optimisation. Achieving a maximum hydrochar yield of 39% and a bio-oil yield of 46%, while obtaining bio-oils with a carbon content of 71–74% and an energy value of approximately 35 MJ/kg, and hydrochar with a carbon content of up to 73.8% and an energy value of 30 MJ/kg, confirms the potential of this technology as an alternative to conventional biomass processing methods.

The exploration of synergistic effects represents another frontier in advancing pyrolysis technologies [18,19]. Kudelytė et al. have investigated the synergistic effects of co-pyrolysis of plastic waste and lignin; they highlighted new opportunities for enhancing hydrocarbon recovery [19]. Their systematic investigations of polypropylene and low- and high-density polyethylene mixtures with lignin in diverse mass ratios exemplify an interdisciplinary methodology that integrates polymer chemistry with biomass technologies. The discovery that the best results are obtained for lignin–LDPE/HDPE mixtures at 600 °C in a mass ratio of 1:3, together with the finding that plastic radicals promote lignin depolymerisation, provides valuable information on synergistic mechanisms in co-pyrolysis processes. The increased recovery of high value-added hydrocarbons confirms the potential of this technology in the circular economy.

The development of high-temperature pyrolysis technologies represents the cutting edge of current research efforts, particularly in applications requiring high-performance characteristics [20,21]. Deutsch et al. have undertaken to determine the lowest thermal treatment temperature for wood that allows a biogenic reducing agent to be obtained that is suitable for use in existing carbon injection systems in blast furnaces (7). Their research under various temperature conditions ranging from 285 to 340 °C is an example of a meticulous strategy to improve chemical and mechanical properties. At the same time, they have retained the necessary transportability in a conventional pneumatic dense-phase transport system. Researchers have indicated that achieving an 18–40% reduction in CO_2 emissions in blast furnaces while maintaining transport properties similar to those of coal represents significant progress in the decarbonisation of the steel sector. This technology is a practical interim solution in the energy transition of heavy industry, offering the possibility of gradually replacing conventional fossil fuels.

An analysis of the presented studies reveals a fascinating mosaic of approaches to the use of biomass, which, despite the diversity of raw materials and applications, shows common themes. The first of these is the pursuit of process parameter optimisation to maximise product yields and quality. All research teams, regardless of the specific nature of their projects, focus precisely on determining the optimal conditions for temperature, retention time, and process atmosphere.

The second common theme is a pragmatic approach to the use of locally available waste raw materials. From rice husks in Sri Lanka to coffee grounds and wine production waste, scientists are constantly looking for ways to transform waste management problems into energy opportunities by using such diverse biomass waste streams that are locally available and respond to local challenges. This method demonstrates a profound comprehension of the tenets of the circular economy and sustainable development. The third key theme is the evolution from simple energy applications of biochar towards advanced industrial applications. Research on the use of biochar in the steel industry, composite materials, and electronics indicates a growing understanding of the potential of this material as a multifunctional component of modern technologies.

Despite promising results, the analysed studies also reveal significant technological and economic challenges. The main problem remains the deterioration of some technological properties when replacing conventional fuels with biochar. Rejdak's research showed a systematic decrease in coking properties, while Kieush and Schenk identified problems with process stability when using biochar alone. These challenges culminate in the assertion that the most advantageous strategy is to employ hybrid mixtures that integrate biochar with conventional fuels. This method facilitates the incremental implementation of sustainable solutions while preserving essential technological attributes, which is vital for industrial acceptance. Another challenge is the economic scaling of processes from the laboratory to the industrial level. Most of the studies presented were conducted on a pilot scale, and the transition to full-scale production requires solving problems related to raw material logistics, product quality control, and cost optimisation.

The analysed studies provide valuable guidance for energy policy and decarbonisation strategies. The results point to the need for an integrated approach to the development of pyrolysis technologies that considers not only technical aspects but also economic and social aspects. Supporting the development of local biomass-based value chains is vital for fostering regional development and mitigating greenhouse gas emissions. Technologies such as the reactor developed by Illankoon et al. illustrate that advanced technological solutions can be customised to meet the specific needs of local communities. Establishing a suitable regulatory framework and quality standards is equally essential for promoting the commercialisation of pyrolysis technologies. The variety of raw materials and applications necessitates a flexible standardisation approach that considers specific local conditions and requirements.

3. Conclusions and Outlook

The presented analysis of the latest research on the pyrolytic use of biomass reveals a dynamically developing field of science and technology that offers real opportunities to accelerate the energy transition. The variety of research methodologies, spanning from fundamental investigations of material properties to practical industrial applications, illustrates the advancement of this field and its preparedness for broader implementation. The primary conclusion of the analysis is that pyrolytic technologies represent not a singular solution but rather a collection of complementary technologies that can be tailored to various feedstocks, local conditions, and end applications. This adaptability represents both the paramount strength and the principal challenge for this domain. The potential for

CO_2 emission reductions of 18–59%, depending on the application [5,8], combined with the possibility of organic waste recovery, makes pyrolysis technologies a key element of decarbonisation strategies. However, realising this potential requires further research on process optimisation, the development of quality standards, and the creation of appropriate support mechanisms. The future of pyrolysis technologies is likely to be characterised by further specialisation and integration with other sustainable technologies. Investigations into composite materials incorporating biochar [4], the synergistic effects of co-pyrolysis [7], and hydrothermal alternatives to conventional pyrolysis [6] indicate the trajectory of this advancement. The success of pyrolysis technologies will ultimately hinge on the integration of scientific rigour with the practical requirements of industry and society. The studies analysed provide a solid scientific basis for this transformation, but its implementation will require further cooperation between the scientific community, industry, and policymakers.

Author Contributions: All authors contributed equally to this work. All authors have read and agreed to the published version of the manuscript.

Acknowledgments: The authors would like to express their sincere gratitude to MDPI for the opportunity to serve as Guest Editors for this Special Issue. We also wish to acknowledge the editorial staff of *Energies* for their cooperation, patience, and dedicated support throughout the editorial process. We are grateful for the opportunity to review and analyse the research presented and to contribute to the advancement of knowledge in the field of biomass pyrolysis and high-value utilisation of pyrolytic carbon. Special thanks are extended to all the research teams whose valuable work formed the foundation of this analysis. Publication supported by the Faculty of Energy and Environmental Engineering, Silesian University of Technology. Publication supported by the Rector's pro-quality grant.

Conflicts of Interest: The authors declare no conflicts of interest.

List of Contributions

1. Rejdak, M.; Wojtaszek-Kalaitzidi, M.; Gałko, G.; Mertas, B.; Radko, T.; Baron, R.; Książek, M.; Yngve Larsen, S.; Sajdak, M.; Kalaitzidis, S. A Study on Bio-Coke Production—The Influence of Bio-Components Addition on Coke-Making Blend Properties. *Energies* **2022**, *15*, 6847. https://doi.org/10.3390/en15186847.
2. Illankoon, W.A.M.A.N.; Milanese, C.; Karunarathna, A.K.; Alahakoon, A.M.Y.W.; Rathnasiri, P.G.; Medina-Llamas, M.; Collivignarelli, M.C.; Sorlini, S. Development of a Dual-Chamber Pyrolizer for Biochar Production from Agricultural Waste in Sri Lanka. *Energies* **2023**, *16*, 1819. https://doi.org/10.3390/en16041819.
3. Kieush, L.; Schenk, J.; Koveria, A.; Hrubiak, A.; Hopfinger, H.; Zheng, H. Evaluation of Slag Foaming Behavior Using Renewable Carbon Sources in Electric Arc Furnace-Based Steel Production. *Energies* **2023**, *16*, 4673. https://doi.org/10.3390/en16124673.
4. Kieush, L.; Schenk, J. Investigation of the Impact of Biochar Application on Foaming Slags with Varied Compositions in Electric Arc Furnace-Based Steel Production. *Energies* **2023**, *16*, 6325. https://doi.org/10.3390/en16176325.
5. D'Eusanio, V.; Lezza, A.; Anderlini, B.; Malferrari, D.; Romagnoli, M.; Roncaglia, F. Technological Prospects of Biochar Derived from Viticulture Waste: Characterization and Application Perspectives. *Energies* **2024**, *17*, 3421. https://doi.org/10.3390/en17143421.
6. Wądrzyk, M.; Katerla, J.; Janus, R.; Lewandowski, M.; Plata, M.; Korzeniowski, Ł. High-Energy-Density Hydrochar and Bio-Oil from Hydrothermal Processing of Spent Coffee Grounds—Experimental Investigation. *Energies* **2024**, *17*, 6446. https://doi.org/10.3390/en17246446.
7. Deutsch, R.; Kienzl, N.; Stocker, H.; Strasser, C.; Krammer, G. Characteristics of High-Temperature Torrefied Wood Pellets for Use in a Blast Furnace Injection System. *Energies* **2025**, *18*, 458. https://doi.org/10.3390/en18030458.

References

1. IEA Report 2020: Iron and Steel Technology Roadmap Towards More Sustainable Steelmaking. Available online: https://iea.blob.core.windows.net/assets/eb0c8ec1-3665-4959-97d0-187ceca189a8/Iron_and_Steel_Technology_Roadmap.pdf (accessed on 11 July 2025).
2. Fu, P.; Hu, S.; Xiang, J.; Sun, L.; Zhang, A.; Yang, T.; Jiang, L. Evolution of pore structure of biomass particles during pyrolysis. *CIESC J.* **2009**, *60*, 1793–1799.
3. Nomura, S.; Arima, T.; Kato, T. Coal blending theory for dry coal charging process. *Fuel* **2004**, *82*, 1771–1776. [CrossRef]
4. Ueki, Y.; Nunome, Y.; Yoshiie, R.; Naruse, I.; Nishibata, Y. Effect of woody biomass addition on coke properties. *ISIJ Int.* **2014**, *54*, 2454–2460. [CrossRef]
5. Schure, J.; Pinta, F.; Cerutti, P.O.; Kasereka-Muvatsi, L. Efficiency of Charcoal Production in Sub-Saharan Africa: Solutions beyond the Kiln. *Bois For. Des. Trop.* **2019**, *340*, 57–70. [CrossRef]
6. Ighalo, J.O.; Eletta, O.A.A.; Adeniyi, A.G. Biomass Carbonisation in Retort Kilns: Process Techniques, Product Quality and Future Perspectives. *Bioresour. Technol. Rep.* **2022**, *17*, 100934. [CrossRef]
7. Mašek, O.; Buss, W.; Roy-Poirier, A.; Lowe, W.; Peters, C.; Brownsort, P.; Mignard, D.; Pritchard, C.; Sohi, S. Consistency of Biochar Properties over Time and Production Scales: A Characterisation of Standard Materials. *J. Anal. Appl. Pyrolysis* **2018**, *132*, 200–210. [CrossRef]
8. Kirch, T.; Medwell, P.R.; Birzer, C.H.; van Eyk, P.J. Small-Scale Autothermal Thermochemical Conversion of Multiple Solid Biomass Feedstock. *Renew Energy* **2020**, *149*, 1261–1270. [CrossRef]
9. Kieush, L.; Rieger, J.; Schenk, J.; Brondi, C.; Rovelli, D.; Echterhof, T.; Cirilli, F.; Thaler, C.; Jaeger, N.; Snaet, D.; et al. A Comprehensive Review of Secondary Carbon Bio-Carriers for Application in Metallurgical Processes: Utilization of Torrefied Biomass in Steel Production. *Metals* **2022**, *12*, 2005. [CrossRef]
10. DiGiovanni, C.; Li, D.; Ng, K.W.; Huang, X. Ranking of Injection Biochar for Slag Foaming Applications in Steelmaking. *Metals* **2023**, *13*, 1003. [CrossRef]
11. Ismail, A.N.; Yunos, N.F.D.M.; Rajeswari, M.S.R.; Idris, M.A. Production of Metallic Iron from Electric Arc Furnace (EAF) Slag Using Palm Shells Char/Coke Blends. In Proceedings of the AIP Conference Proceedings, Arau, Malaysia, 3 May 2021; p. 020202.
12. Sahajwalla, V. Recycling Waste Plastics in EAF Steelmaking: Carbon/Slag Interactions of HDPE-Coke Blends. *Steel Res. Int.* **2009**, *80*, 535–543.
13. Di Fraia, S.; Uddin, M.R. Energy Recovery from Waste Paper and Deinking Sludge to Support the Demand of the Paper Industry: A Numerical Analysis. *Sustainability* **2022**, *14*, 4669. [CrossRef]
14. Chen, Y.H.; Lan Thao Ngo, T.N.; Chiang, K.Y. Enhanced hydrogen production in co-gasification of sewage sludge and industrial wastewater sludge by a pilot-scale fluidized bed gasifier. *Int. J. Hydrogen Energy* **2021**, *46*, 14083–14095. [CrossRef]
15. Carotenuto, A.; Di Fraia, S.; Massarotti, N.; Sobek, S.; Uddin, M.R.; Vanoli, L.; Werle, S. Sewage Sludge Gasification Process Optimization for Combined Heat and Power Generation. *Energies* **2023**, *16*, 4742. [CrossRef]
16. FAOSTAT. Available online: https://www.fao.org/faostat/en/#data (accessed on 11 July 2025).
17. Adan-Mas, A.; Alcaraz, L.; Arévalo-Cid, P.; López-Gómez, F.A.; Montemor, F. Coffee-Derived Activated Carbon from Second Biowaste for Supercapacitor Applications. *Waste Manag.* **2021**, *120*, 280–289. [CrossRef] [PubMed]
18. Akubo, K.; Nahil, M.A.; Williams, P.T. Co-Pyrolysis–Catalytic Steam Reforming of Cellulose/Lignin with Polyethylene/Polystyrene for the Production of Hydrogen. Waste Dispos. *Sustain. Energy* **2020**, *2*, 177–191.
19. Kudelytė, V.; Eimontas, J.; Paulauskas, R.; Striūgas, N. Co-Pyrolysis of Plastic Waste and Lignin: A Pathway for Enhanced Hydrocarbon Recovery. *Energies* **2025**, *18*, 275. [CrossRef]
20. Geldart, D. The Effect of Particle Size and Size Distribution on the Behaviour of Gas-Fluidised Beds. *Powder Technol.* **1972**, *6*, 201–215. [CrossRef]
21. Hilgraf, P. *Pneumatische Förderung: Grundlagen, Auslegung und Betrieb von Anlagen*; Springer: Berlin/Heidelberg, Germany, 2019; ISBN 978-3-662-58407-1.

energies

MDPI

Article

A Study on Bio-Coke Production—The Influence of Bio-Components Addition on Coke-Making Blend Properties

Michał Rejdak [1,*], Małgorzata Wojtaszek-Kalaitzidi [1], Grzegorz Gałko [1], Bartosz Mertas [1], Tomasz Radko [1], Robert Baron [2], Michał Książek [3], Sten Yngve Larsen [4], Marcin Sajdak [5,6] and Stavros Kalaitzidis [7]

1 Institute of Energy and Fuel Processing Technology, 1 Zamkowa Street, 41-803 Zabrze, Poland
2 Koksownia Częstochowa Nowa, 51 Chłodna Street, 00-867 Warszawa, Poland
3 Sintef AS, 3 Richard Birkelands Street, 7034 Trondheim, Norway
4 Eramet Norway AS, CO/O Sintef Industri, Alfred Getz vei 2 B, 7034 Trondheim, Norway
5 Department of Air Protection, Faculty of Energy and Environmental Engineering, Silesian University of Technology in Gliwice, 44-100 Gliwice, Poland
6 School of Chemical Engineering, University of Birmingham, Edgbaston, Birmingham B15 2TT, UK
7 Department of Geology, School of Natural Sciences, University of Patras, GR-265 04 Rio-Patras, Greece
* Correspondence: mrejdak@itpe.pl; Tel.: +48-32-621-63-58

Abstract: Due to global warming, technologies reducing CO_2 emissions in the metallurgical industry are being sought. One possibility is to use bio-coke as a substitute for classic coke made of 100% fossil coal. Bio-coke can be produced on the basis of coal with the addition of substances of biomass origin. Blends for the production of bio-coke should have appropriate coke-making properties to ensure the appropriate quality of bio-coke. The article presents the results of the research on the influence of the addition (up to 20%) of bio-components of different origins to the coke blend on its coke-making properties, i.e., Gieseler Fluidity, Arnu—Audibert Dilatation and Roga Index. The bio-components used in the research were raw and thermally processed waste biomass of different origins (forestry: beech and alder woodchips; sawmill: pine sawdust; and the food industry: hazelnut shells and olive kernels) and commercial charcoal. Studies have shown that both the amount of additive and the type of additive affect the obtained coking properties. There was a decrease in fluidity, dilatation and Roga Index values, with more favorable results obtained for the addition of carbonized biomass and for additives with a higher apparent density. A regressive mathematical model on the influence of the share of the additive and its properties (oxygen content and apparent density) on the percentage decrease in fluidity was also developed.

Keywords: bio-coke; biomass; coke-making; renewable; bio-components; coke; biomass

1. Introduction

Coke is currently an important material for the global economy. In addition to iron ore, it is an crucial raw material in the integrated steel production process (BF-BOF), which currently accounts for over 70% of the world's steel production [1–3]. It is estimated that about 0.6 tons of coke is used to produce 1 ton of crude steel, which corresponds to about 0.77 tons of coking coal used to produce coke [4]. In addition to steel production, coke is also used, inter alia, in the foundry industry, the production of ferroalloys, insulating wool, quicklime, calcium carbide or as a fuel for heating purposes. Depending on the process in which it is used, it can play an energetic role (fuel—energy and heat source), chemical role (elemental carbon source for reduction in ores and carburization) and physical role (support for charge materials, gas permeability of bed, liquid metal drainage). In many processes, due to their technological conditions and the functions performed by coke, it is currently an irreplaceable stock. The use of coke in these processes, due to the fact that it is produced from fossil fuel, is associated with the emission of carbon dioxide, which is a greenhouse gas contributing to global warming. The global policy of many countries, including

European Union countries, to counteract global warming, aims to reduce significantly the CO_2 emissions from all economic sectors. The European Commission adopted the 'fit for 55' package, adapting existing climate and energy legislation to meet the new EU objective of a minimum 55 % reduction in greenhouse gas (GHG) emissions by 2030 [5]. The Iron and Steel sector itself is responsible for the emissions of approx. 7% of carbon dioxide [6], and the vast majority of it is the result of the use of fossil-based materials, such as the aforementioned coke or coal. Carbon dioxide emissions by industrial plants are associated not only with the harmful impact on the environment, but also with higher costs of business activities related to the need to purchase CO_2 emission allowances. Therefore, measures to minimize emissions are extremely important from the point of view of ecological and economic efficiency of production activities. One of the methods of reducing CO_2 emissions from coke production and from processes in which coke is used (as a fuel or reducing agent) would be, to replace elemental carbon of fossil origin with renewable carbon of biomass origin, which is considered to be zero-emission [7]. Moreover, replacing a certain part of the coking coal with biomass is beneficial from the point of view of reducing its consumption—coking coal is a scarce resource and is included in the list of EU critical raw materials. Therefore, in recent years, there has been a great deal of interest in the subject of bio-coke production, i.e., coke produced from blends of coking coals and biomass-derived materials [8–18]. It is not an easy task, because in order to produce coke, it is necessary to use coking coals with appropriate coking properties and the addition of other raw materials, including even other coals (non-coking), usually does not have a positive effect on the above-mentioned coking properties of the prepared blend. Therefore, in the industrial production of coke, it is of key importance to select the right coking blend recipe, the properties of which will guarantee the production of coke of an appropriate quality. The recipes of blend for the production of coke are based on the knowledge of the coking properties of its individual components. A properly selected recipe of the coking blend is key to the production of a well-sintered and durable coke structure. The most frequently assessed coke-making properties are plastic properties according to the Gieseler method, Audibert-Arnu dilatometric properties and sintering properties (Roga Index, Caking Index). Generally, it is considered that in order to produce coke of adequate strength from a coal blend, its fluidity should be in the range of 200–1000 ddpm (min. 100 ddpm) [19]. In the case of dilatation, it is generally recognized that a higher value of dilatation has a positive effect on the quality of coke [20]. According to the literature, the value of the total dilatation should not be lower than 30–50% [21,22]. In the case of the Roga Index, it is considered that the higher the value, the better the caking properties of coal. According to the standard PN-G-97002:2018-11, the RI value for coking coals should be min. 60 for gas-coking coal (equivalent to semi-soft coal) and min. 70 for ortho-coking coals (equivalent to hard coking coal).

Some articles related to the production of bio-coke also raise questions related to the coke-making properties of blends. These publications concern the addition of raw and thermally treated biomass including commercial charcoals, and focus mainly on thermoplastic properties, in particular according to the Gieseler method. Montiano et al. [23] investigated the effect of the addition of three types of sawdust, raw and thermally treated at 250 °C (chestnut, oak and pine) on the thermoplastic properties of coal. It was noticed that the addition of 2% of bio-components reduces the fluidity by about half, while the addition of thermally processed sawdust has a more negative effect than raw sawdust. The authors also suggest that the physical nature of the additive, rather than the chemical one, plays a major role. In another work, Montiano et al. [24] studied the possibility of adding waste sawdust (from chestnut and pine) to two industrial coal blends (up to 5%). They found that biomass caused a decrease in the plastic properties of the coking blend, with reductions in the Gieseler maximum fluidity of around 50% for 3 wt.% additions of the two different sawdusts. Kokonya et al. [25] found that coal fluidity decreased with the addition of biomass but increased with an increase in the heating rate up to 20 K/min (with biomass addition) and an increase in the final temperature for fast heating. There were no significant changes in the fluidity for different particle sizes of biomass <0.5 and >0.5 mm, nor in the case of the addition of torrefied biomass (Scots pine). MacPhee et al. [26] investigated the effect of the addition of commercial hardwood charcoal on the fluidity and dilatation of three coal blends (up to 10%). It was observed that for each of the tested blends, the addition of charcoal reduces dilatations and fluidity, but even with the addition of 10%, they were at an acceptable level. Castro Diaz et al. [27] studied the effect of biomass addition on high-temperature SAOS rheometry. It was found that sugar beet can be added in coking blends up to 5 wt. % without altering the viscoelastic properties of the coal, whereas pine wood and miscanthus reduce the fluidity even with 2 wt. % additions. The authors suggest that during the plastic phase, gas-solid or liquid-solid interactions affect the rheological properties. Fraga et al. [28] investigated the effect of charcoal addition (up to 10%) on the thermoplastic properties of coking coal. They found that even a small addition of charcoal, significantly reduces the fluidity of the coal and plastic range, while the addition of tar raises the fluidity above the level of coal without the addition of charcoal. The total addition of charcoal and tar was 10%. The addition of charcoal and coal tar to the coking coal revealed that it is possible to insert biomass into the coke-making process without depreciation in its thermoplastic behavior. Yustanti et al. [10] investigated the effect of rice husk and coconut shell biochar addition on fluidity and dilatation (up to 25%). It was found that fluidity decreased from 4.1 to 3.3 (log ddpm) for the blends with coconut shells, and from 3.2 to 2.8 for the blends with rice husks. The dilatation of the sample with coconut shells decreased from 78% to 28%, whereas that with the rice husks decreased from 61% to 16%. Solar et al. [29] investigated the effect on the fluidity of charcoal addition from waste woody biomass (Pinus radiata). It was found that the addition of 2 and 5% charcoal reduced the fluidity of the coking blend to the level of 398 and 258 ddpm, respectively (from the base 463 ddpm). Guerrero et al. [30] investigated the effect of the addition of charcoal on thermoplastic properties of three coals (with maximum fluidity of 373 (LF), 541 (MF) and 1891 (HF). Charcoal was added in various grain sizes (<20, 20–80, and 80–212 μm) and in various amounts (2, 5, 10 and 15%). It was found that a charcoal addition to the coals caused a progressive inverse exponential reduction in fluidity. The smallest grains of charcoal caused the greatest reduction in fluidity. A similar trend was observed for the HF and LF coals, which had worn off about 50% fluidity for as little as 5% of the additive. In the case of MF coal, the decrease was smaller. According to the authors, the different susceptibility of the coals to the addition of charcoal is influenced by their inherent properties, including the content of inertinite macerals. The decrease in dilatation, along with the addition of 5 and 10% biomass substances (torrefied pine/spruce woodchips, sawdust and high temperature torrefied pellets) was also reported by El-Tawil et al. [31]. This tendency was also observed for 5% of additives after modification (e.g., washing, kaolin coating) [14].

According to the literature reports, even a small amount of biomass addition can significantly deteriorate the thermoplastic properties of coals or coking blends. However, it should be noted that the vast majority of studies concerned a small amount of the introduced bio-components and the addition of more than 10% was used only in a few studies. From the point of view of the end user of coke, it is crucial to have the largest possible amount of renewable carbon in the bio-coke structure, due to the increasingly stringent requirements to reduce CO_2 emissions and increase the prices of emission allowances. Therefore, the addition of bio-components should be as high as possible. An important task is to find an appropriate bio-component (or the properties it should have) that has the least negative impact on the coking properties, which should make it possible to introduce more of it into the coking blend. Obviously, the conclusions obtained in the tests of coking properties will have to be later verified in comprehensive tests of bio-coke production based on the evaluation of its parameters.

As previously mentioned, most of the studies concerned plastic properties, mainly Gieseler fluidity, and only a few concerned dilatometric properties. Unfortunately, there is no information in the literature on the caking properties (e.g., Roga Index), which are also important because they assess the binding potential of coking blends.

The aim of this study was to assess the effect of bio-components of various origin, properties and share (up to 20%) on the coke-making indices of the coke blend for the purpose of selecting the raw material for the production of bio-coke for the ferroalloy industry. The evaluated parameters were Gieseler plastic properties, Arnu-Audibert dilatometric properties and caking properties by the Roga method. The selected raw materials are four types of waste biomass of different origins (forestry: beech and alder woodchips; sawmill: pine sawdust; and the food industry: hazelnut shells and olive kernels, raw and thermally-processed), commercial charcoal (nine additives in total) and a coke blend used in one of the Polish coking plants for the production of metallurgical coke. The use, in research, of nine types of bio-components with different chemical and physical properties, added to a specific coal blend, allowed us to identify key factors influencing the coke-making properties of the coal/biomass blend and allowed us to develop an equation taking into account the influence of the properties of bio-components on the predicted change in fluidity. In addition, the use of such an amount of additives, including additives not previously studied (beech/alder, olive kernels, hazelnut shells), also allowed for the current state of knowledge in the field of the impact of biomass addition on the coking properties to be extended.

2. Materials and Methods

2.1. Coal Blend and Biocomponents Analysis and Preparation of Biochars

The subject of the research was the coking blend (CB) used for the production of metallurgical coke at one of the Polish coking plants. The selected biomass raw materials—bio-components—were waste biomass of various origins (forest and sawmill waste: pine sawdust (PS), beech and alder chips (BAC); food industry waste: olive kernels (OK) and hazelnut shells (HS)) and commercially available charcoal (CC) produced on the basis of a mixture of trees such as beech, hornbeam, oak, birch, alder and maple. The choice of biomass was based primarily on its availability and diversity of properties—in order to properly identify and demonstrate the impact of various biomass properties on the coke-making indices of the coking blend. All biomass, except for olive kernels, came from Poland. European countries including Poland have well-developed forestry, furniture and food processing industries, which generate significant amounts of organic waste [32–34]. Some of them are used for energy purposes by both professional and industrial power plants, wood pellet producers and individual recipients, e.g., in the form of pellets (domestic boilers). The use of biomass for the production of bio-coke in domestic coking plants may be an alternative to the above-mentioned uses. Olive kernels came from Greece. As European countries are also the largest producers of olive oil [35], the production of which generates a significant amount of waste (including kernels with very favorable properties—very high density), it was also decided to use olive kernels. Poland is also a significant producer of charcoal [36], which is commercially available and has basically more favorable properties than raw biomass (it is already pre-carbonized), which means that among the selected types it has the greatest implementation potential.

Raw biomasses were additionally carbonized in order to minimize the content of volatiles and oxygen. The codes assigned to individual biochars were created by adding the letters BS to the abbreviation assigned to the raw biomass, e.g., PS (pine sawdust), PSBC (pine sawdust biochar). The biomass carbonization was carried out in a 10 kW electric resistance furnace (Figure 1). The furnace is equipped with a cylinder-shaped retort with an internal diameter of 123 mm, a height of 350 mm and a total capacity of 4.16 dm^3. The pyrolysis tests were carried out until the temperature of 500–520 °C was reached in the axis of the pyrolyzed charge. The heating rate was 3 K/min and heating was controlled by an automatic controller. The charge temperature was measured by an independent thermocouple of the furnace automation system, located directly in the central part of the charge. After reaching the set temperature in the batch axis, the biomass was kept for 1 h. The process was carried out under a nitrogen atmosphere. Raw biomass samples and produced biochars were presented in Figure 2.

Figure 1. The electric resistance furnace and the steel retort loaded with biomass.

Figure 2. Raw biomass samples and produced biochars.

The moisture content of raw materials was assessed according to PN-ISO 11722:2009 (coal/biochar) and PN-EN ISO 18134-1:2015-11 (biomass). Ash yield was assessed according to PN-ISO 1171:2002 (coal/biochar) and PN-EN ISO 18122:2016-01 (biomass). Volatile matter content was determined according to ISO 562:2010 (coal) and PN-EN ISO 18123:2016-01 (biomass). Carbon, hydrogen and nitrogen content was assessed according to ISO 29541:2010 (coal/biochar) and PN-EN ISO 16948:2015 (biomass). The total sulfur content was determined according to ISO 19579:2006 (coal/biochar) and PN-EN ISO 16994:2016 (biomass). Oxygen content was assessed according to PN-ISO 1928:2002 (coal/biochar) and PN-EN ISO 16993:2016 (biomass). For the measurement of real density (ρ_t) and total pore volume V_{TOTAL}, an AccuPyc II 1340 helium pycnometer from Micromeritics (Norcross, NA, USA) was used. For the measurement of apparent density (ρ_{app}), a GeoPyc 1350 density analyser from Micromeritics was used. Total porosity (P) was obtained via the Formula (1)–(ρ_t–real density of particle, g/cm^3; ρ_{app}–apparent density of particle g/cm^3):

$$P = \frac{\rho_t - \rho_{app}}{\rho_t} \times 100 \tag{1}$$

The properties of coal blend, raw biomasses and produced biochars are presented in Table 1 and the expanded uncertainty of measured parameters are presented in Table A1 (Appendix A).

Table 1. Properties of used coal blend, raw and carbonized biomass (^{ad}–air dried basis).

	CB	CC	BAC	PS	OK	HS	BACBC	PSBC	OKBC	HSBC
Moisture content, M^{ad},%	1.02	4.5	5.80	8.3	4.4	6.7	6.7	4.7	6.3	5.5
Ash yield, A^{ad}, %	8.35	4.8	0.60	0.60	1.1	0.80	1.5	1.6	2.5	2.2
Volatile matter, VM^{ad}, %	28.73	11.40	79.80	77.50	76.50	73.4	7.81	13.14	6.54	8.51
Carbon content, C^{ad}, %	78.4	82.5	47.0	47.20	48.50	47.9	84.3	83.1	84.7	84.1
Hydrogen content, H^{ad}, %	4.89	2.33	5.58	5.50	5.93	5.74	1.96	2.67	1.75	2.22
Nitrogen content, N^{ad}, %	1.44	0.52	0.23	0.17	0.20	0.22	0.53	0.36	0.39	0.46
Oxygen content, O^{ad}, %	5.4	5.3	40.8	38.2	39.9	38.6	5.0	7.6	4.4	5.5
Total sulfur content, $S_T{}^{ad}$, %	0.50	0.02	<0.02	<0.02	<0.02	<0.02	0.02	0.01	0.01	0.01
Fixed carbon, %	61.9	79.3	13.8	13.6	18.0	19.1	84.0	80.6	84.7	83.8
Real density, ρ_t, g/cm^3	1.357	1.502	1.446	1.431	1.433	1.434	1.494	1.410	1.529	1.470
Apparent density, ρ_{app}, g/cm^3	1.316	0.460	0.531	0.335	1.217	0.986	0.377	0.157	0.862	0.759
Total pore volume, V_{TOTAL}, cm^3/g	0.023	1.510	1.192	2.282	0.124	0.317	1.982	5.668	0.506	0.640
Total porosity, P, %	3.0	69.4	63.3	76.6	15.1	31.2	74.8	88.9	43.6	48.3

2.2. *Thermogravimetric Analysis*

For raw materials, a thermogravimetric analysis of the carbonization process was conducted. The range of temperatures was equal to industrial-scale coking temperatures, which are up to 1000 °C. The analysis was conducted by a thermogravimetry analyzer STA 409 PG Netsch in the range of temperatures from 40 to 1000 °C degrees with a heating rate of 10 K/min. On the basis of the obtained data, the percentage of volatiles obtained in a specific temperature range (e.g., VM100-415), the yield of the solid residue (CY) and the thermogravimetric curve (DTG curve) were obtained. Thermogravimetric curves were developed for data averaged from two measurements. The T_{max}—temperature for which the amount of released volatile parts is the highest was determined on the basis of the curve.

2.3. *Analysis of Coke-Making Properties*

The coking properties of coal and biomass blends were assessed using the standards dedicated to coal. The determination of plastic properties was carried out by the Gieseler method based on PN-G-04565:1994 which is consistent with ISO 10329:2009. The method consists of measuring the coal fluidity with a constant torque stirrer immersed in a coal sample. The heating rate is 3 K/min and the sample is heated up to 550 °C. On the basis of the measurement, the following parameters are determined: t_1—the initial softening temperature—is the first change of coal state, which is expressed by the moving of the plastometric dial by 1 degree; t_{max}—maximum fluidity temperature—is the temperature at which the coal sample achieves the highest fluidity which is expressed by the greatest number of dial rotations per 1 minute; t_3—solidification temperature—is the temperature in which coal resolidify and turns into the semicoke, the dial rotation at this temperature stops; all temperatures are measured in °C; F_{max}—maximum fluidity—the greatest number of rotations of the stirrer achieved at the temperature of maximum fluidity. The fluidity result is expressed as total dial divisions turned by the stirrer in a 1 min time period (ddpm); t_3–t_1—plastic range—is the range of temperatures between the beginning and the end of fluidity.

Determination of dilatometric properties (Arnu–Audibert dilatometer test) was carried out based on ISO 349:1975. The standard specifies a method for determining the swelling properties of coal when heated under standard conditions in a dilatometer (5 °C/min; up to 550 °C). In performing the method, a pencil made of powdered coal formed under pressure is inserted into a precisely calibrated narrow tube and topped by a calibrated steel rod, which slides in the bore of the tube. The pencil is heated at a constant and definite rate, making regular readings of the displacement of the rod as a function of the temperature and expressing the displacements observed as percentages of the original length of the pencil; a curve is plotted. The results of the analysis are the contraction and dilatation coefficients. The contraction coefficient is calculated from a proportion of a difference between the initial length and shortest length of the coal pencil to the initial length (a). The dilatation coefficient is calculated from a proportion of a difference between the initial and final length of the coal pencil to the initial length (b).

The determination of Roga Index RI was carried out based on PN-G-04508:2020-05. The RI expresses the mechanical strength of the crucible of coke obtained by carbonization under standard conditions of a blend of 1 g of coal and 5 g of standard anthracite. The crucible of coke obtained is submitted to a drum test in a strictly prescribed manner and the Roga Index is calculated from the results.

3. Results and Discussion

The raw materials used in the research are significantly diversified in terms of physical and chemical properties (Table 1). Raw biomasses were characterized, as reasonably expected, by a much lower ash yield (0.6–1.1) than the coal blend (8.35) and a much higher content of volatile matter (73.4–79.8) and oxygen (38.2–40.8)—the content of volatile matters and oxygen in the coal mixture was 28.7 and 5.4%, respectively. They were also advantageously low in total sulfur content, which was below the detection level, i.e., <0.02% which is advantageous for the coke production process. Sulfur is an undesirable element. In a blast furnace, only sulfate sulfur is completely tied to the slag. Other forms of sulfur are bounded with the lime (flux) and transformed to CaS. Sulfur bounding by alkali components of slag requires an adequate amount of lime which increases coke usage. Based on industrial experiences, it was noticed that 0.1% of sulfur content causes an increase in coke usage in a blast furnace from 0.3 to 1.1% of unit consumption. On the other hand, the blast furnace efficiency is decreasing by 2% [37]. Due to the high content of volatiles, fixed carbon was at a much lower level (13.6–19.1%) than in the case of the coal blend (61.9%). Raw biomasses were also characterized by a significantly lower content of elemental carbon (47.0–48.5%) in relation to the coal blend (78.4%). Carrying out the carbonization process resulted in a significant decrease in the volatile matter (6.54–13.14%) and oxygen content (4.4–7.6%). Fixed carbon increased significantly, which was at a higher level (80.6–84.0%) than for coal blend and commercial charcoal (79.3%). Therefore, the addition of charcoal or carbonized biomass (biochar) will leave a much larger residue in the bio-coke structure than the addition of raw biomass, which is beneficial. Total sulfur content remained at a very low level (0.01–0.02%). Along with the release of volatile parts, the ash content increased (1.5–2.5%), but despite this, its level in biochars was lower than for the coal blend. The raw materials were also equal in terms of physical properties, which was particularly noticeable in the case of apparent density and porosity. The coal blend had an apparent density of 1.316 g/cm^3. In the case of raw biomass, the differences were significant—for pine sawdust and beech and alder woodchips, the apparent density of their particles was 0.460 and 0.531 g/cm^3, respectively, while in the case of olive kernels, the apparent density was at a very high level of 1.217 g/cm^3, respectively. The apparent density of hazelnut shells was somewhat lower and amounted to 0.986 g/cm^3. Similar results were found for the total volume of pores and, consequently, for the total porosity. In the case of the coal blend, the lowest values of the total pore volume and total porosity were noted, which were 0.02 cm^3/g and 3.0%, respectively. In the case of raw biomass, the highest pore volume and total porosity were noticed for pine sawdust. The total pore volume, in this case, was 2.282 cm^3/g and the total porosity was 76.6%. Lightly lower values were noticed for beech and alder chips. The total pore volume was 1.192 cm^3/g and the total porosity was 63.3%. In the case of olive pits, a very low pore volume (0.124 cm^3/g) and a total porosity (15.1%) were noted. In the case of hazelnut shells, these values were about twice as high and amounted to 0.317 cm^3/g and 31.2%, respectively. For commercial charcoal, the total pore volume was 1.510 cm^3/g and the total porosity was 69.4%. In the case of the biochar produced from raw biomass, the values of the total pore volume were about twice as high as initially: beech and alder chips, 1.982 cm^3/g; pine sawdust, 5.668 cm^3/g; olive kernel, 0.506 cm^3/g; and hazelnut shell, 0.640 cm^3/g. It should be noted that the values obtained for pine sawdust biochar are approx. 10 times higher than for olive kernel biochar. The differences in the total porosity are not as great as in the case of the total pore volume but are also significant. Pine sawdust biochar has a total porosity of 88.9%, and beech/alder biochar has a porosity of 74.8%. Biochars made from olive kernels and hazelnut shells have much lower porosities of 43.6 and 48.3%, respectively. It should be observed that the properties of raw biomass have a decisive influence on the properties of biochar produced from them. Biochars made of raw materials with lower porosity and higher apparent density, also had lower porosity and apparent density e.g., the porosity and apparent density of OK were 15.1% and 1.217 g/cm^3, respectively, and OKBC was 43.6% and 0.862 g/cm^3, respectively, while the porosity of PS was 76.6% and 0.335 g/cm^3, respectively, and PSBC was 88.9% and 0.157 g/cm^3, respectively.

The results of the thermogravimetric analysis are presented in Table 2 and Figures A1–A10 (Appendix A). The obtained results indicate that the individual raw materials are characterized by different dynamics of the carbonization process. In the case of raw biomass, the greatest weight loss related to the separation of volatile parts was observed for the temperature below 415 °C, i.e., below the initial softening temperature of the tested coal blend. In this temperature range, it degasses more than 60% of volatile matter (in the case of beech alder chips, even more than 70%). For the raw biomass, the temperature for which the maximum temporary loss of volatile matter (T_{max}) was recorded, was also well below the aforementioned temperature and was in the range of approx. 340–360 °C. Some differences in the dynamics of the carbonization process are the result of differences in the chemical and structural composition of individual types of biomass, including different contents of hemicellulose, cellulose and lignin. For example, according to the data from the Phyllis 2 database [38], the content of cellulose, hemicellulose and ligin for the types of biomass assessed in our study may be, respectively: olive kernels: 28.1%, 37.2%, 25.3%; beech: 49.0%, 22.0%, 24.0%; alder: 38.6%, 33.1%, 28.1%; hazelnut shells: 28.9%, 29.9%, 42.5%; pine sawdust: 23.7%, 24.9%, 50%. In general, however, individual species may be characterized by a fairly wide variability of the above-mentioned components [39]. In the case of the coal blend, the greatest weight loss occurred in the temperature range of 350–500 °C, i.e., in the range corresponding to the plastic state of coal. The highest dynamics of the volatile matter releasing took place at 451 °C, and then it gradually decreased. In the case of carbonized biomass, the highest degassing dynamics were in the range of 750–1000 °C, except for commercial charcoal where a slightly higher loss of volatile matter was recorded in the range of 500–750 °C. This is most likely due to the fact that the industrial production conditions (temperature, residence time) were slightly different than in the case of biomass carbonized in the laboratory, i.e., lower temperature or shorter residence time. Small amounts of volatile matter released in the temperature range below 500 °C results from the prior degassing of the biomass during the biochar production process. Small amounts of volatile matter released in the temperature range below 500 °C results from the prior degassing of the biomass during the biochar production process and thus the release of volatile components. In addition to the differences in the dynamics and course of the process, a very important factor is the amount of solid post-process residue—char yield CY, which ultimately affects the amount of renewable elemental carbon in the structure of the produced bio-coke. The data in Table 2 show that in the case of raw biomass, the yield of char is over three times lower than in the case of carbonized biomass. This means that in the case of adding raw biomass, the content of renewable carbon in bio-coke will be several times lower than in the case of adding carbonized biomass. In the case of carbonized biomass for which the solid residue amount is greater than in the case of the coal blend, the yield of renewable carbon in the bio-coke structure will be higher than it would appear from the amount of its addition to the mentioned coking blend.

Table 2. Results of TG analysis of investigated raw materials.

	T_{max} [°C]	DTG_{max} [%/min]	VM 100–415 [%]	VM 415–500 [%]	VM 500–750 [%]	VM 750–1000 [%]	CY [%]
CB	451.5	2.08	5.39	13.38	9.69	4.09	67.45
BAC	366.2	10.19	71.63	3.63	3.63	2.91	18.13
PS	352.9	9.35	65.14	3.77	4.25	2.43	24.42
OK	344.9	8.3	63.49	4.8	4.33	2.26	25.12
HS	361.1	7.64	60.34	5.36	5.36	2.96	26.92
CC	664.6	0.64	2.62	1.41	8.24	5.08	82.66
BACBC	739.0	0.63	2.41	0.67	7.51	19.8	90.79
PSBC	694.6	0.68	2.49	1.71	10.89	24.07	91.03
OKBC	695.0	0.42	2.56	0.57	5.31	15.84	92.6
HSBC	778.6	0.88	1.46	0.57	6.9	21.78	87.14

The key feature of coking coals, which in principle allows the formation of a porous sintered structure of coke, is the ability to convert these coals to a plastic state in a certain temperature range (conventionally at a temperature of approx. 350–500 °C). The properties of the plastic mass existing in the temperature range mentioned, directly affect the process of forming the porous structure of the coke, and thus its properties. The fluidity of the plastic mass enables the reorganization of structural units of coal and the creation of a unique coke microtexture, it also affects the morphology of pores formed during the release of pyrolytic gases and solidification of the plastic mass [40]. The rheological properties of the plastic mass also determine other properties of coal including dilatation and sintering ability. The transformation of coal into a plastic state enables the molecular reorganization of structural units and the development of an ordered (more or less) coke texture, which is known to be an important factor influencing its quality parameters [41–43]. The addition of non-plasticizing substances to coal blends, as a rule, significantly affects the rheological properties of plastic mass, therefore the assessment of thermoplastic properties of coal with the addition of biomass substances is extremely important from the point of view of the bio-coke production process.

The results of the influence of biomass additives on plastometric properties analyzed by the Gieseler method are shown in Figures 3 and 4 (and in Tables A2 and A3—Appendix A) and as it results from the presented data, all of the parameters assessed with the Gieseler method changed under the influence of the addition of biomass. With the addition of bio-components, an increase in the initial softening temperature t_1 was observed. In the case of 5% of the additive, basically regardless of the type of biomass, an increase of approx. 5 °C was recorded (t_1 for the mixture without additive: 415 °C). Some differences became noticeable for 10% of the additive, and for 20% of the additive they were more visible, i.e., in the case of the thermally processed biomass, the t_1 values were higher and oscillated around 430 ± 3°C, while for raw biomass they were a few degrees lower. No significant differences were observed when it comes to the t_{max} maximum fluidity temperature value. Regardless of the type and amount of the additive, the t_{max} value was about 460 °C with slight fluctuations of 2–3 degrees, which can be considered as the uncertainty of determination.

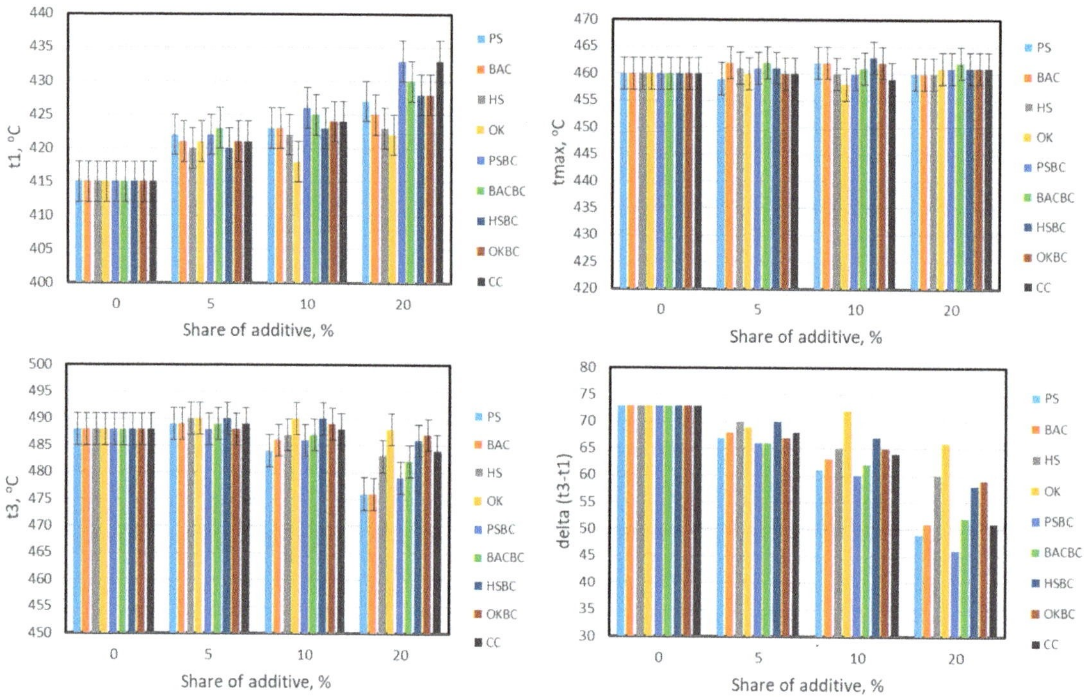

Figure 3. Influence of type and share of bio-components on characteristic temperatures of fluidity.

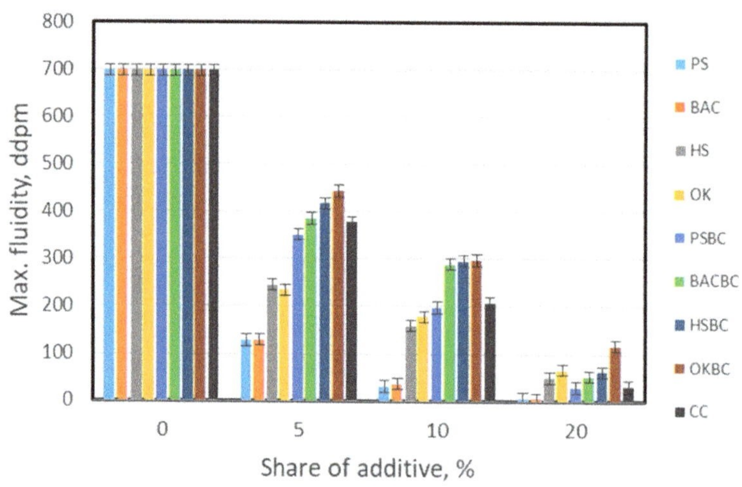

Figure 4. Influence of type and share of bio-components on maximum fluidity.

A slight decrease in t_3 solidification temperature was also observed. For 5 and 10% additive, the decrease is very slight, while in the case of 20% additive, it is more visible, especially for pine sawdust and beech and alder woodchips. The changes in the mentioned temperatures t_1 and t_3 resulted in a narrowing of the plastic range defined as t_3–t_1, which decreased from 73 °C (for the blend without any additive) to the level of 45–66 °C depending on the additive. The decrease was observed to be lower for raw and carbonized biomass from olive kernels and hazelnut shells. In the case of the initial fluidity temperature t_1, the observed tendencies are small (an increase) while for the temperature of maximum fluidity t_{max} and the final fluidity temperature t_3, tendencies are more visible, i.e., both mentioned temperatures decrease. As a consequence, the plastic range defined as the difference between t_3 and t_1, was unfavorably narrowed.

As expected, the most noticeable changes were for the maximum fluidity F_{max}, depending on the type of additive and its amount. The most unfavorable changes were recorded for raw biomass, i.e., pine sawdust and beech/alder woodchips. In this case, already for 5% of their addition, F_{max} decreased by approx. 82% (from the level of 699 ddpm to approx. 130 ddpm). For the 10% additive, the decrease was in the order of 500% to approx. 30 ddpm, while for the 20% additive, the plasticity value was only a few ddpm (5 and 4, respectively). Better results were observed in the case of raw olive seeds and hazelnut shells, for the 5% additive, the decrease in fluidity was about 65% (to the level of 245 and 234 ddpm), which is a twice higher result. For the addition of 10%, there was a further decline to around 159 and 178 ddpm. for 20% addition, the fluidity value dropped below 100 ddpm and amounted to 49 and 66 ddpm. Much better results were obtained for the addition of the carbonized biomass. For the 5% additive, the decrease was in the order of about 36–49% (from 352 to 444 ddpm). For the 10% additive, the fluidity value was from 200 to 300 and for the 20% additive except for the olive kernels char (117 ddpm) it was well below 100 ddpm. On this basis, it should be noted that the addition of carbonized biomass is much more advantageous than raw biomass. It is particularly visible for the 5 and 10% additive levels where the fluidity level is about three times higher for carbonized than for raw biomass. It should also be noted, that the type of biomass also significantly influences the obtained results. In general, the obtained results are consistent with the literature data on the deterioration of thermoplastic properties with the addition of biomass [23,25,27,30]. The reduction in fluidity with the addition of biomass to the coal blend can be attributed to the chemical and physical effects. The decrease of fluidity can be attributed to the harmful effect of oxygen groups present in volatile parts released during the pyrolysis process but the reduction in fluidity may also result from the sorption of plasticizing coal components by the porous structure of the biomass char [27,28].

In our case, we are dealing with additives that differ significantly in terms of volatile matter and oxygen content (raw and carbonized biomass) as well as in terms of physical properties (apparent density, total porosity). In general, the oxygen content in the examined raw biomass is at a relatively similar level (38.2–40.8%), while in the case of carbonized biomass this level is much lower and amounts to 4.4–7.6%, which may partially explain the more favorable effect of the addition of carbonized biomass. The amount of volatile matter released in the temperature range close to the plastic state range of coal (415–500 °C) is also significantly different for raw and carbonized biomass. A significant part of the volatile matter is released before the commencement of the plasticization process (below the initial softening temperature), so theoretically it should not much affect the reduction in fluidity, but according to literature reports, the decomposition/reaction of the hydroxyl- and carboxyl-group generates cross-linking prior to the initial softening temperature of coal, and thus affect the thermoplastic properties of coal [19,44]. In addition, the coking process in the coking chamber is layered (due to the temperature gradient the plastic layer moves from the chamber wall to the axis of the coal charge), so the volatile parts released in the inner part of the charge must pass through the plastic layer before they leave the coking chamber, so they can also react with components of plastic mass. This phenomenon may be important during the industrial coking process.

Therefore, taking into account the relatively similar content of oxygen and a similar amount of volatiles released during the plastic state, such large differences in the fluidity of raw biomass most likely result from physical interaction. In the work [23], it was found that the physical effect may be more important than the chemical one. As previously mentioned, raw biomass particles differ significantly in physical properties. Olive kernels and hazelnut shells have a much higher apparent density and much lower total porosity and total pore volume than pine sawdust and beech/alder woodchips particles. The differences in the apparent density mean that in the case of a specific mass addition of biomass to the coal blend, the volume fraction of biomass particles is different. This results in their different numbers and total area of introduced particles. Both raw and carbonized biomass acts as inert during the transition of the coal through the plastic state, so the greater surface area is associated with greater adsorption of plasticizing components (adhesion on the surface of solid particles), which in turn causes an increase in metaplastic viscosity, and, thus, a decrease in fluidity. It is also related to the increase in internal friction caused by a greater amount of solid particles suspended in the plastic mass, which causes an increase in viscosity and, consequently, a decrease in fluidity. A more developed porous structure manifested by a higher total pore volume and total porosity may also exacerbate this phenomenon by absorbing the plastic components into the char structure. This is consistent with the research of Fernandez et al. [45], where it was found that the addition of inert substances with higher porosity causes a greater degree of reduction in fluidity. It may also be responsible for an increase in initial softening temperature and a decrease in solidification temperature due to the sorption of small amounts of plastic (fluid) components that appear at the beginning of the plastic phase and the sorption of disappearing plastic components at the end of the plastic phase. In the case of the addition of raw biomass, the porous structure also develops during the process, which may also contribute to increased sorption of plastic components. This may confirm the obtaining of the most negative results for pine sawdust and beech/alder woodchips, which are characterized by the highest values of total porosity and pore volume and lowest values of apparent density. On the other hand, the best results were obtained with the addition of biomass with the highest apparent density and the lowest porosity values—olive kernels. In the case of raw biomass, there is also the phenomenon of loss of a significant part of the volatile matter before the pre-plastic state and during the plastic state of the coal blend, which is also related to the decrease in mass and volume (shrinkage) of biomass particles. Therefore, the real amount of char from raw biomass is much lower than that from carbonized biomass (much lower than was added to the blend), which, assuming only a physical impact, theoretically should benefit from the addition of raw biomass. Only with the addition of 20% can it be observed that for two types of raw biomass (olive kernel and hazelnut shell) the maximum fluidity value was achieved at a level similar to that of carbonized biomass, which may suggest that the char mass reduction is a result of the devolatilization process that also affects fluidity in some way. Probably, in the case of a larger amount of additive, the negative effect is also partially compensated by the higher apparent density of the added raw biomass particles and hence their lower actual amount and total surface area (for the same mass addition). Generally, it is considered that in order to produce coke of adequate strength from a coal blend, its fluidity should be in the range of 200–1000 ddpm (min. 100 ddpm) [46]. For some bio-components, the level below 200 was already observed for 10% of their addition. As previously mentioned, for the 20% additive, the fluidity level dropped below 100 ddpm (except from OKBC). It seems, therefore, that with such significant additions above 10%, for some types of bio-components, obtaining a strong coke structure will be very difficult to achieve. Therefore, it is crucial to select the additive with the least possible negative effect on fluidity. In addition, the rheological properties can be improved by adding to the blend coals with higher fluidity or by adding substances with a proven positive influence on fluidity (e.g., coal tar or pitch) [28,47]. On the other hand, the study [48] found that despite a significant decrease in fluidity as a result of coal weathering (from 272 to 26 ddpm), coke was obtained with a similar CSR (Coke Reactivity Index). The influence of bio-components on the properties of bio-coke is planned as part of further work.

In order to identify the quantitative influence of the properties of the added biomass on the decrease in fluidity, it was decided to perform a statistical analysis using polynomial regression with backward elimination of irrelevant regression coefficients. Statistica 13.3 software by TIBCO (Palo Alto, NA, USA) was used for the analysis. Physical properties (apparent density and porosity), chemical properties (oxygen content) and the amount of additive were selected as independent variables (predictors), and the dependent variable (expected) was the percentage decrease in F_{max} (ΔF_{max}).

On the basis of the analysis, Equation (2) was obtained that describes the impact of changes in the above-mentioned factors on the value of the decrease in the maximum fluidity.

$$\Delta F_{max} = 42.36 + 0.0157 \times O^2 - 13.47 \times \rho_{app}^2 + 2.38 \times S \quad (2)$$

The value of the coefficient of determination is $R^2 = 0.943$, the value of F-Test is $F = 61.29$, and the critical value of the $F_{CR}(0.01; 3; 27) = 4.76$, which proves the statistical significance of the obtained equation at the significance level $\alpha = 0.01$ (99% confidence level). Figure 5 shows graphically the correlation between the value obtained in the research and the value predicted using the obtained regression model. The obtained equation allows us to predict the decrease in fluidity quite well—the average forecasting error is 8.3% (±5.7 percentage points) and the coefficient of linear determination is $R^2 = 0.89$. As can be seen from Equation (2), the variables that significantly affect fluidity changes are primarily the amount of additive (S), oxygen content (O) and apparent density (ρ_{app}), which confirms both the chemical and physical type of influence of biomass addition on the fluidity value.

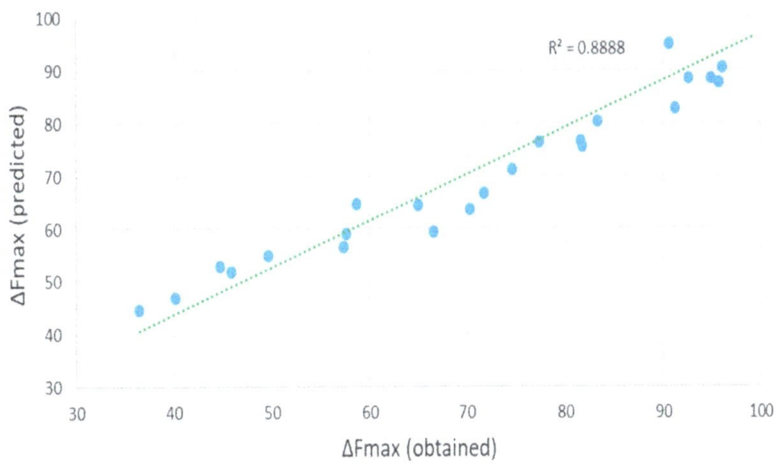

Figure 5. Predicted vs. obtained values of maximum fluidity F_{max}.

The addition of bio-components and their influence on the rheological properties of the plastic mass also directly affected other coke-making indices, such as dilatometric properties and sintering properties (Roga Index). In the case of dilatation, a similar trend was observed as in the case of Gieseler fluidity (Figure 6, Tables A4 and A5). Dilatometric properties are very important for the formation of strong coke. According to Nomura et al. [38], insufficient dilatation of the coal grain (low expansion volume) causes the swelling grains to be unable to fill the inter-particle voids in the charge bed located in the coking chamber—not all of the particles stick to each other properly. This results in weak bonds and, consequently, weak coke. If the volume of coal expansion is larger than the inter-particle voids volume, the fused coal particles come into contact with each other, stick together and, finally, form strong bonds. Nomura [22] noticed that the minimum value of total dilatation ensuring the obtaining of coke of appropriate quality also depends on the conditions of the coking process (i.a. bulk density) and should be min. 30–50% (higher density partially compensates for lower dilatation). Lower values of dilatation can be the reason for an inadequately-sintered coke structure of high porosity. On the other hand, in the work of MacPhee et al. [26], where with the addition of charcoal, the value of dilatation fell significantly to the level of a few points (but remained at a positive level), it was considered by the authors as workable. In our case, the worst results were obtained for the addition of raw biomass with the lowest apparent density (pine sawdust and beech/alder chips). In the case of the mentioned components, for the content of 20%, a complete lack of dilatation properties was noted. In the case of the addition of olive kernels and hazelnut shells, significantly higher dilatations were noted, with values comparable to the addition of the carbonized biomass. Generally, for 20% of the additive, a very low level of expansion was recorded. Only the values for olive kernel and hazelnut shells biochars stood out from the rest. However, taking into account that the value of total dilatation also takes into account contraction (total dilatation = contraction + dilatation), which for all examined cases was in the range of 30–40%, it seems that all bio-components provide the appropriate level of dilatation up to 10% of their addition. In the case of 20%, only the carbonized additives with the highest apparent density provide the appropriate level, i.e., biochars from olive kernels and hazelnut shells.

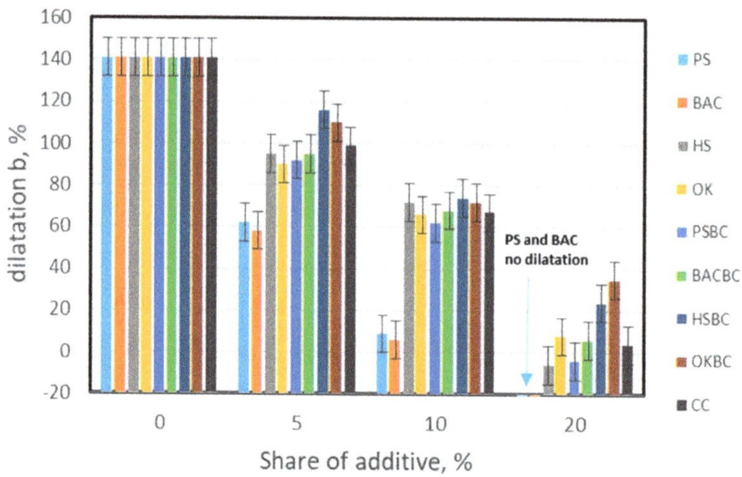

Figure 6. Influence of type and share of bio-components on dilatation.

The effect of dilatation reduction with the addition of inert substances, such as raw or carbonized biomass, is associated with the modification of the gas permeability of the plastic mass. The higher the permeability, the easier the gases released during pyrolysis are released and the value of the dilatation should be lower. Literature data confirm that the addition of inert substances to the coal charge reduces the permeability of the plastic phase [49,50]. This is because the inert substance does not become plastic during the pyrolysis process, therefore its structure does not add much resistance against the gases released during pyrolysis. Moreover, the porous structure of the inert allows these gases to escape more easily. The above may explain the different behavior of sawdust and beech/alder biomass, olive kernels and hazelnut shells. As previously mentioned, as a result of large differences in the apparent density of particles of these types of biomass, at the assumed constant mass, their volume share in the mixture with coal is different. The low density of pine sawdust and beech/alder results in more particles compared to olive kernels and hazelnut shells and, thus, more zones/points being a pathway for gas release. The easier release of these gases results in a lower expansion of the plastic mass (heterogeneous: a mixture of metaplastic and inert additive particles) resulting in a decrease in the dilatation value. The ease of gas release also depends on the nature of the porous structure of the additive (the number of pores and their morphology). The charcoals made of pine and beech/alder are characterized by a much higher porosity and total pore volume than those made of olive kernels and hazelnut shells, which significantly facilitates the escape of the emerging gases; thus, reducing the value of dilatation. In addition, during the pyrolysis process, raw biomass changes much more in volume than previously carbonized biomass, i.e., it shrinks more, leaving voids, which also facilitates the escape of pyrolysis gases, which also explains why the value of dilatation for the addition of raw biomass is lower than for carbonized biomass.

The addition of bio-components also caused a decrease in the sintering properties of the coking blend expressed by the Roga Index (Figure 7, Table A6). In general, no significant differences were noted for carbonized biomass, while for the addition of raw biomass they were clearly noticeable. Moreover, the RI values for the addition of raw biomass were lower, and the greater their share, the more noticeable the difference in relation to carbonized biomass. In the PN-G-97002:2018-11 standard, setting out the requirements for certain types of coal, the minimum RI value for coals used in the production of coke, i.e., gas-coking coal (semi-soft equivalent) and ortho-coking coal (hard coking coal equivalent), was, respectively, 60 and 70. Therefore, it can be concluded that for the tested coal blend, the addition of bio-components up to 5% resulted in a decrease in the RI value, but generally, the obtained values corresponded to the values required for ortho-coking coals. In the case of 10%, only the addition of carbonized biomass would meet this condition, but raw biomass would also allow a result to be obtained corresponding to gas-coking coals. For the 20% addition, the RI value corresponding to gas-coking coals was obtained only in the case of the addition of carbonized biomass.

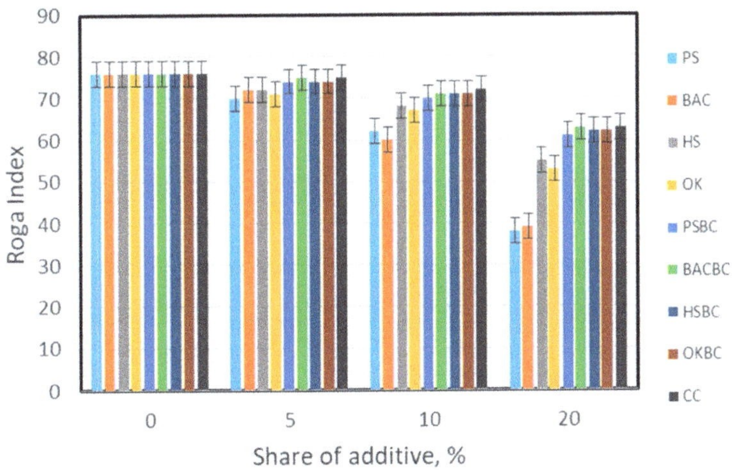

Figure 7. Influence of type and share of bio-components on Roga Index.

The principle of the Roga Index determination method is to determine the mechanical strength of a char made of 5 g of anthracite (an inert substance) and 1 g of tested coal. It determines the bonding properties of coal. Thus, the addition of the particles of the subsequent inert substance reduces the sintering properties because the real content of the metaplast binder in relation to the substance requiring binding is reduced. In this case, the properties of the additive can also play a role. As in the case of the previously discussed indicators, the addition of biomass with a lower apparent density means that the coal must stick together more inert particles than in the case of the addition of a substance with a higher apparent density, which is related to a larger surface area. Metaplast also penetrates the porous structure which also affects the final result, because there is less of it in the space between the particles that must be bounded. Therefore, adding a greater amount of additive and an additive with a lower apparent density and higher porosity density to the blend can result in the deterioration of the strength of the char due to the need of binding a larger surface (external and internal) of the inert additive. According to the authors [51], in the case of the addition of biomass, the contact area is of great importance in terms of the strength of the char (coke). An additional reason that is particularly apparent for raw biomass may be the different shrinkage of the additive during the plastic phase during both the fluidity development period (before maximum fluidity) and the semi-coke formation (after maximum fluidity). Different types of biomass undergo different shrinkage during pyrolysis [52], which causes the formation of voids that must be filled in order to properly sinter the particles that make up the coking blend. Thus, in order for the coal particles to be tightly bound together, the appropriate expansion potential is necessary [22,51]. In the study of Ueki et al. [51], it was stated that the release of volatile substances at lower carbonization temperatures and the related shrinkage of biomass particles inhibits connectivity between coal particles. Too low metaplastic expansion capacity (confirmed by reduced dilatation) is not able to compensate for the voids formed (due to shrinkage), which worsens the sintering conditions and lowers the strength of the resulting char, which is reflected by the lower Roga Index. Lower values of dilatation observed for a larger amount of additive (especially visible for raw biomass—mainly PS and BAC) may also be associated with lower values of internal pressure, which also does not favor the formation of a strong structure of the char. In the case of coke production, the low value of dilatation can be partially compensated by the use of a higher coal charge density which significantly reduces the voids between the grains of the blend and contributes to the improvement of sintering conditions and a stronger coke structure.

Summing up, as results from the obtained results, the addition of bio-components has a negative effect on all the analyzed coking properties. The key point is to select the right additive with the least possible adverse effects in order to obtain a strong bio-coke. It is also possible to further optimize the coking blend by, i.e., the addition of higher fluidity coals, the addition of plasticizers, increasing the density of the coal charge and grain size optimization. This will be verified in future bio-coke manufacturing and quality assessment tests.

4. Conclusions

The aim of this study was to assess the effect of bio-components of various origins, properties and share (up to 20%) on the coke-making indices of the coking blend. Based on the research carried out, the following conclusions can be drawn:

- The addition of raw and carbonized biomass deteriorates the coking properties of the coal blend. The properties measured by the Gieseler, Arnu–Audibert and Roga methods significantly deteriorated. A decrease in the value of key indicator characteristics for the above-mentioned methods, i.e., decrease in the maximum fluidity F_{max}, decrease in the value of dilatation b and decrease in RI. Both the share of the added bio-component and its type contributed to the deterioration of the coke-making properties.
- It was observed that the addition of raw biomass causes more negative effects than carbonized biomass but the magnitude of this influence depends on the initial biomass properties. In the case of raw biomass, more negative effects were noted for raw materials characterized by low apparent density and high porosity. The differences between carbonized biomass are not as clear as in the case of raw biomass, but they are noticeable, especially in the case of maximum fluidity and dilatation.

Therefore, it seems that, apart from the chemical properties (oxygen groups content), physical factors play an important role in shaping the coke-making properties—both those related to the properties of the samples (apparent density, porosity) and the behavior during the pyrolysis process—as well as volumetric shrinkage and pore structure development. From the point of view of bio-coke production, regardless of the type of added biomass (raw or carbonized), it is preferable to add the one with the highest possible apparent density and the lowest porosity. It will be verified during further tests of bio-coke production and evaluation of its quality parameters.

Author Contributions: Conceptualization, M.R. and B.M.; methodology, M.R., M.W.-K. and G.G.; investigation, M.R., T.R., M.W.-K. and G.G.; writing-original draft preparation, M.R.; writing-review and editing, M.R., M.W.-K., B.M., G.G., T.R., M.K., R.B., S.Y.L., M.S. and S.K. All authors have read and agreed to the published version of the manuscript.

Funding: The research leading to these results has received funding from the Norway Grants 2014-2021 via the National Centre for Research and Development. Contract no.: NOR/POLNOR/BioCoke4FAI/0070/2019-00.

Data Availability Statement: The data is contained in the article in Appendix A.

Acknowledgments: The authors would like to express their thanks to all employees of the Laboratory of Analytical Chemistry, Department of Circular Economy and Department of Coke-making Technologies of the Institute of Energy and Fuel Processing Technology involved in the research. The authors would like to express their thanks to Aris Kapsampelis from Eleourgia Prevezis S.A (Greece) for providing the olive kernels and for companies: OZEN (Poland) for providing the charcoal and Tree Nuts (Poland) for providing the nut shells.

Conflicts of Interest: The authors declare no conflict of interest.

Appendix A

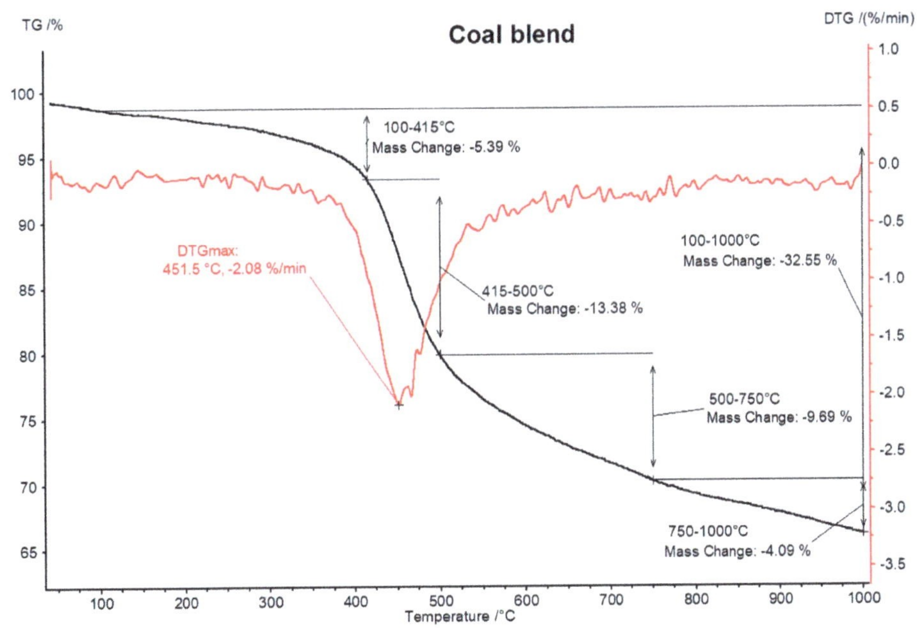

Figure A1. TG/DTG curves obtained for coal blend.

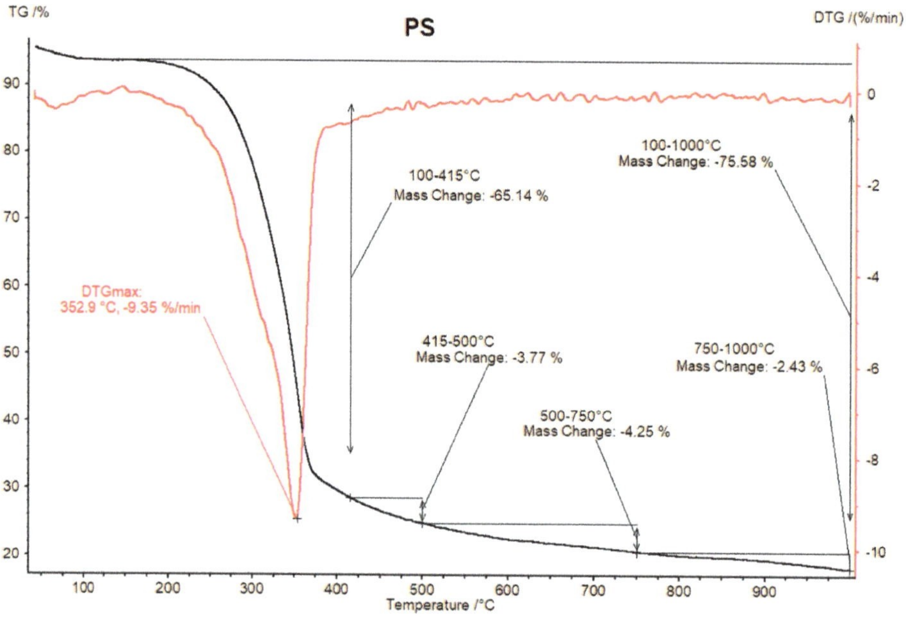

Figure A2. TG/DTG curves obtained for pine sawdust.

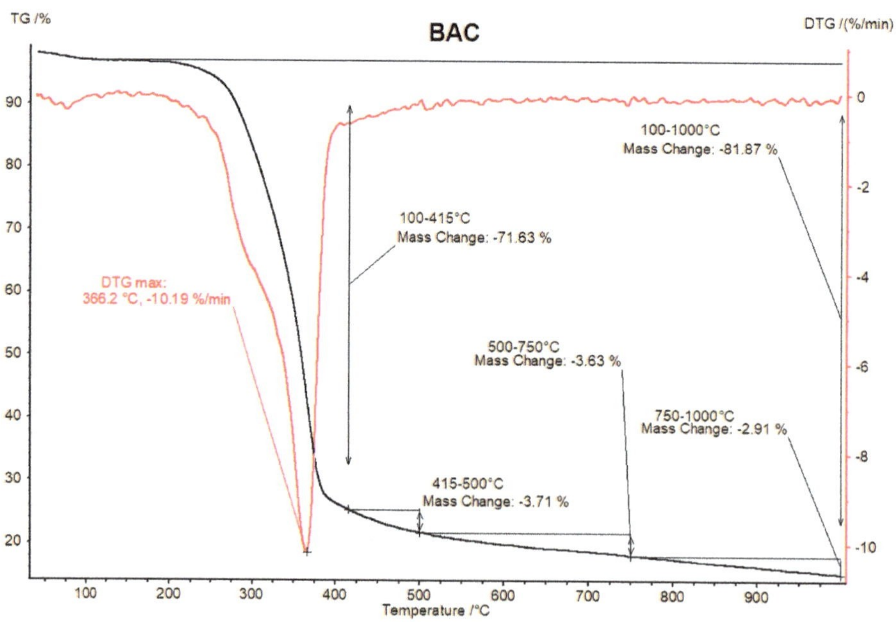

Figure A3. TG/DTG curves obtained for beech/alder chips.

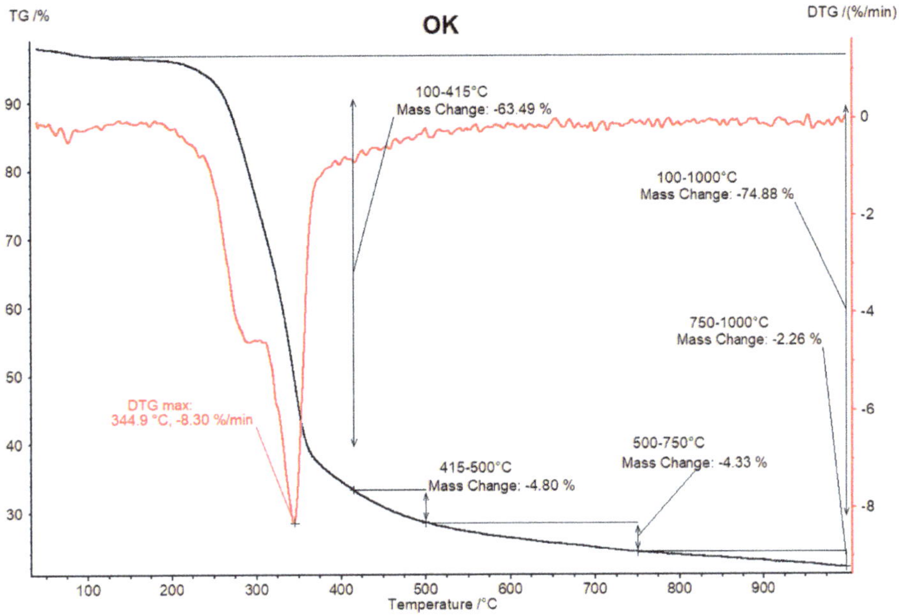

Figure A4. TG/DTG curves obtained for olive kernels.

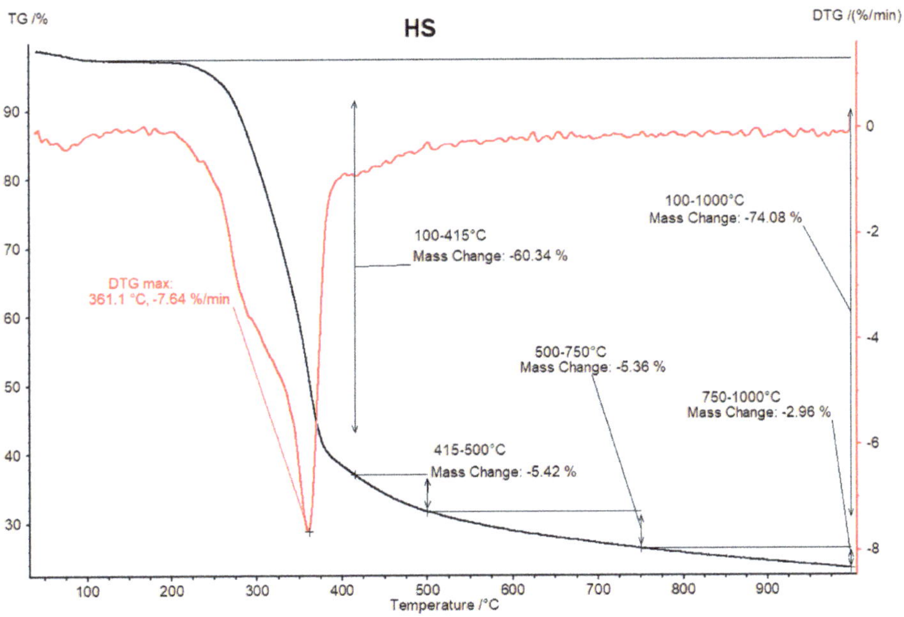

Figure A5. TG/DTG curves obtained for hazelnut shells.

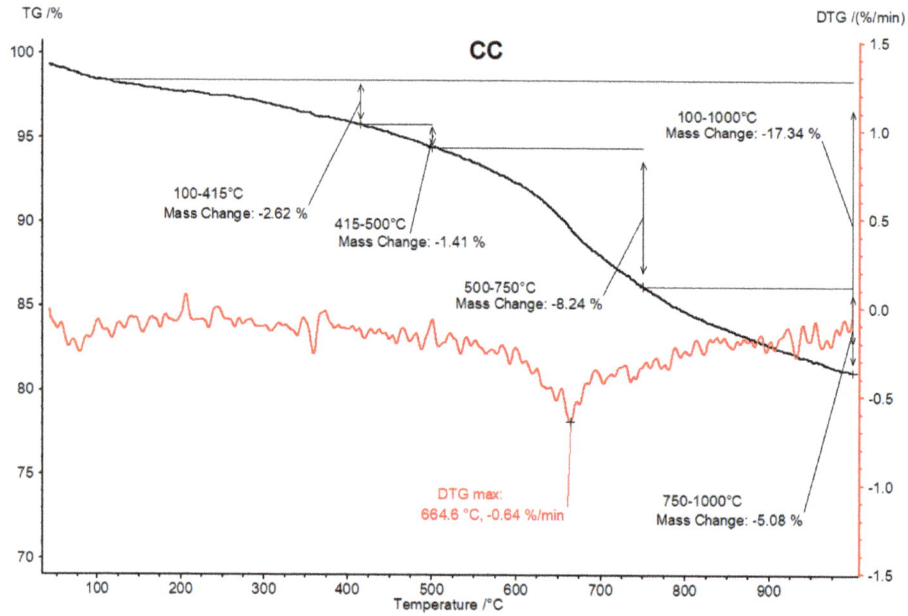

Figure A6. TG/DTG curves obtained for commercial charcoal.

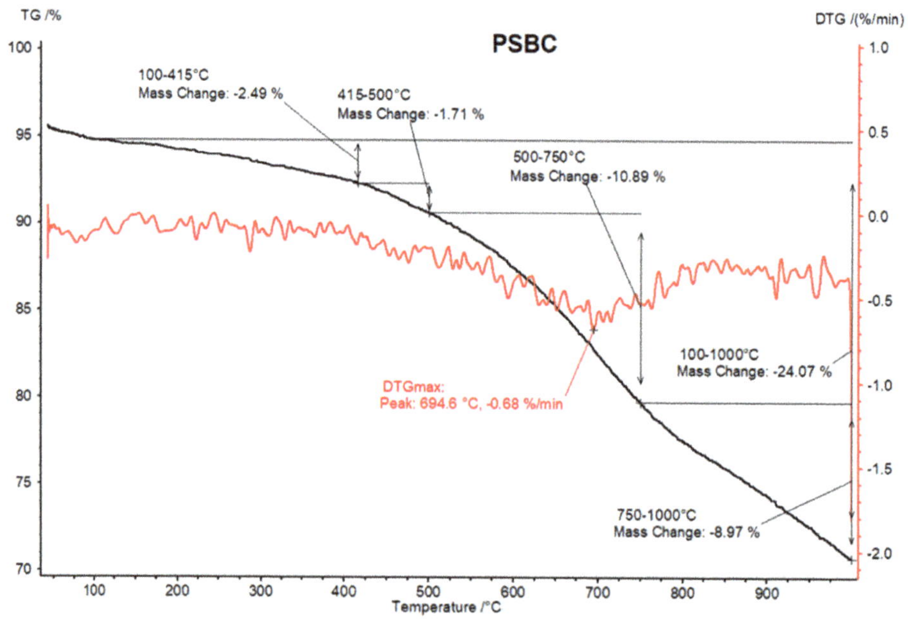

Figure A7. TG/DTG curves obtained for pine sawdust biochar.

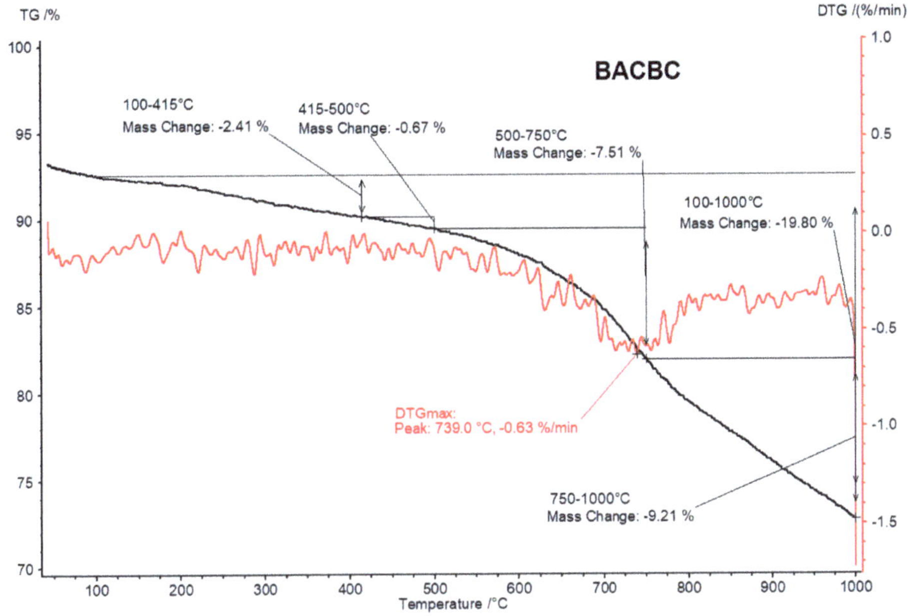

Figure A8. TG/DTG curves obtained for beech/alder chips biochar.

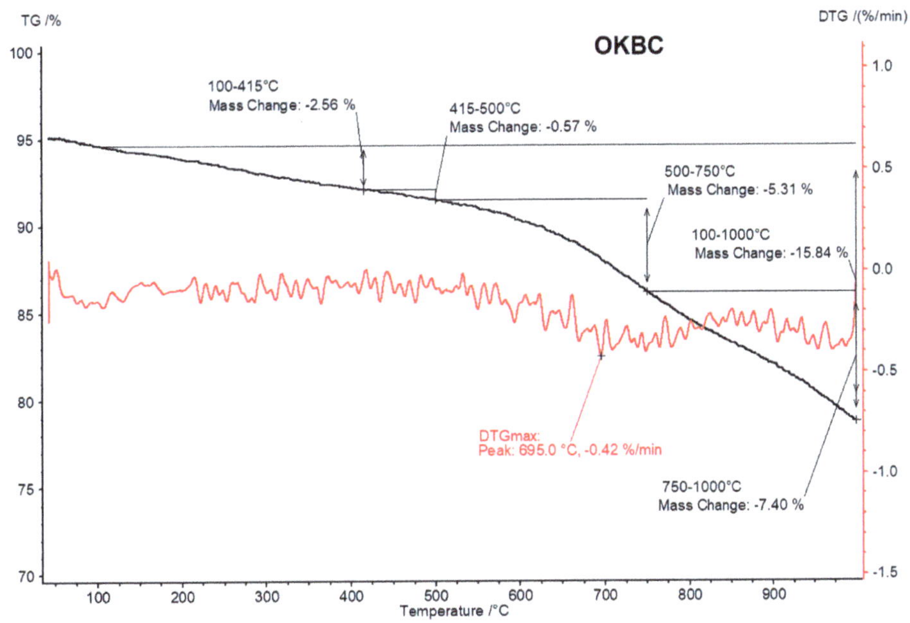

Figure A9. TG/DTG curves obtained for olive kernel biochar.

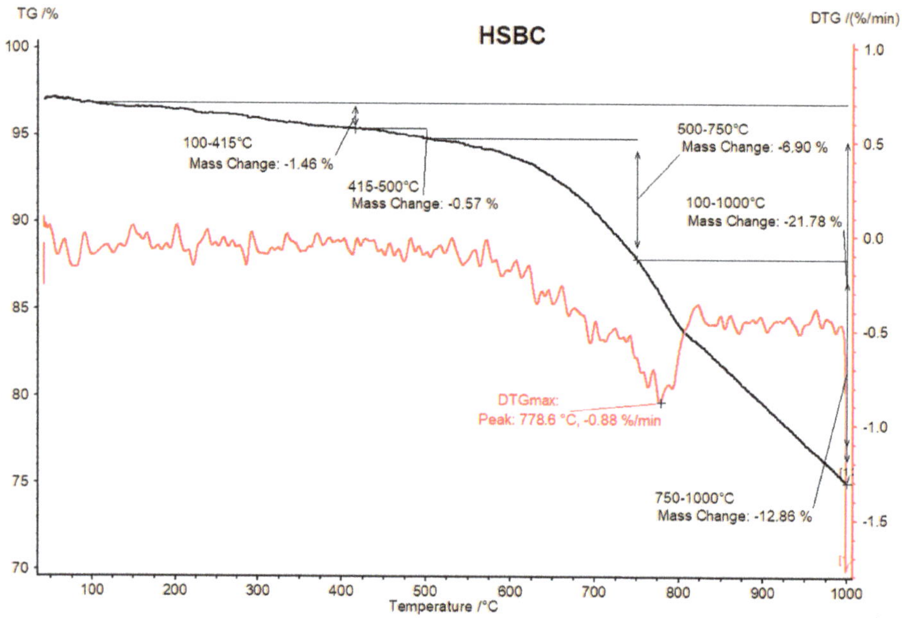

Figure A10. TG/DTG curves obtained for hazelnut shells biochar.

Table A1. The values of expanded uncertainty (α = 0.05) of performed determinations (^{ad}–air dried basis).

	Value of Expanded Uncertainty, u_e ± (pp)	
	Coal/Biochar	**Biomass**
Moisture content, M^{ad}	0.2	0.2
Ash yield, A^{ad}	0.2	0.2
Volatile matter, VM^{ad}	0.19	0.5
Carbon content, C^{ad}	0.4	0.6
Hydrogen content, H^{ad}	0.10	0.32
Nitrogen content, N^{ad}	0.03	0.11
Total sulfur content, S_T^{ad}	0.05	0.03
	Value of expanded uncertainty, u_e ± (%)	
Real density, ρ_t	0.1	0.1
Apparent density, ρ_{app}	0.8	0.8
Total pore volume, V_{TOTAL}	2.2	2.2
Total porosity, P	1.4	1.4

Table A2. The influence of raw biomass supplementation on Gieseler plastometric properties.

Plastometric Properties by Gieseler Method					
Additive	**Additive Content, %**				**Expandend Uncertainty**
	0%	**5%**	**10%**	**20%**	
	Coal Blend				
t_1, °C	415	-	-	-	±3
t_{max}, °C	460	-	-	-	±3
t_3, °C	488	-	-	-	±3
F_{max}, ddpm	699	-	-	-	±12

Table A2. *Cont.*

Plastometric Properties by Gieseler Method					
		Pine Sawdust			
t_1, °C	-	422	423	427	±3
t_{max}, °C	-	459	462	460	±3
t_3, °C	-	489	484	476	±3
F_{max}, ddpm	-	128	30	5	±12
		Beech and Alder Woodchips			
t_1, °C	-	421	423	414	±3
t_{max}, °C	-	462	462	460	±3
t_3, °C	-	489	486	476	±3
F_{max}, ddpm	-	129	36	4	±12
		Hazelnut shells			
t_1, °C	-	420	422	423	±3
t_{max}, °C	-	461	460	460	±3
t_3, °C	-	490	487	483	±3
F_{max}, ddpm	-	245	159	49	±12
		Olive kernels			
t_1, °C	-	421	418	422	±3
t_{max}, °C	-	460	458	461	±3
t_3, °C	-	490	490	488	±3
F_{max}, ddpm	-	234	178	66	±12

Table A3. The influence of carbonized biomass supplementation on Gieseler plastometric properties.

Plastometric Properties by Gieseler Method					
Additive	Additive Content, %				Expandend Uncertainty
	0%	5%	10%	20%	
		Coal Blend			
t_1, °C	415	-	-	-	±3
t_{max}, °C	460	-	-	-	±3
t_3, °C	488	-	-	-	±3
F_{max}, ddpm	699	-	-	-	±12
		Biochar—Pine Sawdust			
t_1, °C	-	422	426	433	±3
t_{max}, °C	-	461	460	461	±3
t_3, °C	-	488	486	479	±3
F_{max}, ddpm	-	352	198	28	±12
		Biochar—Beech and Alder Woodchips			
t_1, °C	-	423	425	430	±3
t_{max}, °C	-	462	461	462	±3
t_3, °C	-	489	487	482	±3
F_{max}, ddpm	-	386	289	52	±12
		Biochar—Hazelnut Shells			
t_1, °C	-	420	423	428	±3
t_{max}, °C	-	461	463	461	±3
t_3, °C	-	490	490	486	±3
F_{max}, ddpm	-	418	296	62	±12
		Biochar—Olive Kernels			
t_1, °C	-	421	424	428	±3
t_{max}, °C	-	460	462	461	±3
t_3, °C	-	488	489	487	±3
F_{max}, ddpm	-	444	298	117	±12

Table A3. *Cont.*

Plastometric Properties by Gieseler Method					
Additive	Additive Content, %				Expandend Uncertainty
	0%	5%	10%	20%	
		Biochar—Charcoal			
t_I, °C	-	421	424	433	±3
t_{max}, °C	-	460	459	461	±3
t_3, °C	-	489	488	484	±3
F_{max}, ddpm	-	378	208	30	±12

Table A4. The influence of raw biomass supplementation on dilatometric properties by Arnu–Audibert method.

Dilatometric Properties by Arnu-Audibert Method					
Additive	Additive Content, %				Expandend Uncertainty
	0%	5%	10%	20%	
		Coal Blend			
t_I, °C	395	-	-	-	±6
t_{II}, °C	435	-	-	-	±5
t_{III}, °C	475	-	-	-	±6
a, %	34	-	-	-	±6
b, %	141	-	-	-	±9
		Pine Sawdust			
t_I, °C	-	375	330	330	±6
t_{II}, °C	-	440	445	485	±5
t_{III}, °C	-	475	475	485	±6
a, %	-	40	35	37	±6
b, %	-	62	9	lack	±9
		Beech and Alder Woodchips			
t_I, °C	-	335	330	330	±6
t_{II}, °C	-	445	450	500	±5
t_{III}, °C	-	475	475	500	±6
a, %	-	35	35	40	±6
b, %	-	58	6	lack	±9
		Hazelnut Shells			
t_I, °C	-	390	390	345	±6
t_{II}, °C	-	440	440	450	±5
t_{III}, °C	-	475	475	475	±6
a, %	-	38	38	31	±6
b, %	-	95	72	-6	±9
		Olive Kernels			
t_I, °C	-	395	390	385	±6
t_{II}, °C	-	440	440	450	±5
t_{III}, °C	-	475	475	475	±6
a, %	-	37	38	32	±6
b, %	-	90	66	8	±9

Table A5. Influence of carbonized biomass supplementation on dilatometric properties by Arnu–Audibet method.

Additive	Dilatometric Coefficients Value by Arnu-Audibert Method				
	Additive Content, %				Expandend Uncertainty
	0%	5%	10%	20%	
Coal Blend					
t_I, °C	395	-	-	-	±6
t_{II}, °C	435	-	-	-	±5
t_{III}, °C	475	-	-	-	±6
a, %	34	-	-	-	±6
b, %	141	-	-	-	±9
Biochar—Pine Sawdust					
t_I, °C	-	395	395	395	±6
t_{II}, °C	-	440	445	450	±5
t_{III}, °C	-	480	475	475	±6
a, %	-	34	32	30	±6
b, %	-	62	9	lack	±9
Biochar—Beech and Alder Woodchips					
t_I, °C	-	395	395	400	±6
t_{II}, °C	-	440	445	450	±5
t_{III}, °C	-	470	475	470	±6
a, %	-	35	36	30	±6
b, %	-	95	68	6	±9
Biochar—Hazelnut Shells					
t_I, °C	-	400	400	400	±6
t_{II}, °C	-	440	445	450	±5
t_{III}, °C	-	480	475	475	±6
a, %	-	36	35	32	±6
b, %	-	116	74	24	±9
Biochar—Olive Kernels					
t_I, °C	-	395	400	400	±6
t_{II}, °C	-	440	445	450	±5
t_{III}, °C	-	480	475	475	±6
a, %	-	35	30	32	±6
b, %	-	110	72	35	±9
Biochar—Charcoal					
t_I, °C	-	400	400	400	±6
t_{II}, °C	-	440	440	450	±5
t_{III}, °C	-	475	475	470	±6
a, %	-	33	34	30	±6
b, %	-	99	67	4	±9

Table A6. The influence of biomass additive on RI values.

Additive	Roga Index RI				
	Additive Content,%				Expandend Uncertainty
	0%	5%	10%	20%	
Coal blend	76	-	-	-	±3
Pine sawdust	-	70	62	38	±3
Beech and alder woodchips	-	72	60	39	±3
Hazelnut shells	-	72	68	55	±3
Olive kernels	-	71	67	53	±3

Table A6. *Cont.*

Additive	Roga Index RI				
	Additive Content,%				Expandend Uncertainty
	0%	5%	10%	20%	
Biochar—pine sawdust	-	74	70	61	±3
Biochar—beech and alder woodchips	-	75	71	63	±3
Biochar—hazelnut shells	-	74	71	62	±3
Biochar—olive kernels	-	74	71	62	±3
Biochar—charcoal	-	75	72	63	±3

References

1. World Steel in Figures 2022. Available online: https://worldsteel.org/wp-content/uploads/World-Steel-in-Figures-2022.pdf (accessed on 26 July 2022).
2. Niesler, M.; Stecko, J.; Stelmach, S.; Kwiecińska-Mydlak, A. Biochars in Iron Ores Sintering Process: Effect on Sinter Quality and Emission. *Energies* **2021**, *14*, 3479. [CrossRef]
3. Duda, A.; Fidalgo Valverde, G. The Economics of Coking Coal Mining: A Fossil Fuel Still Needed for Steel Production. *Energies* **2021**, *14*, 7682. [CrossRef]
4. Dincer, I. *Comprehensive Energy Systems*; Elsevier: Amsterdam, The Netherlands, 2018; ISBN 978-0-12-814925-6.
5. Communication from the Commission to the European Parliament, the Council, the European Economic and Social Committee and the Committee of the Regions "Brussels, 14.7.2021 (COM 2021) Final". Available online: https://ec.europa.eu/info/sites/default/files/chapeau_communication.pdf (accessed on 26 July 2022).
6. IEA Report 2020: Iron and Steel Technology Roadmap towards More Sustainable Steelmaking. Available online: https://iea.blob.core.windows.net/assets/eb0c8ec1-3665-4959-97d0-187ceca189a8/Iron_and_Steel_Technology_Roadmap.pdf (accessed on 26 July 2022).
7. Cavaliere, P. *Clean Ironmaking and Steelmaking Processes Efficient Technologies for Greenhouse Emissions Abatement*; Springer Nature Switzerland AG: Cham, Switzerland, 2019. [CrossRef]
8. Koveria, A.; Kieush, L.; Svietkina, O.; Perkov, Y. Metallurgical coke production with biomass additives. Part 1. A review of existing practices. *Can. Metall. Q.* **2020**, *59*, 417–429. [CrossRef]
9. Koskela, A.; Suopajärvi, H.; Mattila, O.; Uusitalo, J.; Fabritius, T. Lignin from Bioethanol Production as a Part of a Raw Material Blend of a Metallurgical Coke. *Energies* **2019**, *12*, 1533. [CrossRef]
10. Yustanti, E.; Wardhono, E.Y.; Mursito, A.T.; Alhamidi, A. Types and Composition of Biomass in Biocoke Synthesis with the Coal Blending Method. *Energies* **2021**, *14*, 6570. [CrossRef]
11. Suopajärvi, H.; Dahl, E.; Kemppainen, A.; Gornostayev, S.; Koskela, A.; Fabritius, T. Effect of Charcoal and Kraft-Lignin Addition on Coke Compression Strength and Reactivity. *Energies* **2017**, *10*, 1850. [CrossRef]
12. Matsumura, T.; Ichida, M.; Nagasaka, T.; Kato, K. Carbonization Behaviour Of Woody Biomass and Resulting Metallurgical Coke Properties. *ISIJ Int.* **2008**, *48*, 572–577. [CrossRef]
13. Flores, B.D.; Flores, I.V.; Guerrero, A.; Orellana, D.R.; Pohlmann, J.G.; Diez, M.A.; Borrego, A.G.; Osório, E.; Vilela, A.C.F. Effect of charcoal blending with a vitrinite rich coking coal on coke reactivity. *Fuel Processing Technol.* **2017**, *155*, 97–105. [CrossRef]
14. El-Tawil, A.A.; Björkman, B.; Lundgren, M.; Bäck, F.; Ökvist, L.S. Influence of Modified Bio-Coals on Carbonization and Bio-Coke Reactivity. *Metals* **2022**, *12*, 61. [CrossRef]
15. Bazaluk, O.; Kieush, L.; Koveria, A.; Schenk, J.; Pfeiffer, A.; Zheng, H.; Lozynskyi, V. Metallurgical Coke Production with Biomass Additives: Study of Biocoke Properties for Blast Furnace and Submerged Arc Furnace Purposes. *Materials* **2022**, *15*, 1147. [CrossRef]
16. Kieush, L.; Schenk, J.; Pfeiffer, A.; Koveria, A.; Rantitsch, G.; Hopfinger, H. Investigation on the influence of wood pellets on the reactivity of coke with CO_2 and its microstructure properties. *Fuel* **2022**, *309*, 122151. [CrossRef]
17. Xing, X.; Rogers, H.; Zhang, G.; Hockings, K.; Zulli, P.; Deev, A.; Mathieson, J.; Ostrovski, O. Effect of charcoal addition on the properties of a coke subjected to simulated blast furnace conditions. *Fuel Process. Technol.* **2017**, *157*, 42–51. [CrossRef]
18. Montiano, M.G.; Díaz-Faes, M.G.; Barriocanal, C. Partial briquetting vs direct addition of biomass in coking blends. *Fuel* **2014**, *137*, 313–320. [CrossRef]
19. Mochizuki, Y.; Naganuma, R.; Uebo, K.; Tsubouchi, N. Some factors influencing the fluidity of coal blends: Particle size, blend ratio and inherent oxygen species. *Fuel Process. Technol.* **2017**, *159*, 67–75. [CrossRef]
20. North, L.; Blackmore, K.; Nesbitt, K.; Mahoney, M.R. Models of coke quality prediction and the relationships to input variables: A review. *Fuel* **2018**, *219*, 446–466. [CrossRef]
21. Walker, A. The Carbon Texture of Metallurgical Coke and Its Bearing on Coke Quality Prediction. Doctoral Thesis, Loughborough University, Loughborough, UK, 2012. Available online: https://hdl.handle.net/2134/10950 (accessed on 7 September 2022).
22. Nomura, S.; Arima, T.; Kato, T. Coal blending theory for dry coal charging process. *Fuel* **2004**, *82*, 1771–1776. [CrossRef]

23. Montiano, M.G.; Barriocanal, C.; Alvarez, R. Effect of the addition of waste sawdust on thermoplastic properties of a coal. *Fuel* **2013**, *106*, 537–543. [CrossRef]
24. Montiano, M.G.; Diaz-Faes, E.; Barriocanal, C.; Alvarez, R. Influence of biomass on metallurgical coke quality. *Fuel* **2014**, *116*, 175–182. [CrossRef]
25. Kokonya, S.; Castro-Diaz, M.; Barriocanal, C.; Snape, C.E. An investigation into the effect of fast heating on fluidity development and coke quality for blends of coal and biomass. *Biomass Bioenergy* **2013**, *56*, 295–306. [CrossRef]
26. MacPhee, J.A.; Gransden, J.F.; Giroux, L.; Price, J.T. Possible CO_2 mitigation via addition of charcoal to coking coal blends. *Fuel Process. Technol.* **2009**, *20*, 16–20. [CrossRef]
27. Castro Diaz, M.; Zhao, H.; Kokonya, S.; Dufour, A.; Snape, C.E. The Effect of Biomass on Fluidity Development in Coking Blends Using High-Temperature SAOS Rheometry. *Energy Fuels* **2012**, *26*, 1767–1775. [CrossRef]
28. Fraga, M.; Flores, B.; Osório, E.; Vilela, A. Evaluation of the thermoplastic behavior of charcoal, coal tar and coking coal blends. *J. Mater. Res. Technol.* **2020**, *9*, 3406–3410. [CrossRef]
29. Solar, J.; Caballero, B.M.; Barriocanal, C.; Lopez-Urionabarrenechea, A.; Acha, E. Impact of the Addition of Pyrolysed Forestry-Waste to the Coking Process on the Resulting Green Biocoke. *Metals* **2021**, *11*, 613. [CrossRef]
30. Guerrero, A.; Diez, M.A.; Borrego, A.G. Influence of charcoal fines on the thermoplastic properties of coking coals and the optical properties of the semicoke. *Int. J. Coal Geol.* **2015**, *147–148*, 105–114. [CrossRef]
31. El-Tawil, A.A.; Björkman, B.; Lundgren, M.; Robles, A.; Sundqvist Ökvist, L. Influence of Bio-Coal Properties on Carbonization and Bio-Coke Reactivity. *Metals* **2021**, *11*, 1752. [CrossRef]
32. Wieruszewski, M.; Górna, A.; Mydlarz, K.; Adamowicz, K. Wood Biomass Resources in Poland Depending on Forest Structure and Industrial Processing of Wood Raw Material. *Energies* **2022**, *15*, 4897. [CrossRef]
33. Rejdak, M.; Czardybon, A.; Ignasiak, K.; Robak, J. Utilization of waste forest biomass: Pelletization studies of torrefied sawmill wood chips. *E3S Web Conf.* **2019**, *100*, 68. [CrossRef]
34. Sevket Guney, M. Utilization of hazelnut husk as biomass. *Sustain. Energy Technol. Assess.* **2013**, *4*, 72–77. [CrossRef]
35. Di Giacomo, G.; Romano, P. Evolution of the Olive Oil Industry along the Entire Production Chain and Related Waste Management. *Energies* **2022**, *15*, 465. [CrossRef]
36. Eurostat Data. Available online: https://ec.europa.eu/eurostat/en/data/database (accessed on 7 September 2022).
37. Ściążko, M.; Sobolewski, A. *Coke Quality Forecasting*; IChPW: Zabrze, Poland, 2015. (In Polish)
38. Phyllis 2 Database. Available online: https://phyllis.nl/ (accessed on 7 September 2022).
39. Díez, D.; Urueña, A.; Piñero, R.; Barrio, A.; Tamminen, T. Determination of Hemicellulose, Cellulose, and Lignin Content in Different Types of Biomasses by Thermogravimetric Analysis and Pseudocomponent Kinetic Model (TGA-PKM Method). *Processes* **2020**, *8*, 1048. [CrossRef]
40. Loison, R.; Foch, P.; Boyer, A. *Coke: Quality and Production*; Butterworths: London, UK, 1989.
41. Pusz, S.; Kwiecińska, B.; Koszorek, A.; Krzesińska, M.; Pilawa, B. Relationships between the optical reflectance of coal blends and the microscopic characteristics of their cokes. *Int. J. Coal Geol.* **2009**, *77*, 356–362. [CrossRef]
42. Flores, B.D.; Borrego, A.G.; Diez, M.A.; da Silva, G.L.R.; Zymla, V.; Vilela, A.C.F.; Osório, E. How coke optical texture became a relevant tool for understanding coal blending and coke quality. *Fuel Process. Technol.* **2017**, *164*, 13–23. [CrossRef]
43. Rejdak, M.; Strugała, A.; Sobolewski, A. Stamp-Charged Coke-Making Technology—The Effect of Charge Density and the Addition of Semi-Soft Coals on the Structural, Textural and Quality Parameters of Coke. *Energies* **2021**, *14*, 3401. [CrossRef]
44. Solomon, P.R.; Serio, M.A.; Girish, V.; Kroo, D.; Kroo, E. Cross-linking reactions during coal conversion. *Energy Fuels* **1990**, *4*, 42–54. [CrossRef]
45. Fernández, A.M.; Barriocanal, C.; Díez, M.A.; Alvarez, R. Importance of the textural characteristics of inert additives in the reduction of coal thermoplastic properties. *Fuel* **2010**, *89*, 3388. [CrossRef]
46. Diez, M.A.; Alvarez, R.; Barriocanal, C. Coal for metallurgical coke production: Predictions of coke quality and future requirements for cokemaking. *Int. J. Coal Geol.* **2002**, *50*, 389–412. [CrossRef]
47. Sakurovs, R. Some factors controlling the thermoplastic behaviour of coals. *Fuel* **2000**, *79*, 379–389. [CrossRef]
48. Smedowski, L.; Piechaczek, M. Impact of weathering on coal properties and evolution of coke quality described by optical and mechanical parameters. *Int. J. Coal Geol.* **2016**, *168*, 119–130. [CrossRef]
49. Florentino-Madiedo, L.; Casal, D.; Díaz-Faes, E.; Barriocanal, C. Effect of sawdust addition on coking pressure produced by two low vol bituminous coals. *J. Anal. Appl. Pyrolysis* **2017**, *127*, 369–376. [CrossRef]
50. Nomura, S.; Mahoney, M.; Fukuda, K.; Kato, K.; Le Bas, A.; McGuire, S. The mechanism of coking pressure generation I: Effect of high volatile coking coal, semi-anthracite and coke breeze on coking pressure and plastic coal layer permeability. *Fuel* **2010**, *89*, 1549–1556. [CrossRef]
51. Ueki, Y.; Nunome, Y.; Yoshiie, R.; Naruse, I.; Nishibata, Y. Effect of woody biomass addition on coke properties. *ISIJ Int.* **2014**, *54*, 2454–2460. [CrossRef]
52. Fu, P.; Hu, S.; Xiang, J.; Sun, L.; Zhang, A.; Yang, T.; Jiang, L. Evolution of pore structure of biomass particles during pyrolysis. *CIESC J.* **2009**, *60*, 1793–1799.

energies

MDPI

Article

Development of a Dual-Chamber Pyrolizer for Biochar Production from Agricultural Waste in Sri Lanka

W. A. M. A. N. Illankoon [1,*], Chiara Milanese [2], Anurudda Karunarathna Karunarathna [3], A. M. Y. W. Alahakoon [3], Puhulwella G. Rathnasiri [4], Maria Medina-Llamas [2,5], Maria Cristina Collivignarelli [6] and Sabrina Sorlini [1,*]

1 Department of Civil, Environmental, Architectural Engineering, and Mathematics (DICATAM), University of Brescia, Via Branze 43, 25123 Brescia, Italy
2 Department of Chemistry & Center for Colloid and Surface Science, University of Pavia, Viale Taramelli 16, 27100 Pavia, Italy
3 Department of Agricultural Engineering, Faculty of Agriculture, University of Peradeniya, Peradeniya 20400, Sri Lanka
4 Department of Chemical and Process Engineering, University of Moratuwa, Bandaranayake Mawatha, Moratuwa 10400, Sri Lanka
5 Unidad Académica Preparatoria, Plantel II, Universidad Autónoma de Zacatecas, Zacatecas 98068, Mexico
6 Department of Civil Engineering and Architecture, University of Pavia, Via Ferrata 3, 27100 Pavia, Italy
* Correspondence: a.wijepalaabeysi@unibs.it (W.A.M.A.N.I.); sabrina.sorlini@unibs.it (S.S.)

Abstract: This study investigates the design and development of a pyrolysis reactor for batch-type biochar production from rice husks. The main objective is to develop an appropriate technology to regulate pyrolysis temperature and biomass residence time that can be easily operated under field and household conditions with minimal operational and technical requirements. The designed novel dual-chamber reactor comprises two concentrical metal cylinders and a syngas circulation system. The outer cylinder is for energy generation and the inner one is for pyrolysis. Temperature profiles, energy exchanges, syngas production, and the physicochemical characteristics of biochar were obtained to determine the performance of the reactor. Different trials were carried out to obtain different pyrolysis temperatures under constant amounts of feedstock and fuel. The temperature was monitored continuously at three predetermined reactor heights, the temperature profile varied from 380 °C to 1000 °C. The biochar yield was 49% with an average production rate of 1.8 ± 0.2 kg h^{-1}. The reactor consumed 11 ± 0.1 kg of rice husk as feedstock and 6 ± 1 kg h^{-1} of wood as fuel. The gaseous products from the pyrolysis were CH_4, CO_2, H_2, CO, and C_nH_m, which contributed 23.3 ± 2.3 MJ m^{-3} of energy as fuel for the pyrolysis process. The specific surface area of the biochar was 182 m^2 g^{-1}. The achieved operational capacity and thermal efficiency of the reactor show biochar production is a suitable option to convert discarded biomass into a value-added product that can potentially be used in several environmental applications.

Keywords: rice husk; pyrolysis reactor; pyrolysis; energy; agricultural residues; biochar; Sri Lanka; decarbonization; green technologies

1. Introduction

Biochar production methods are widely used to enhance soil quality and address environmental problems [1–3]. Due to rapid urbanization and industrialization, large quantities of pollutants are released into freshwater bodies. When the concentration of these pollutants in drinking water or wastewater reaches a certain level, the water is no longer safe to drink or use for human purposes. At this stage, there is a need for an effective treatment method that addresses these environmental problems, that can be inexpensive and suitable for large-scale applications. In this case, biochar adsorption technology is considered a suitable option. Many scholars have found it to be an ideal method for removing various types of pollutants from water [4–10]. Biochar is a low-density carbon substance generated by the thermal degradation of biomass with or without oxygen [11–13], it has multiple mineral components and is distinguished by its porous structure, high specific surface area, and many surfaces functional groups [14,15]. Due to its physicochemical characteristics, biochar has recently attracted great attention for a wide range of applications (Figure 1) [16–19].

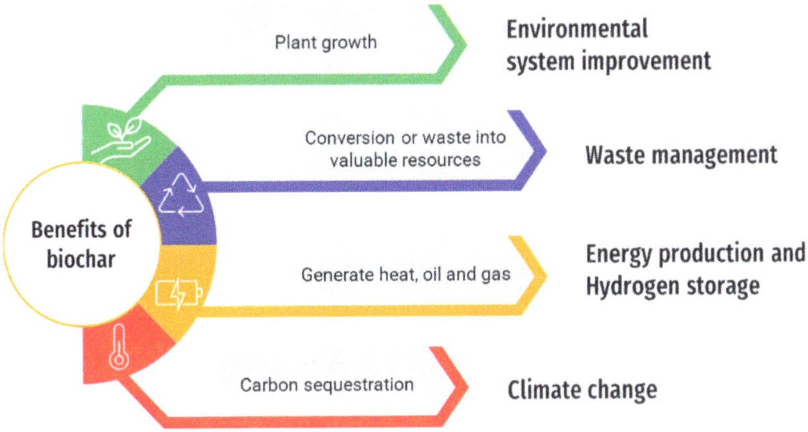

Figure 1. Utilization of biochar in various environmental sectors [16–19].

Biochar System for Global Smallholder Agriculture

A significant proportion of farms in the world are smallholding [20,21]. Small farms comprise the vast majority (562 million) of the world's agricultural activities. However, about 25% of the entire agricultural land is occupied by farms of typical size (average size is 1260 M ha) [20,22,23]. It is estimated that 2.6 billion individuals working on small farms provide food to meet more than 70% of the global demand [24]. To lessen reliance on commercial fertilizers and reduce food production costs, biochar produced from agricultural waste from smallholder farms might be a sustainable solution.

Large amounts of agricultural leftovers such as cereal straw and rice husk, bagasse, and tree limbs might be converted into resources rather than waste. Currently, many stakeholders engage in agricultural sectors practice the open burning of agricultural waste, leading to air pollution [25]. Reports on sustainable biochar production in South India found that it could increase the systemic resilience of farming-based communities in Karnataka; small-scale biochar systems might be beneficial for accomplishing this goal [26]. Therefore, we expect the results of this research will provide a suitable solution for agricultural waste, without capitalizing on its potential uses, as mentioned in Figure 1. Small-scale biochar production is claimed to be more cost-effective than industrial production [27–29]. Several types of small-scale batch-type and industrial scale continuous type biochar production technologies have been reported worldwide, such as earth pit kiln, earth mound kiln, Casamance earth mound kiln, metal kiln, brick kiln, Reichert kiln, retort kiln, multiple hearth reactors, screw type pyrolizers, paddle drum type reactors, and the carbonex kiln [29–32]. The usage of each of these reactors is accompanied by its own set of benefits and drawbacks, which are mentioned in Table 1. Although there are several initiatives to promote biochar production worldwide, conventional biochar production techniques at the small scale and industrial production methods must be improved. As a matter of fact, only a few research studies are available that describe the kiln's design with data on biochar's production and quality. The studies outlined that it is important to control the amount of oxygen (O_2) the biomass is exposed to in order to produce enough heat to carry out the process. Therefore, taking into account these aspects is crucial to produce high-quality biochar. Table 1 shows the challenges associated with pyrolyzing biomass at small and industrial scales.

Table 1. Comparison of various types of reactors for biochar production [29–32].

Production Type	Reactor Type	Advantages	Disadvantages
Small-scale batch-type processes	Earth pits and mounds, brick, concrete and double metal kilns, retorts	Portable, low-cost, and simple technological solution.	Inefficient, low yield, no heat recovery, hence, much of the feedstock is burned up. Pyrolysis gas and vapors are released into the air, polluting the environment. No control of the pyrolysis temperature, homogeneity of the final product. Some methods are unsafe and require high volumes of water to cool down biochar.
Industrial-scale continuous processes	Retort, multiple hearth reactors, screw type pyrolizers	Higher yields; feedstock flexibility; heat integration; possible co-generation of char and energy; technology that is easy to use and has been available for a long time; co-generation of char and energy; the unit that can be moved or be stationary (depending on size).	Systems that are more sophisticated and have higher costs than batch operations. There are no useful byproducts.
	Paddle drum type reactors	Feedstock flexibility; proven technology; integrated char and energy generation; available as a portable or fixed system (depending on size); greater yields, heat integration, and potential co-production of char and energy.	Complex system; higher cost than batch processes.

Our investigation is focused on rice husk, the most common agricultural waste from the rice sector in Sri Lanka and other Southeast Asian nations. Its production is increasing with growing paddy production patterns [25,33,34]. However, in many developing countries, direct burning is the most common approach to managing this agricultural waste, but it is not an environmentally friendly solution [25,33,35]. Although, during the parboiling process, the rice husk produced in the rice mills is used to generate steam [25]. Producing biochar is a sustainable alternative to direct burning, as described in Figure 1. Therefore, in this work, a dual-chamber pyrolizer was designed with the intention of promoting high-quality biochar production in small and medium-scale paddy farmers and householders due to its appropriateness as a technology that requires minimal specialized operations and it is easy to use. This research uses rice husk as a useful resource to produce biochar without disrupting the conventional parboiling process. This work analyzes the effect of using the produced syngas as an additional heat source during pyrolysis. The productivity and efficiency of the dual-chamber biochar furnace is evaluated by varying the O_2 flow. In addition, evaluate the physicochemical characteristics of the produced biochar under different O_2 inputs. Thereby, this study is designed to test the hypothesis that, "Performance efficiency of a simplified pyrolysis reactor can be improved by controlling the input volumetric air flowrate".

2. Materials and Methods

2.1. The Feedstocks

Rice husk was selected as the feedstock for biochar production, and Gliricidia (*Gliricidia sepium*) wood was chosen as the primary energy source for the pyrolizer. Rice husk was collected from medium-sized rice mills in the north-central province of Sri Lanka, air dried, and suitably stored till its use in the reactor.

2.2. Dual-Chamber Pyrolizer

The introduced dual-chamber pyrolizer is based on the basic principle of indirect heating of biomass with minimum oxygen content. The reactor was introduced as a small-scale and household-level biochar production technique because most of the existing small-scale reactors do not have the ability to control the operational parameters such as temperature and biomass retention time. The novelty of the idea is mainly its ability to control fuel: air ratio and the recirculation of the generated syngas within the reactor to control the pyrolysis temperature and biomass retention time. These modifications enable the production of biochar from different biomass types with different pyrolysis conditions. The dual-chamber biochar reactor was developed using an empty oil barrel as the outer chamber. The reactor consisted of six parts (Figures 2 and 3) and the specifications of each part of the reactor are described in Table 2.

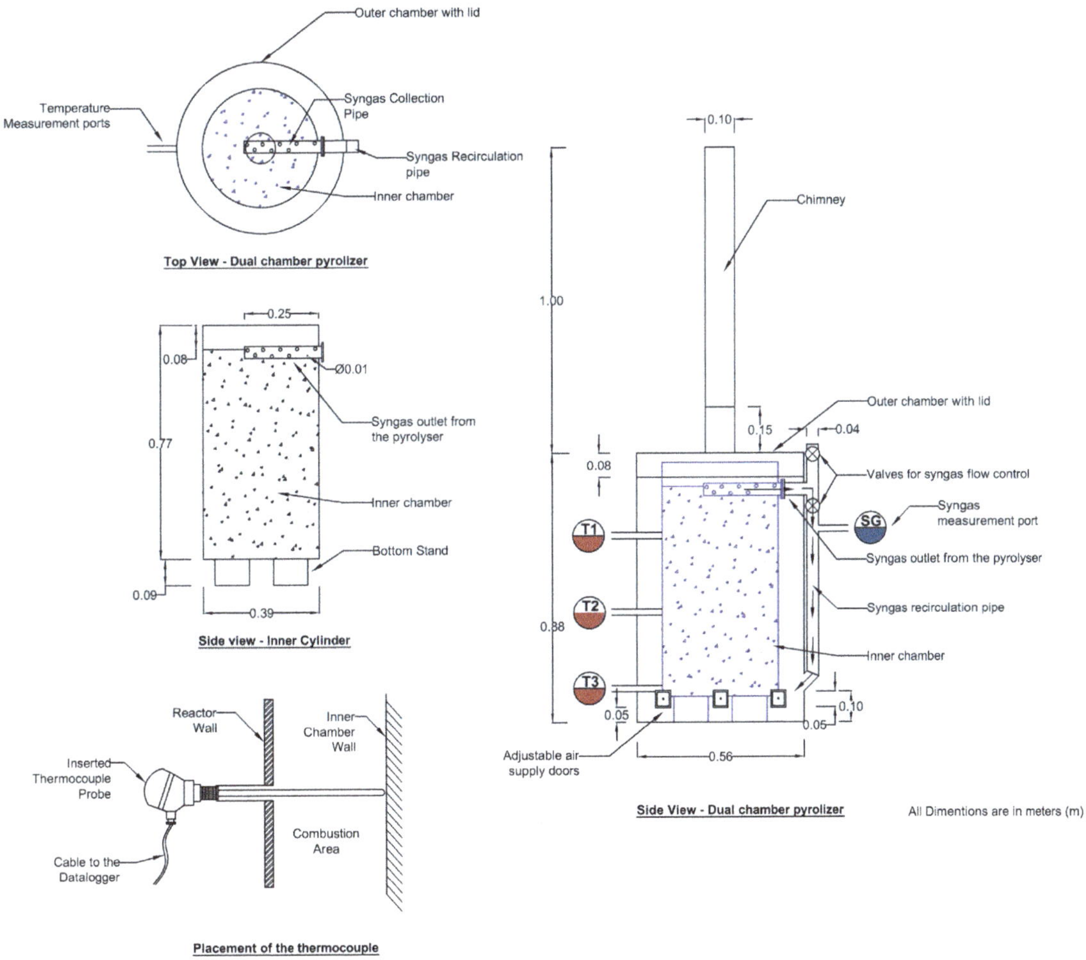

Figure 2. Schemes of the dual-chamber pyrolizer.

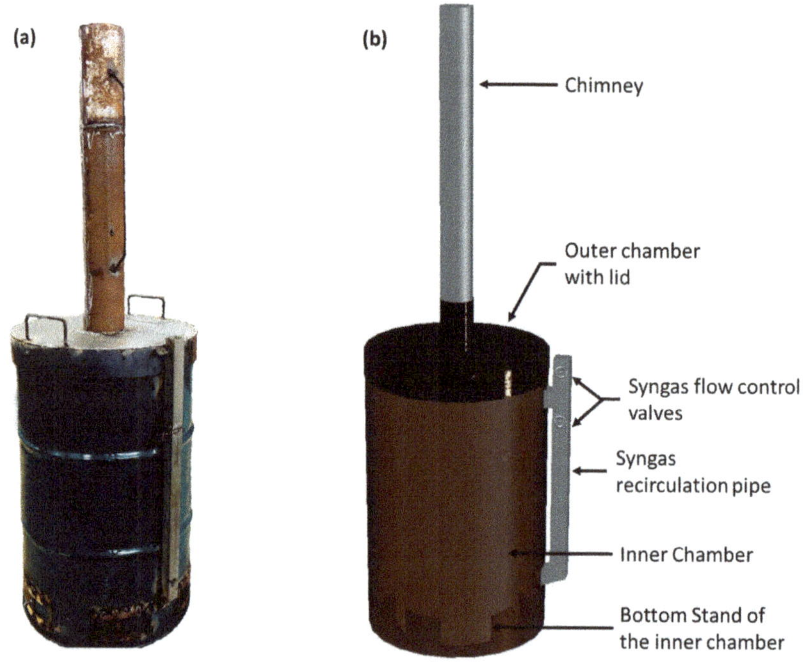

Figure 3. Image of the fabricated dual-chamber reactor: (**a**) actual reactor, (**b**) 3D model of the reactor.

Table 2. Specifications of the dual-chamber reactor.

System Component	Description	Dimensions (Diameter × Height)
Outer chamber	Material—Used oil barrel. Modifications were made in the oil barrel to achieve the design requirements.	0.56 m × 0.88 m
Inner chamber	Material—Gauge 14 (0.016 m) thick steel metal sheet. A 0.01 m high stand was fabricated at the bottom of the cylinder to facilitate a proper heat supply from the bottom of the reactor.	0.39 m × 0.76 m
Lid of the outer chamber	Material—Gauge 14 (0.016 m) thick steel metal sheet. The lid comprised an exhaust gas opening to connect the chimney.	0.56 m × 0.08 m
Lid of the inner chamber	Material—Gauge 14 (0.016 m) thick steel metal sheet. Tightly fixed to minimize oxygen supply.	0.39 m × 0.08 m
Chimney	Material—Gauge 14 (0.016 m) thick steel metal sheet. Fixed to the outer chamber lid to facilitate the updraft of the exhaust gas.	0.10 m × 1.0 m
Syngas circulation pipe	Material—1 mm thick galvanized square tube. Fabricated in two parts for easy dissembling during cleaning, 1 cm holes were drilled in the tube section inside the pyrolizer.	0.032 m × 0.032 m (tube cross section)

Figure 3 shows an image and 3D model of the fabricated dual-chamber reactor. Figure 4 shows the steps to set up the reactor. The inner chamber, with a 39 cm diameter, was designed to store rice husk and produce biochar by using the heat coming from the outer chamber. The bottom of the outer chamber was sealed, and the top was left open. The open top was used to insert the inner chamber into the outer chamber. Wood chips were used to generate heat by combustion. After inserting the inner chamber into the outer chamber, the syngas outlet valve from the inner chamber was connected to the syngas circulation pump mounted on the surface of the outer chamber (Figure 4a). Subsequently, the rice husk was fed into the inner chamber, which was closed with its lid. At the same time, wood chips were arranged in the outer chamber space between the inner and outer chamber (Figure 4b–d). Six air supply doors were designed on the bottom surface of the outer chamber. Four tests were conducted to analyze the temperature profiles and biochar characteristics. The air supply was controlled by adjusting the air supply doors during each experiment, and the syngas supply was controlled by using syngas flow control valves. Four different tests were performed:

- Trial 1: all air supply doors were fully opened, and syngas was supplied as fuel from the beginning of syngas generation.
- Trial 2: air supply for the combustion was reduced by half by controlling the size of the air supply doors, and syngas was supplied as fuel from the beginning of syngas generation.
- Trial 3: all air supply doors were opened, and the produced syngas was taken out from the reactor.
- Trial 4: all air supply doors were closed, and the produced syngas was taken out from the reactor.

4a: Syngas outlet valve that connects to the syngas circulation pump on the surface of the outer chamber.

4b: Rice husk and split wood feeding to the chambers.

4c: The inner chamber closed with its lid.

4d: The system before starting the process.

Figure 4. Dual-chamber pyrolizer start-up process.

2.3. *Performance Evaluation of Dual-Chamber Reactor*

The reactor's performance was assessed during each trial of batch operation by varying volumetric air flow rate and the syngas recirculation flow rate was kept constant. The temperature was monitored and mass balances were performed for both wood and rice husk. The reactor's performance was assessed during each batch operation by monitoring its temperature, mass balance, and volumetric flow rate. These findings aided in characterizing the reactor performance and in making comparisons among each trial. In addition, the behavior of the combustion chamber was analyzed by calculating the heat loss that occurred throughout the process. The basic energy balance for the reactor was performed by applying the first law of thermodynamics where energy can be neither created nor destroyed; thus, for any system, the following Equation (1) applies [36]. This helped to find out the theoretical heat loss during the batch operation, and thereby the efficiency of the reactor. The energy content of each biomass material was calculated using Equation (2) and the energy balance (Equation (1)) was used to calculate the theoretical heat losses:

$$E_{fs} + E_{fuel} = E_{biochar} + E_{sg} + E_{Exh} + E_{Losses} \quad (1)$$

$$Q = C \cdot M \quad (2)$$

where E_{fs} is the energy content of feedstock (MJ), E_{fuel} is the energy content of fuel wood (MJ), $E_{biochar}$ is the energy content of produced biochar (MJ), E_{sg} is the energy content of the produced syngas (MJ), E_{Exh} is the energy content of the exhaust gas produced from the combustion (MJ), and E_{Losses} is the heat losses from the reactor wall (MJ). In Equation (2), Q is the energy content of the biomass material (MJ), C is the heating value of the material (MJ/m^3), and M is the mass of the material (kg).

2.4. Quality of the Produced Biochar

The biochar produced from rice husk in the dual-chamber pyrolizer was analyzed physically, chemically, and proximately. Since the synthesized biochar was intended for use in various environmental applications. ASTM D3172-13, the standard procedure for proximate analysis of coal and coke, was used to describe the biochar recovered using the following Equation (3). The percentage recovery of the biomass was calculated on a wet basis by using Equation (3), where $W_{biochar}$ is the weight of biochar obtained per batch and $W_{feedstock}$ is the weight of wet feedstock biomass used per batch.

$$\% \text{ Recovery} = \frac{W_{biochar}}{W_{feedstock}} \cdot 100 \quad (3)$$

2.5. Characterization

Analyses were performed on the raw husk and on the pyrolyzed samples by various instrumental techniques. The thermal behavior of the samples was described using thermo-analytical methods. The thermogravimetric curves were acquired by a thermogravimetric analyzer (TGA Q5000, TA Instruments Inc., Lindon, UT, USA) interfaced with a TA5000 data station by heating about 5 mg of powder in a Pt crucible under N_2 flux (50 mL min^{-1}) from 25 °C to 1000 °C at 10 °C/min. Differential scanning calorimetry (DSC) was performed by a Q2000 apparatus (TA Instruments, New Castle, DE, USA) interfaced with a TA5000 data station by heating about 3 mg of powder in an open aluminum crucible from −50 °C to 350 °C and then cooling down to −50 °C (heating and cooling rate = 5 °C min^{-1}) under nitrogen flux (50 mL min^{-1}). Three independent measurements were taken on each sample. The temperature accuracy of the instrument is ±0.1 °C, the precision is ±0.01 °C, and the calorimetric reproducibility is ±0.05%. DSC data were analyzed by the Universal Analysis software by TA Instruments. X-ray diffraction patterns of each sample were obtained (Bruker D2 diffractometer, Ettlingen, Germany) using a zero-background sample holder in the 2θ range 5°–50° with CuKα radiation, step scan of 0.02° and a counting time of 2 s/step at 40 KV and 40 mA. The structural properties of the samples were analyzed at room temperature using the Nicolet FTIR iS10 spectrometer (Nicolet, Madison, WI, USA) equipped with Smart iTR with a diamond plate. The biochar was dried at 45 °C before analysis. Thirty-two scans in the 4000–600 cm^{-1} range at 4 cm^{-1} resolution were coadded. The elemental compositions of the pyrolyzed samples were analyzed by energy dispersive X-ray spectrometry (EDX) using a Zeiss EVO MA10 (Carl Zeiss, Oberkochen, Germany) scanning electron microscope equipped with an Oxford Xmax 50 mm^2 detector following the standard procedure (ASTM E 1508). The tests were carried out at a working distance of 8.5 mm and a voltage of 20 kV for electron generation in an ultra-high vacuum environment.

3. Results and Discussion

3.1. Characteristics of Feedstock

The heating of the reactor walls and system components caused an initial high thermal inertia for the reactor. Therefore, energy had to be supplied at the commencement from the heat source in a quantity larger than required to maintain a constant temperature. Due to this, a small quantity (less than 1 kg) of coconut shells, which is a high-energy biomass source, was added to *Gliricidia* wood as an additional energy source at the pyrolizer's combustion area at the beginning of the process. The energetic and compositional parameters of raw rice husk, *Gliricidia* wood, and coconut are shown in Table 3.

Table 3. Characterization of the feedstocks [34,37–49].

Parameters	Feedstocks		
	Rice Husk	*Gliricidia* Wood	Coconut Shells
Moisture content (%)	7.7 [a]	62.3 [b] [42]	8.14–10.53 [b] [37,39]
Specific gravity (kg/m^3)	140 [a]	670 [b] [42]	1738 [b] [47]
Calorific value (MJ/kg)	12. 85–14.02 [b] [44,49]	19.0–20.5 [b] [42,43,48]	34.1 [b] [40]
Fuelwood value index	-	1255 [b] [42]	-
Lignin content (%)	20–25 [b] [45,46]	26.26–26.8 [b] [41,42]	27 [b] [38]
Volatile matter (%)	67.7 [a]	73.1–82.9 [b] [41,43]	67.7–78.3 [b] [37,39]
Fixed carbon (%)	17.0 [a]	10.3 [b] [43]	17.6–20.96 [b] [37,39]
Ash (%)	15.2 [a]	1.11–1.28 [b] [41,43]	0.74–0.73 [b] [39]

[a] Results from laboratory tests conducted in this work. [b] Results obtained from the literature.

3.2. Mass Balance per Batch

Table 4 shows the input and output data for the reactor operational conditions. For each trial, an approximately constant amount of rice husk (about 11 kg) and wood as fuel was used (about 28 kg). The yield of biochar production varied from 42.37% to 49.28%. However, bio-oil was not extracted throughout the process since it was not intended to be a byproduct. As mentioned above, coconut shell (calorific value 8143 kcal/g) was selected as a high-energy supplementary biomass source; since it has high flammability and heat retention capacity. As the amount of coconut shells (supplied in the form of small pieces) is less than 1 kg for each experiment, it was not considered in the calculations.

Table 4. Mass of reactants and products in the reactor and biochar yield (input and output data).

Trial	Wood		Rice Husk		
	Weight of Wood (kg)	Weight of Ash Residue (kg)	Weight of Rice Husk (kg)	Weight of Biochar (kg)	Biochar Yield (%)
1	28.6	0.5	10.8	4.7	43.52
2	28.1	1.3	11	4.96	45.05
3	28.2	1.13	11	4.66	42.37
4	28	1.58	11	5.42	49.28

Figure 5a shows the TGA curve of rice husk, which can be divided into four distinct mass losses. Between 45 °C and 120 °C, adsorbed water (about 5% in mass) is released. Next, a weight loss of 16.8% occurs between 150 °C and 310 °C due to the evaporation of the fixed water content. Carbonaceous components such as lignin and cellulose are destroyed between 310 °C and 400 °C, resulting in a mass loss of 36.30%. Additional elimination of various carbonaceous components leads to the last stage of weight loss (10.52%). The total mass loss at 1000 °C is equal to 68.95%. In the dual-chamber reactor, experiments are performed with a higher mass of rice husk 11 kg, compared to 5 mg in the TGA analysis, the mass loss after each trial is 56.48% (Trial 1), 54.95% (Trial 2), 57.63% (Trial 3), and 50.72% (Trial 4). Figure 5b shows the calorimetric profile of rice husk obtained by DSC. The analysis shows the presence of the two endothermic events due to adsorbed water release and the first decomposition step starting at around 150 °C is confirmed by the TGA analysis.

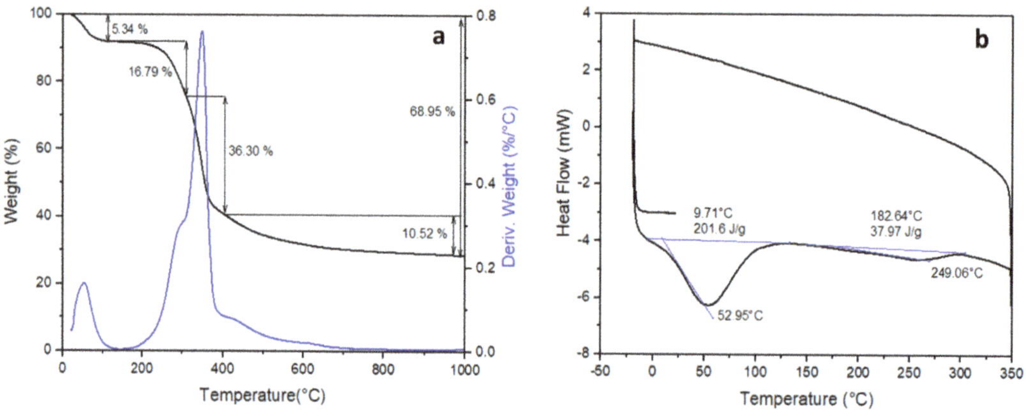

Figure 5. TGA and DTG curves (**a**) and calorimetric profile (**b**) for rice husk (exo up).

3.3. Temperature Profiles

Three temperature measurement ports were fabricated in the outer cylinder wall at a 25 cm distance each. The bottom temperature port was located 10 cm from the bottom edge (Figure 2). During the pyrolysis three continuous temperature profiles of the outer surface of the inner chamber were acquired for each trial, the results are shown in Figure 6. The results show noticeable temperature variations in the reactor's lower zone (T3) compared to the top zones (T1). When the fire front moves down, the reactor employs thermochemical conversion reactions in stratified layers inside the pyrolizer. Reaction completion time changes by varying airflow (O_2) and syngas supply: the times taken to complete the reaction were 176 min (Trial 1), 209 min (Trial 2), 241 min (Trial 3), and 401 min (Trial 4). Trial 1 showed the lowest reaction time; the average temperature varied from 400 °C to 800 °C. According to the data of Trial 4, a long time was needed to complete the reaction when all air supply doors were closed, and syngas was not supplied. Furthermore, the average temperature of T1 and T2 varied between 400 °C and 600 °C. The T3 temperature range was wider than T1 and T2 because the burning charcoals went down to the bottom and induce a temperature rise. It is to underline that the suggested method is more suitable for producing biochar in a lower temperature range but, in this condition, it takes a longer time to complete the reaction, which is a disadvantage. Trials 2 and 3 showed no significant temperature difference or difference in the reaction completion time.

When the rice husk started to become heated, sequential thermochemical reactions started. Rice husk drying takes place initially, in the temperature range of 20–180 °C. The dried rice husk is then subjected to a temperature of 180–250 °C which helps for the torrefaction reactions. Then exothermic pyrolysis takes place at stable temperatures around 250–400 °C. The reactor was designed with the intention of pyrolyzing rice husk at a specific temperature range by varying the airflow (O_2) and syngas supply. However, the temperature profiles fluctuate with the wood-burning rate in the outer chamber. The thermal inertia of the internal pyrolysis cylinder worked as a heat reservoir; thus, temperature changes in the combustion zone did not affect the temperature of the inner pyrolysis cylinder. Therefore, when the temperature in the combustion region decreases, the exothermic reactions may disperse the heat to the biomass, allowing the pyrolysis process to continue [34,50].

Thermochemically, the pyrolysis process is subjected to both endothermic and exothermic reactions, depending on the operating temperature. The exothermic processes of pyrolysis create enough energy to balance out the endothermic reactions [51–53]. Said et al. [54] performed TGA and DSC tests on rice husk (moisture 9%) and found that the devolatilization process would provide 4.437 kJ/g of energy through fragmentation, reforming, cracking polymerization, and dehydration. This energy might be taken to utilize in pyrolysis' endothermic processes.

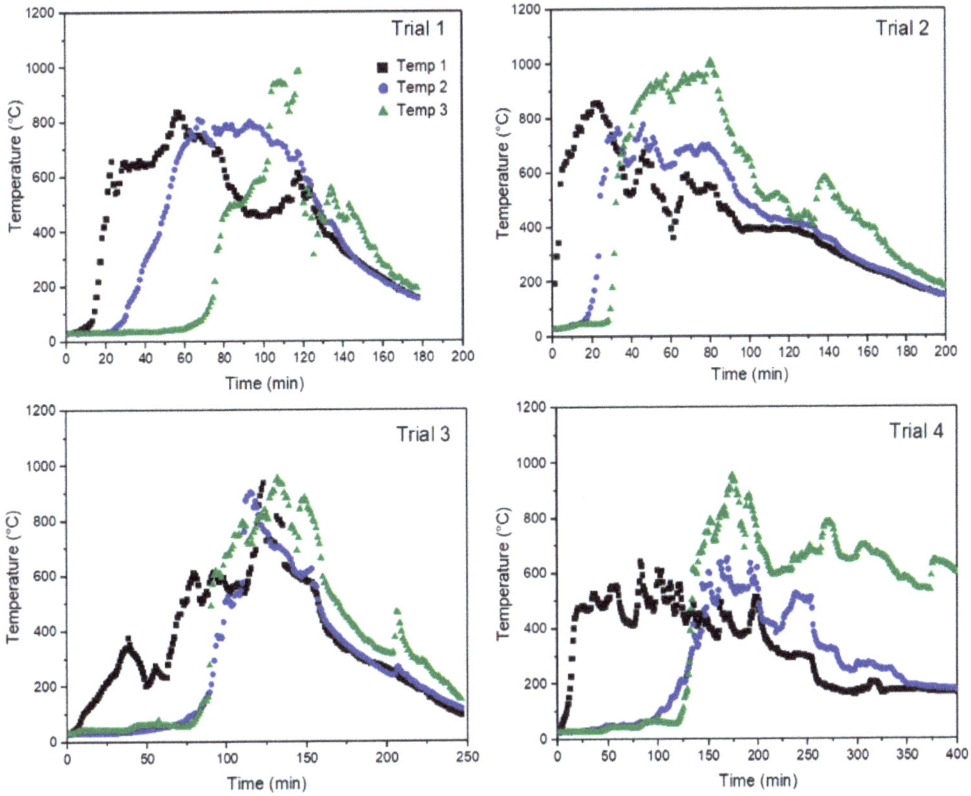

Figure 6. Temperature profiles for the three thermocouples placed in the dual-chamber pyrolizer at different trials.

3.4. Syngas Composition and Energy Value

Figure 7 depicts the chemical composition of the syngas produced by pyrolyzing rice husks and their higher heating values (HHV). Two syngas samples were collected at constant intervals for each trial. The main gaseous products obtained from rice husk pyrolysis are CH_4 (up to 14.29%), CO_2 (up to 33.06%), H_2 (up to 26.38%), CO (up to 25.53%), and C_nH_m (up to 0.94%). From Figure 7, the sequence of the gases generated is essentially consistent across all samples, albeit Trial 2 and Trail 3 exhibit a higher release of syngas and have a maximum HHV.

The catalytic effect of the minerals in pyrolyzed rice husk (ash concentration of about 15.2%) may account for the high hydrogen (H_2) level in the syngas mixture. The literature reports found that the percentage of hydrogen (H_2) released during the pyrolysis of rice husks is proportional to the amount of ash in the husk, suggesting that the ash has a catalytic influence during thermal cracking and, more specifically, in the dehydrogenation process, which in turn increases H_2 generation [55–57]. In addition, oxidized group breakdowns such as decarboxylation and decarbonylating may produce CO and CO_2 [58].

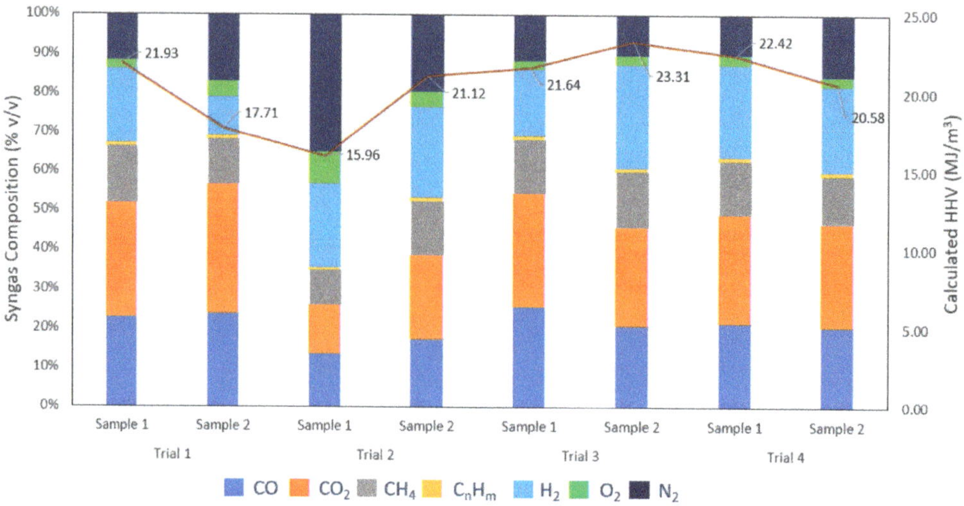

Figure 7. Chemical composition of the syngas and HHV.

Since only H_2, CO, and CH_4 are combustible, the HHV of syngas is the calorific value of these three gases. A mass of 1 kg of biomass is expected to provide 2 m^3 syngas [59–61], hence Equation (4) [59] was used to calculate the HHV of the generated gas combination using the information on fuel HHV and syngas composition:

$$\Delta H\ (2\ m^3\ \text{syngas}) = (HHV_{H2} \times 2\ m^3 \times H_2\%) + (HHV_{CO} \times 2\ m^3 \times CO\%) + (HHV_{CH4} \times 2\ m^3 \times CH_4\%) \quad (4)$$

where HHV_{H2}, HHV_{CO}, and HHV_{CH4} are standard HHV for: H_2 = 12.76 MJ m^{-3}; CO = 12.63 MJ m^{-3}; and CH_4 = 39.76 MJ m^{-3} [62,63]. This investigation achieves a maximum of 23.3 MJ m^{-3} across all pyrolysis settings, suggesting a rich supply of combustible gases. The gas mixture's relatively high heating value could provide supplementary heat for the pyrolysis reactor. The calculated heat capacities of gases fall within the range of those reported in prior research. For example, pyrolysis of solar-dried sewage sludge resulted in gases with an HHV of around 25 MJ m^{-3}, according to reference [64].

3.5. Heat Balance of the Reactor

According to Equation (2), the energy contents of each input and output of the process were calculated and theoretical heat loss for the process was calculated using Equation (1). Table 5 shows the different energy values for the feedstock, fuel, and products yielded from each trial. By using the data, a simple energy balance was conducted. The calculated heat losses are 77.61% (Trial 1), 77.98% (Trial 2), 75.15% (Trial 3), and 74.41% (Trial 4). The results revealed that there is a significant heat loss from the reactor surface due to the absence of an insulation layer. The study suggests the construction of a simple insulation layer such as a brick or clay wall.

3.6. Characteristics of Produced Biochar

3.6.1. Elemental Composition

Biochar's' qualities may be determined from their analytic constituents and H/C and O/C ratios [65,66]. Carbon (C), oxygen (O), silicon (Si), aluminum (Al), potassium (K), and chlorine (Cl) were detected throughout an EDX analysis for each trial (Figure 8). It is to be considered that H is not visible by this technique. As the temperature rises, hydrogen and oxygen are released at a far greater rate than carbon [65]. Based on that, it can be concluded that the waste in Trial 1 has been subjected to a higher temperature than other trials, as it has lower O content than the other three tests. Thermally-induced CH_4 dehydration is an indication of a shift in biochar recalcitrance [67,68].

Table 5. Energy values for each biomass material in the pyrolysis process for the different trials.

	Trial	Rice Husk	*Gliricidia* Wood	Rice Husk Biochar	Rice Husk Biochar	Exhaust Gas (m^3) *
Mass (kg)	1	10.8	28.6	4.7	30.12	4.88
	2	11.0	28.1	4.96	26.8	4.83
	3	11.0	28.2	4.66	26.97	5.07
	4	11.0	28	5.42	26.42	5.07
Calorific value (MJ/kg)	1	12.85	20.5	12.35	0.27	19.82
	2	12.85	20.5	12.35	0.27	18.53
	3	12.85	20.5	12.35	0.27	22.47
	4	12.85	20.5	12.35	0.27	21.49
Energy content (MJ/batch)	1	138.78	586.3	58.05	8.13	96.72
	2	141.35	576.05	61.26	7.24	89.5
	3	141.35	578.1	57.55	7.28	113.92
	4	141.35	574	66.94	7.13	108.95

* For the exhaust gas, the tabulated values express the volume (m^3) and the calorific value is expressed as MJ m^{-3}.

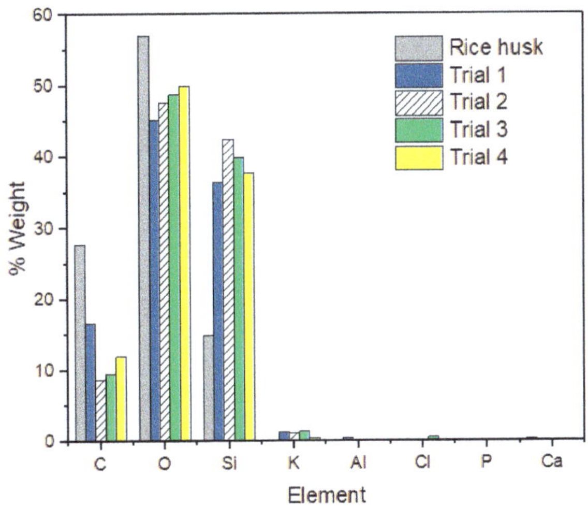

Figure 8. Elemental analysis of the produced biochar.

Furthermore, a biomass material typically contains both labile and recalcitrant O fractions, with the former lost after first heating and preserved in the char of the final product [69]. Charring occurs at high temperatures, leading to dehydration and decarboxylation processes, which reduces the O/C ratios. The O/C ratios in each experiment varied in the following sequence: 5.52 (Trial 2) > 5.15 (Trial 3) > 4.18 (Trial 4) > 2.71 (Trial 1).

To the same extent, a reduced C/O ratio at higher temperatures implies a structural arrangement of the aromatic rings [70], which create highly stable crystal graphite-like structures [71,72]. Since biochar from Trial 2 has a lower C/O ratio (0.18), it is expected to have a stronger graphite-like structure than other samples. Compared to biochar generated from agricultural wastes at lower temperatures, biochar generated from wood at higher temperatures is less biologically labile because it includes a proportionally greater quantity of aromatic-organic matter [73,74]. According to some researchers, the presence of a higher amount of silicon in rice husk biochar creates a dense carbon structure with Si-encapsulated carbon [75]. Figure 8 displays a higher Si concentration in the Trial 2 sample; therefore, Trial 2 biochar may have a dense carbon structure than the other three samples.

3.6.2. Thermal Analysis (TGA and DSC) and Surface Area Results

Thermal analysis is an effective tool for investigating biochar's composition and structure [76,77]. To assess the relative amount of moisture, volatile matter, ash, and fixed carbon in pyrolyzed samples, thermogravimetric analysis was performed. Figure 9 shows the TGA analysis of the biochar obtained in the four trials. At a first glance, it is evident that the TGA profiles of the obtained biochar (Figure 9) are strongly different from the one of the raw rice husk (Figure 5), which is more structured and well-defined, with steps occurring in a sharper way in smaller temperature ranges. First, all the biochar samples followed a similar thermal degradation pattern, with a first step of humidity release (between 4 and 5% for each sample) ending at around 120 °C followed by a huge decomposition step taking place from 150 °C to 1000 °C.

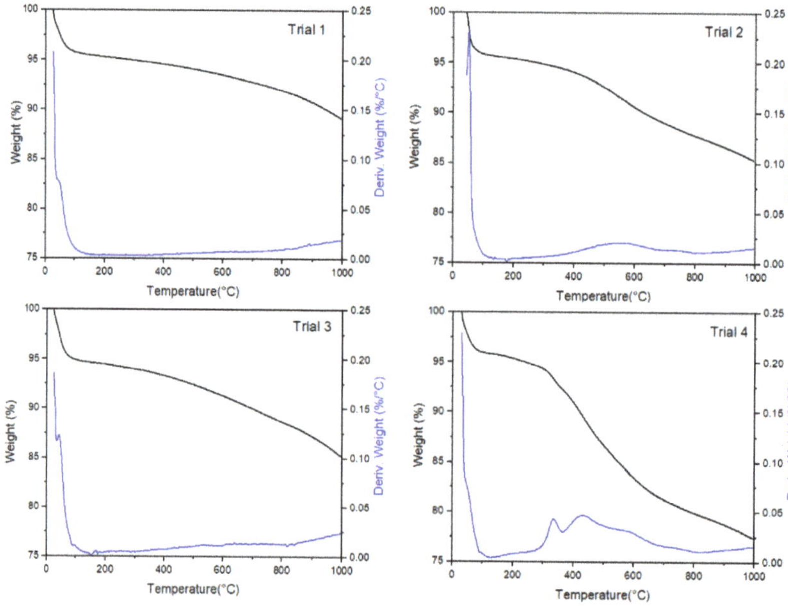

Figure 9. TGA and DTG curves for biochar samples obtained in each trial.

However, some differences can be noticed by observing the shape of this last mass loss step and the evolution of the derivative curve DTG (blue line). In particular, for Trial 1 the degradation rate is slow up to around 600°C, and then gradually increases, probably due to the decomposition of some residual carbonaceous materials such as lignin, cellulose, and hemicellulose [25,78,79]. A similar trend, even if with a higher volatile release is evident for the Trial 3 sample (6% wt. for Trial 1 and 9% wt. for Trial 3). On the contrary, for biochar obtained from Trials 2 and 4, the decomposition is more structured. The DTG curve shows at least two subsequent decomposition processes between 250 °C and 650 °C for Trial 2 and at least three for Trial 4, for which the evolution of processes with different rates is evident also from the TGA curve. From 650 °C to 1000 °C, two other steps of mass release are evident. The mass loss under these steps is 9.6% wt. for Trial 2 and up to 18%, for Trial 4. Therefore, it can be concluded that Trial 4 biochar contains more residual carbonaceous components than the other samples, followed by Trial 2: this points out that for these samples the pyrolysis reaction in the reactor has proceeded to a lower extent.

The release of adsorbed water at a temperature lower than 100 °C and the beginning of a huge and slow decomposition process starting from 150 °C is evident also in the calorimetric profiles (Figure 10), showing a clear endothermic peak ending at around 100 °C and an increasing endothermic variation from the baseline starting at around 150 °C and more evident in Trial 4 curve. Furthermore, in this case, the profile is different from that of the raw husk (Figure 5b), since the amount of adsorbed water is lower in the pyrolyzed samples. Compare an enthalpy of 200 J/g for the raw material vs. a maximum value of 70 J/g for Trial 2 and 4 and the decomposition steps are slower and less evident (the second peak present in the calorimetric profile of raw husk is a gradual detachment from the baseline in these samples). The lower degradation degree in Trial 4 is evident in its lower surface area as obtained by BET analysis: the surface area of raw rice husk is as low as 2.8 m^2/g and it increases to 182 m^2/g for Trial 1, 141 m^2/g for Trial 2, 168 m^2/g for Trial 3, and 132 m^2/g for Trial 4.

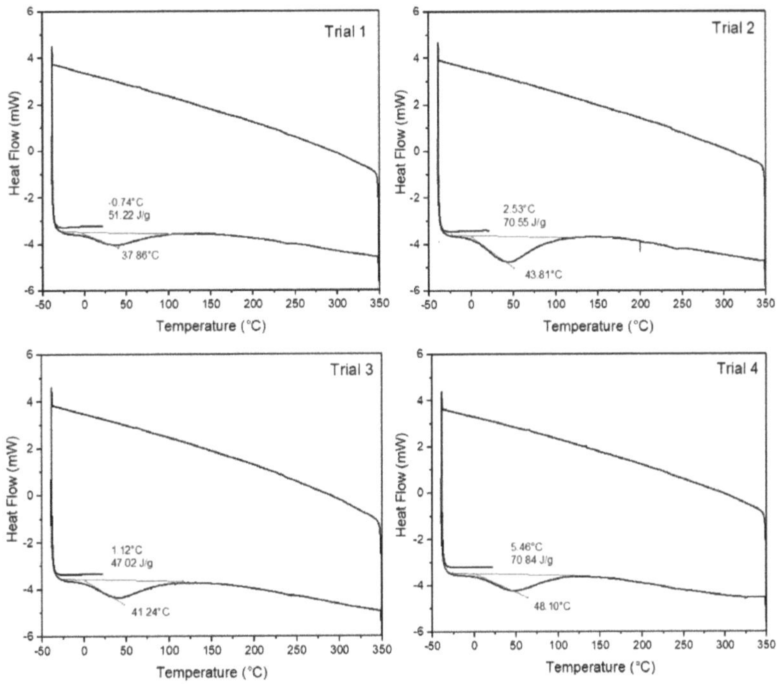

Figure 10. DSC profiles of the pyrolyzed samples (exo up).

3.6.3. Fourier Transform Infrared (FTIR) Spectroscopy

To detect the variation in chemical composition, FT-IR spectroscopy is an effective tool (Figure 11). The C-H (750–900 cm^{-1}), C=C (1380–1450 cm^{-1}) and C-O stretching (1580–1700 cm^{-1}) signals can be seen in all samples, and C=C bonds in the Trial 4 sample caused the most prominent representative peaks for aromatic carbon. Pyrolyzing temperature alters the functional group, as the FTIR spectrum shows a reduction in aliphatic C groups and an increase in aromatic C [80].

According to the literature, the strength of bands such as those corresponding to hydroxyl groups (3200–3400 cm^{-1}) and aromatic groups (1580–1600 and 3050–3000 cm^{-1}) decreases, when the pyrolyzing temperature range is between 700 and 800 °C, while biochar obtained at lower temperatures (300 and 500 °C) include a more significant number of bonds indicating functional groups than those made at 700 °C [81]. Figure 8 and the cited works support the conclusion that all feedstocks were heated to temperatures higher than 700 °C during the biochar synthesis process.

Bands around 800 cm^{-1} and 1040–1100 cm^{-1} were attributed to SiO_2, reflecting the feedstock's composition. These spectral lines were detected in all the trials. Silica, an essential component of phytoliths, prevents carbon in plants from degrading, according to the field of plant physiology [82,83]. Among the many chemical components of rice, silicon dioxide plays a significant role. The shoulder seen in the FTIR spectrum about 1600 cm^{-1}, which was attributed to aromatic compounds, is still present at temperatures as high as 800 °C during biochar synthesis.

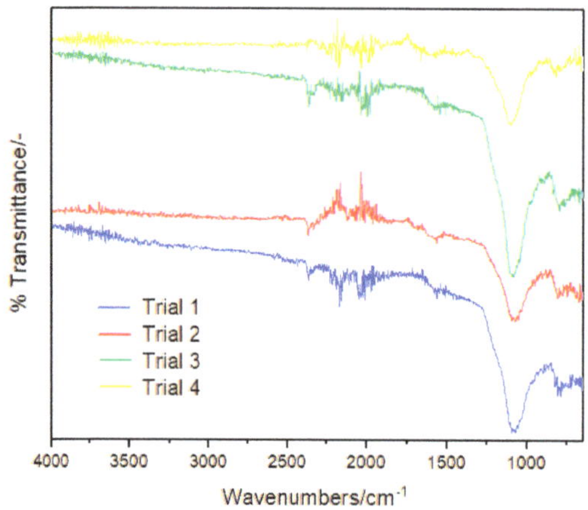

Figure 11. ATR-FTIR spectra of pyrolyzed samples.

3.6.4. XRD

Figure 12 shows the XRD patterns for rice husk (before pyrolysis) and biochar pyrolyzed at the different trial conditions. The XRD patterns of rice husk show the distinctive peaks at 22.4°, 34.1°, and 47.8° that can be assigned to the presence of silica (SiO_2) from rice husk (JCPDS 33-1161). The wide diffraction peaks of the cellulose polymorphs are located around 15°, 17°, and 20°: these wide peaks can be observed only in the rice husk [84]. After pyrolysis, only the peak at 22° due to the C presence is evident and it becomes broader due to the amorphous characteristics of biochar.

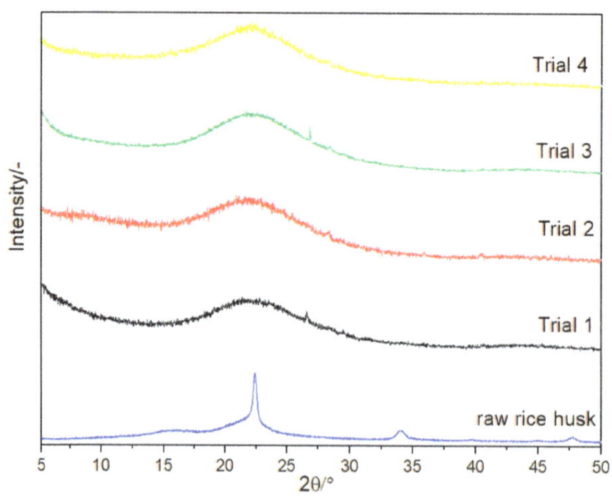

Figure 12. XRD patterns of the rice husk and the pyrolyzed samples.

3.7. Challenges and Future Perspectives

According to the results obtained from four trials, the weight of the biochar produced is nearly the same. However, in Trial 3, the energy content of syngas is higher compared to all other trials. If a farmer uses this syngas to produce heat for cooking or other purposes,

Trial 3 would be beneficial. The main challenge of introducing a new technology is the knowledge dissemination process. There are several circumstances in introducing new technologies to the farmer level with the very low adaptability of farmers to these new technologies. Furthermore, the demand for biomass sources for other purposes such as combustion, animal feed, and incineration also makes it competitive to promote biochar production for agricultural wastes.

The central notion of this investigation is that bio-carbon may be used to create a sustainable, circular bioeconomy, which presents a practical response to the system's dominant administration. Thermochemical technologies employed in biochar synthesis, particularly in rural regions, will aid regional development by allowing small and medium-scale firms to promote industrial symbiosis, boosting farmer income and facilitating waste management in the agricultural sector. Thus, with this method, small-scale production systems may be linked with other systems to create closed system models, where waste from one process can be used as an input for another, resulting in positive social, economic, and ecological consequences in a circular economy. The application of the main product in different environmental applications is also another important aspect to be addressed. Biochar can be used as an adsorbent to remove color and different pollutants in wastewater treatment. In addition, depending on characteristics, heavy metal can also be removed. However, it has to be contained in packed bed columns or other appropriate reactor configurations.

4. Conclusions

The dual-chamber pyrolizer reactor developed in this work was designed to valorize rice industry waste products in rice-growing regions for small-scale applications and batch-type operations. The design specifically has the ability to control the fuel-to-air ratio for combustion by controlling the airflow, as well as it has the ability to utilize produced syngas as needed. With the careful control of airflow and syngas flow, the reactor's capacity to work at temperatures ranging from 400 to 1000 °C was observed. Biochar yielding percentage and quality proved that the dual-chamber reactor can produce biochar from rice husk in the above temperature range at varying biomass residence times (from 200 min up to 400 min). Furthermore, the varying resident times influence the chemical composition of the final product, resulting in a broader range of product quality with varying chemical characteristics. The main gaseous products obtained from rice husk pyrolysis were CH_4 (up to 14.27 ± 0.08%), CO_2 (up to 33.06 ± 2.67%), H_2 (up to 26.38 ± 0.75%), CO (up to 25.53 ± 0.45%), and C_nH_m (up to 0.94 ± 0.11%), that also contributed with 23.3 ± 2.3 MJm^{-3} of energy as a fuel for the pyrolysis process. Recirculating the generated syngas as an energy source for the reactor can minimize the fuel requirement for the process while reducing harmful emissions to the atmosphere. With the above reactor characteristics, small-scale biochar manufacturers will have many advantages from this reactor over other conventional methods, where they can produce biochar at a predetermined temperature and residence time combination. Furthermore, the use of simple and banal fabrication materials such as oil barrels and steel barrels would make it more implementable within the local community. Thereby, the local community can also manipulate the pyrolysis process parameters without in-depth technical knowledge of the pyrolysis process. In the south Asian region, the main use of biochar is for soil application, thus the physiochemical and surface characteristics of produced biochar proved its suitability in soil application. Rice husk biochar produced from this reactor is already being utilized to make a composite fertilizer for various crops, including rice. Additionally, the adaptability of the reactor to different biomass types also requires further investigation. Furthermore, it is necessary to evaluate reactor performance by adding a thermal insulation layer to the outer chamber using locally available materials.

Author Contributions: Conceptualization, W.A.M.A.N.I. and S.S.; methodology, W.A.M.A.N.I., A.M.Y.W.A., A.K.K., P.G.R. and S.S.; software, W.A.M.A.N.I., C.M., A.M.Y.W.A. and M.M.-L.; validation, W.A.M.A.N.I., A.M.Y.W.A., A.K.K. and C.M.; formal analysis, W.A.M.A.N.I.; investigation, W.A.M.A.N.I., A.M.Y.W.A. and A.K.K.; data curation, W.A.M.A.N.I. and A.M.Y.W.A.; writing—original draft preparation, W.A.M.A.N.I.; writing—review and editing, W.A.M.A.N.I., C.M., A.M.Y.W.A., A.K.K., P.G.R., M.M.-L., M.C.C. and S.S.; supervision, C.M., A.K.K., P.G.R., M.C.C. and S.S. All authors have read and agreed to the published version of the manuscript.

Funding: This research received no external funding.

Data Availability Statement: The data presented in this study are available on request from the corresponding author.

Acknowledgments: The authors are grateful to the Department of Agricultural Engineering in the University of Peradeniya, Sri Lanka, and the Postgraduate Institute of Agriculture, University of Peradeniya, Sri Lanka for providing the needful facilities.

Conflicts of Interest: The authors declare no conflict of interest.

References

1. Lehmann, J.; Gaunt, J.; Rondon, M. Bio-Char Sequestration in Terrestrial Ecosystems—A Review. *Mitig. Adapt. Strat. Glob. Chang.* **2006**, *11*, 403–427. [CrossRef]
2. Wu, P.; Ata-Ul-Karim, S.T.; Singh, B.P.; Wang, H.; Wu, T.; Liu, C.; Fang, G.; Zhou, D.; Wang, Y.; Chen, W. A Scientometric Review of Biochar Research in the Past 20 Years (1998–2018). *Biochar* **2019**, *1*, 23–43. [CrossRef]
3. Lehmann, J.; Bossio, D.A.; Kögel-Knabner, I.; Rillig, M.C. The Concept and Future Prospects of Soil Health. *Nat. Rev. Earth Environ.* **2020**, *1*, 544–553. [CrossRef]
4. Xu, Y.; Chen, J.; Chen, R.; Yu, P.; Guo, S.; Wang, X. Adsorption and Reduction of Chromium(VI) from Aqueous Solution Using Polypyrrole/Calcium Rectorite Composite Adsorbent. *Water Res.* **2019**, *160*, 148–157. [CrossRef]
5. Khataee, A.R.; Kasiri, M.B. Photocatalytic Degradation of Organic Dyes in the Presence of Nanostructured Titanium Dioxide: Influence of the Chemical Structure of Dyes. *J. Mol. Catal. A Chem.* **2010**, *328*, 8–26. [CrossRef]
6. Haddad, M.Y.; Alharbi, H.F. Enhancement of Heavy Metal Ion Adsorption Using Electrospun Polyacrylonitrile Nanofibers Loaded with ZnO Nanoparticles. *J. Appl. Polym. Sci.* **2019**, *136*, 47209. [CrossRef]
7. Haddad, M.; Oie, C.; Vo Duy, S.; Sauvé, S.; Barbeau, B. Adsorption of Micropollutants Present in Surface Waters onto Polymeric Resins: Impact of Resin Type and Water Matrix on Performance. *Sci. Total Environ.* **2019**, *660*, 1449–1458. [CrossRef]
8. Mahvi, A.H. Application of Agricultural Fibers in Pollution Removal from Aqueous Solution. *Int. J. Environ. Sci. Technol.* **2008**, *5*, 275–285. [CrossRef]
9. Dehghani, M.H.; Mahvi, A.H.; Rastkari, N.; Saeedi, R.; Nazmara, S.; Iravani, E. Adsorption of Bisphenol A (BPA) from Aqueous Solutions by Carbon Nanotubes: Kinetic and Equilibrium Studies. *Desalination Water Treat.* **2015**, *54*, 84–92. [CrossRef]
10. Roy, H.; Islam, M.S.; Arifin, M.T.; Firoz, S.H. Synthesis, Characterization and Sorption Properties of Biochar, Chitosan and ZnO-Based Binary Composites towards a Cationic Dye. *Sustainability* **2022**, *14*, 14571. [CrossRef]
11. Yuan, Y.; Chesnutt, B.M.; Haggard, W.O.; Bumgardner, J.D. Deacetylation of Chitosan: Material Characterization and in Vitro Evaluation via Albumin Adsorption and Pre-Osteoblastic Cell Cultures. *Materials* **2011**, *4*, 1399–1416. [CrossRef]
12. Regmi, P.; Garcia Moscoso, J.L.; Kumar, S.; Cao, X.; Mao, J.; Schafran, G. Removal of Copper and Cadmium from Aqueous Solution Using Switchgrass Biochar Produced via Hydrothermal Carbonization Process. *J. Environ. Manag.* **2012**, *109*, 61–69. [CrossRef]
13. Beesley, L.; Marmiroli, M. The Immobilisation and Retention of Soluble Arsenic, Cadmium and Zinc by Biochar. *Environ. Pollut.* **2011**, *159*, 474–480. [CrossRef]
14. Tan, K.B.; Vakili, M.; Horri, B.A.; Poh, P.E.; Abdullah, A.Z.; Salamatinia, B. Adsorption of Dyes by Nanomaterials: Recent Developments and Adsorption Mechanisms. *Sep. Purif. Technol.* **2015**, *150*, 229–242. [CrossRef]
15. Maleki, A.; Mahvi, A.H.; Zazouli, M.A.; Izanloo, H.; Barati, A.H. Aqueous Cadmium Removal by Adsorption on Barley Hull and Barley Hull Ash. *Asian J. Chem.* **2011**, *23*, 1373–1376.
16. Palansooriya, K.N.; Ok, Y.S.; Awad, Y.M.; Lee, S.S.; Sung, J.-K.; Koutsospyros, A.; Moon, D.H. Impacts of Biochar Application on Upland Agriculture: A Review. *J. Environ. Manag.* **2019**, *234*, 52–64. [CrossRef] [PubMed]
17. Wani, I.; Ramola, S.; Garg, A.; Kushvaha, V. Critical Review of Biochar Applications in Geoengineering Infrastructure: Moving beyond Agricultural and Environmental Perspectives. *Biomass Convers. Biorefin.* **2021**, 1–29. [CrossRef]
18. Xu, G.; Lv, Y.; Sun, J.; Shao, H.; Wei, L. Recent Advances in Biochar Applications in Agricultural Soils: Benefits and Environmental Implications. *Clean* **2012**, *40*, 1093–1098. [CrossRef]
19. Roy, H.; Prantika, T.R.; Riyad, M.H.; Paul, S.; Islam, M.S. Synthesis, Characterizations, and RSM Analysis of Citrus Macroptera Peel Derived Biochar for Textile Dye Treatment. *S. Afr. J. Chem. Eng.* **2022**, *41*, 129–139. [CrossRef]
20. Yrjälä, K.; Ramakrishnan, M.; Salo, E. Agricultural Waste Streams as Resource in Circular Economy for Biochar Production towards Carbon Neutrality. *Curr. Opin. Environ. Sci. Health* **2022**, *26*, 100339. [CrossRef]

21. Nair, V.D.; Nair, P.K.R.; Dari, B.; Freitas, A.M.; Chatterjee, N.; Pinheiro, F.M. Biochar in the Agroecosystem–Climate-Change–Sustainability Nexus. *Front. Plant Sci.* **2017**, *8*, 02051. [CrossRef] [PubMed]
22. Seeman, T.; Šrédl, K.; Prášilová, M.; Svoboda, R. The Price of Farmland as a Factor in the Sustainable Development of Czech Agriculture (A Case Study). *Sustainability* **2020**, *12*, 5622. [CrossRef]
23. Lowder, S.K.; Skoet, J.; Raney, T. The Number, Size, and Distribution of Farms, Smallholder Farms, and Family Farms Worldwide. *World Dev.* **2016**, *87*, 16–29. [CrossRef]
24. Small Family Farmers Produce a Third of the World's Food. Available online: https://www.fao.org/newsroom/detail/Small-family-farmers-produce-a-third-of-the-world-s-food/en (accessed on 2 January 2023).
25. Illankoon, W.A.M.A.N.; Milanese, C.; Girella, A.; Rathnasiri, P.G.; Sudesh, K.H.M.; Llamas, M.M.; Collivignarelli, M.C.; Sorlini, S. Agricultural Biomass-Based Power Generation Potential in Sri Lanka: A Techno-Economic Analysis. *Energies* **2022**, *15*, 8984. [CrossRef]
26. Müller, S.; Backhaus, N.; Nagabovanalli, P.; Abiven, S. A Social-Ecological System Evaluation to Implement Sustainably a Biochar System in South India. *Agron. Sustain. Dev.* **2019**, *39*, 43. [CrossRef]
27. Clare, A.; Shackley, S.; Joseph, S.; Hammond, J.; Pan, G.; Bloom, A. Competing Uses for China's Straw: The Economic and Carbon Abatement Potential of Biochar. *Bioenergy* **2015**, *7*, 1272–1282. [CrossRef]
28. Leach, M.; Fairhead, J.; Fraser, J. Green Grabs and Biochar: Revaluing African Soils and Farming in the New Carbon Economy. *J. Peasant. Stud.* **2012**, *39*, 285–307. [CrossRef]
29. Schure, J.; Pinta, F.; Cerutti, P.O.; Kasereka-Muvatsi, L. Efficiency of Charcoal Production in Sub-Saharan Africa: Solutions beyond the Kiln. *Bois For. Des. Trop.* **2019**, *340*, 57–70. [CrossRef]
30. Ighalo, J.O.; Eletta, O.A.A.; Adeniyi, A.G. Biomass Carbonisation in Retort Kilns: Process Techniques, Product Quality and Future Perspectives. *Bioresour. Technol. Rep.* **2022**, *17*, 100934. [CrossRef]
31. Mašek, O.; Buss, W.; Roy-Poirier, A.; Lowe, W.; Peters, C.; Brownsort, P.; Mignard, D.; Pritchard, C.; Sohi, S. Consistency of Biochar Properties over Time and Production Scales: A Characterisation of Standard Materials. *J. Anal. Appl. Pyrolysis* **2018**, *132*, 200–210. [CrossRef]
32. Kirch, T.; Medwell, P.R.; Birzer, C.H.; van Eyk, P.J. Small-Scale Autothermal Thermochemical Conversion of Multiple Solid Biomass Feedstock. *Renew Energy* **2020**, *149*, 1261–1270. [CrossRef]
33. Rodrigo, A.S.; Perera, S. Potential and Viability of Rice Husk Based Power Generation in Sri Lanka. *Eng. J. Inst. Eng. Sri Lanka* **2013**, *46*, 9. [CrossRef]
34. Alahakoon, A.M.Y.W.; Karunarathna, A.K.; Dharmakeerthi, R.S.; Silva, F.H.C.A. Design and Development of a Double-Chamber Down Draft (DcDD) Pyrolyzer for Biochar Production from Rice Husk. *J. Biosyst. Eng.* **2022**, *47*, 458–467. [CrossRef]
35. Qian, X.; Xue, J.; Yang, Y.; Lee, S.W. Thermal Properties and Combustion-Related Problems Prediction of Agricultural Crop Residues. *Energies* **2021**, *14*, 4619. [CrossRef]
36. Schoeller, D.A. The Energy Balance Equation: Looking Back and Looking Forward Are Two Very Different Views. *Nutr. Rev.* **2009**, *67*, 249–254. [CrossRef] [PubMed]
37. Hoque, M.M.; Bhattacharya, S.C. Fuel Characteristics of Gasified Coconut Shell in a Fluidized and a Spouted Bed Reactor. *Energy* **2001**, *26*, 101–110. [CrossRef]
38. Jaya Prithika, A.; Sekar, S.K. Mechanical and Fracture Characteristics of Eco-Friendly Concrete Produced Using Coconut Shell, Ground Granulated Blast Furnace Slag and Manufactured Sand. *Constr. Build. Mater.* **2016**, *103*, 1–7. [CrossRef]
39. Li, W.; Yang, K.; Peng, J.; Zhang, L.; Guo, S.; Xia, H. Effects of Carbonization Temperatures on Characteristics of Porosity in Coconut Shell Chars and Activated Carbons Derived from Carbonized Coconut Shell Chars. *Ind. Crops Prod.* **2008**, *28*, 190–198. [CrossRef]
40. Nuriana, W.; Anisa, N. Martana Synthesis Preliminary Studies Durian Peel Bio Briquettes as an Alternative Fuels. *Energy Procedia* **2014**, *47*, 295–302. [CrossRef]
41. Oyelere, A.T.; Oluwadare, A.O. Studies on Physical, Thermal and Chemical Properties of Wood Gliricidia Sepium for Potential Bioenergy Production. *Int. J. Biomass Renew.* **2019**, *8*, 28–38.
42. Mainoo, A.A.; Ulzen-Appiah, F. Growth, Wood Yield and Energy Characteristics of Leucaena Leucocephala, Gliricidia Sepium and Senna Siamea at Age Four Years. *Ghana J. For.* **1996**, *3*, 69–79.
43. Rahayu, S.; Hilyana, S.; Suryani, E.; Nazmi; Sari, H.; Ali, M. Analysis of Wood Pellet Quality from Calliandra Callothyrsus, Gliricida Sepium, and Sawdust as New and Renewable Energy. In *Proceedings of the International Conference on Science and Technology (ICST), Mataram, Indonesia, 16–17 October 2021*; Institute for Research and Research and Community Service: Kota Surakarta, Indonesia, 2021; pp. 110–115.
44. Suryaningsih, S.; Nurhilal, O.; Yuliah, Y.; Mulyana, C. Combustion Quality Analysis of Briquettes from Variety of Agricultural Waste as Source of Alternative Fuels. *IOP Conf. Ser. Earth Environ. Sci.* **2017**, *65*, 012012. [CrossRef]
45. Kumar, P.; Barrett, D.M.; Delwiche, M.J.; Stroeve, P. Methods for Pretreatment of Lignocellulosic Biomass for Efficient Hydrolysis and Biofuel Production. *Ind. Eng. Chem. Res.* **2009**, *48*, 3713–3729. [CrossRef]
46. Gao, Y.; Guo, X.; Liu, Y.; Fang, Z.; Zhang, M.; Zhang, R.; You, L.; Li, T.; Liu, R.H. A Full Utilization of Rice Husk to Evaluate Phytochemical Bioactivities and Prepare Cellulose Nanocrystals. *Sci. Rep.* **2018**, *8*, 10482. [CrossRef]
47. Olanipekun, E.A.; Olusola, K.O.; Ata, O. A Comparative Study of Concrete Properties Using Coconut Shell and Palm Kernel Shell as Coarse Aggregates. *Build Environ.* **2006**, *41*, 297–301. [CrossRef]

48. Ranaraja, C.D.; Devasurendra, J.W.; Maduwantha, M.I.P.; Madhuwantha, G.A.L.; Hansa, R.Y.D. Optimization of an Industrial Boiler Operation. *Ournal Res. Technol. Eng.* **2020**, *1*, 126–134.
49. Zhongqing Ma; Jiewang Ye; Chao Zhao; Qisheng Zhang Gasification of Rice Husk in a Downdraft Gasifier: The Effect of Equivalence Ratio on the Gasification Performance, Properties, and Utilization Analysis of Byproducts of Char and Tar. *Bioresources* **2015**, *10*, 2888–2902.
50. Bhatia, B.E.A. *Overview of Refractory Materials*; Meadow Estates Drive Fairfax: Virginia, GA, USA, 2020.
51. Gong, K.; Cao, Y.; Feng, Y.; Liu, S.; Qin, J. Influence of Secondary Reactions on Heat Transfer Process during Pyrolysis of Hydrocarbon Fuel under Supercritical Conditions. *Appl. Therm. Eng.* **2019**, *159*, 113912. [CrossRef]
52. Strezov, V.; Evans, T.J.; Hayman, C. Thermal Conversion of Elephant Grass (Pennisetum Purpureum Schum) to Bio-Gas, Bio-Oil and Charcoal. *Bioresour. Technol.* **2008**, *99*, 8394–8399. [CrossRef] [PubMed]
53. Cheung, K.-Y.; Lee, K.-L.; Lam, K.-L.; Chan, T.-Y.; Lee, C.-W.; Hui, C.-W. Operation Strategy for Multi-Stage Pyrolysis. *J. Anal. Appl. Pyrolysis* **2011**, *91*, 165–182. [CrossRef]
54. Said, M.M.; Mhilu, C.F.; John, G.R. Thermal Characteristics And Kinetics Of Rice Husk For Pyrolysis Process. *Int. J. Renew. Energy Res.* **2014**, *4*, 275–278.
55. Fonts, I.; Azuara, M.; Gea, G.; Murillo, M.B. Study of the Pyrolysis Liquids Obtained from Different Sewage Sludge. *J. Anal. Appl. Pyrolysis* **2009**, *85*, 184–191. [CrossRef]
56. Domínguez, A.; Menéndez, J.A.; Inguanzo, M.; Pís, J.J. Production of Bio-Fuels by High Temperature Pyrolysis of Sewage Sludge Using Conventional and Microwave Heating. *Bioresour. Technol.* **2006**, *97*, 1185–1193. [CrossRef] [PubMed]
57. ben Hassen Trabelsi, A.; Ghrib, A.; Zaafouri, K.; Friaa, A.; Ouerghi, A.; Naoui, S.; Belayouni, H. Hydrogen-Rich Syngas Production from Gasification and Pyrolysis of Solar Dried Sewage Sludge: Experimental and Modeling Investigations. *Biomed Res. Int.* **2017**, *2017*, 7831470. [CrossRef]
58. Fan, H.; Zhou, H.; Wang, J. Pyrolysis of Municipal Sewage Sludges in a Slowly Heating and Gas Sweeping Fixed-Bed Reactor. *Energy Convers. Manag.* **2014**, *88*, 1151–1158. [CrossRef]
59. Biomass Gasification—Biomass (Solid Waste/Sludge) Gasification. Available online: https://blog.nus.edu.sg/chbewch/gasification-technology/biomass-gasification/ (accessed on 23 December 2022).
60. Wang, W. A Thermal Conversion Efficiency Study on Biomass Gasification of Arundo Donax and Woodchips. Master's Thesis, Eastern Illinois University, Charleston, IL, USA, 2013.
61. Biomass Gasification Research—HHV of Syn-Gas. Available online: https://www.eiu.edu/energy/hhv_of_syn_gas.pdf (accessed on 23 December 2022).
62. Rahmat, N.F.H.; Rasid, R.A. Gasification of Empty Fruit Bunch with Carbon Dioxide in an Entrained Flow Gasifier for Syngas Production. *IOP Conf. Ser. Mater. Sci. Eng.* **2017**, *206*, 012013. [CrossRef]
63. Sultan, S.H.; Palamanit, A.; Techato, K.-A.; Amin, M.; Baloch, K.A. Physiochemical Characterization and Potential of Synthesis Gas Production from Rubber Wood Biomass by Using Downdraft Gasifier. *Mehran Univ. Res. J. Eng. Technol.* **2021**, *40*, 1–15. [CrossRef]
64. Inguanzo, M.; Domínguez, A.; Menéndez, J.A.; Blanco, C.G.; Pis, J.J. On the Pyrolysis of Sewage Sludge: The Influence of Pyrolysis Conditions on Solid, Liquid and Gas Fractions. *J. Anal. Appl. Pyrolysis* **2002**, *63*, 209–222. [CrossRef]
65. Jindo, K.; Mizumoto, H.; Sawada, Y.; Sanchez-Monedero, M.A.; Sonoki, T. Physical and Chemical Characterization of Biochars Derived from Different Agricultural Residues. *Biogeosciences* **2014**, *11*, 6613–6621. [CrossRef]
66. Nguyen, B.T.; Lehmann, J. Black Carbon Decomposition under Varying Water Regimes. *Org. Geochem.* **2009**, *40*, 846–853. [CrossRef]
67. Harvey, O.R.; Herbert, B.E.; Kuo, L.-J.; Louchouarn, P. Generalized Two-Dimensional Perturbation Correlation Infrared Spectroscopy Reveals Mechanisms for the Development of Surface Charge and Recalcitrance in Plant-Derived Biochars. *Environ. Sci. Technol.* **2012**, *46*, 10641–10650. [CrossRef]
68. Harvey, O.R.; Kuo, L.-J.; Zimmerman, A.R.; Louchouarn, P.; Amonette, J.E.; Herbert, B.E. An Index-Based Approach to Assessing Recalcitrance and Soil Carbon Sequestration Potential of Engineered Black Carbons (Biochars). *Environ. Sci. Technol.* **2012**, *46*, 1415–1421. [CrossRef] [PubMed]
69. Rutherford, D.W.; Wershaw, R.L.; Rostad, C.E.; Kelly, C.N. Effect of Formation Conditions on Biochars: Compositional and Structural Properties of Cellulose, Lignin, and Pine Biochars. *Biomass Bioenergy* **2012**, *46*, 693–701. [CrossRef]
70. Spokas, K.A. Review of the Stability of Biochar in Soils: Predictability of O:C Molar Ratios. *Carbon Manag.* **2010**, *1*, 289–303. [CrossRef]
71. Wu, W.; Yang, M.; Feng, Q.; McGrouther, K.; Wang, H.; Lu, H.; Chen, Y. Chemical Characterization of Rice Straw-Derived Biochar for Soil Amendment. *Biomass Bioenergy* **2012**, *47*, 268–276. [CrossRef]
72. Dong, X.; Ma, L.Q.; Zhu, Y.; Li, Y.; Gu, B. Mechanistic Investigation of Mercury Sorption by Brazilian Pepper Biochars of Different Pyrolytic Temperatures Based on X-Ray Photoelectron Spectroscopy and Flow Calorimetry. *Environ. Sci. Technol.* **2013**, *47*, 12156–12164. [CrossRef]
73. Yang, H.; Yan, R.; Chen, H.; Lee, D.H.; Zheng, C. Characteristics of Hemicellulose, Cellulose and Lignin Pyrolysis. *Fuel* **2007**, *86*, 1781–1788. [CrossRef]
74. Khodadad, C.L.M.; Zimmerman, A.R.; Green, S.J.; Uthandi, S.; Foster, J.S. Taxa-Specific Changes in Soil Microbial Community Composition Induced by Pyrogenic Carbon Amendments. *Soil. Biol. Biochem.* **2011**, *43*, 385–392. [CrossRef]

75. Guo, J.; Chen, B. Insights on the Molecular Mechanism for the Recalcitrance of Biochars: Interactive Effects of Carbon and Silicon Components. *Environ. Sci. Technol.* **2014**, *48*, 9103–9112. [CrossRef] [PubMed]
76. Mimmo, T.; Panzacchi, P.; Baratieri, M.; Davies, C.A.; Tonon, G. Effect of Pyrolysis Temperature on Miscanthus (Miscanthus × Giganteus) Biochar Physical, Chemical and Functional Properties. *Biomass Bioenergy* **2014**, *62*, 149–157. [CrossRef]
77. Kalderis, D.; Kotti, M.S.; Méndez, A.; Gascó, G. Characterization of Hydrochars Produced by Hydrothermal Carbonization of Rice Husk. *Solid Earth* **2014**, *5*, 477–483. [CrossRef]
78. Illankoon, W.A.M.A.N.; Milanese, C.; Girella, A.; Medina-Llamas, M.; Magnani, G.; Pontiroli, D.; Ricco, M.; Collivignarelli, M.C. Sabrina Sorlini Biochar derived from the rice industry by-products as sustainable energy storage material. In *Proceedings of the 30th European Biomass Conference and Exhibition (EUBCE), Online, 9–12 May 2022*; Chevet, P.-F., Scarlat, N., Grassi, A., Eds.; ETA-Florence Renewable Energies: Florence, Italy, 2022.
79. Weber, K.; Quicker, P. Properties of Biochar. *Fuel* **2018**, *217*, 240–261. [CrossRef]
80. Lee, J.W.; Kidder, M.; Evans, B.R.; Paik, S.; Buchanan, A.C., III; Garten, C.T.; Brown, R.C. Characterization of Biochars Produced from Cornstovers for Soil Amendment. *Environ. Sci. Technol.* **2010**, *44*, 7970–7974. [CrossRef]
81. Yuan, J.-H.; Xu, R.-K.; Zhang, H. The Forms of Alkalis in the Biochar Produced from Crop Residues at Different Temperatures. *Bioresour. Technol.* **2011**, *102*, 3488–3497. [CrossRef] [PubMed]
82. Parr, J.F. Effect of Fire on Phytolith Coloration. *Geoarchaeology* **2006**, *21*, 171–185. [CrossRef]
83. Wilding, L.P.; Brown, R.E.; Holowaychuk, N. Accessibility and Properties of occluded carbon in biogenetic opal. *Soil Sci.* **1967**, *103*, 56–61. [CrossRef]
84. Gong, J.; Li, J.; Xu, J.; Xiang, Z.; Mo, L. Research on Cellulose Nanocrystals Produced from Cellulose Sources with Various Polymorphs. *RSC Adv.* **2017**, *7*, 33486–33493. [CrossRef]

energies

MDPI

Article

Evaluation of Slag Foaming Behavior Using Renewable Carbon Sources in Electric Arc Furnace-Based Steel Production

Lina Kieush [1,*], Johannes Schenk [1], Andrii Koveria [2], Andrii Hrubiak [3], Horst Hopfinger [4] and Heng Zheng [1]

1 Chair of Ferrous Metallurgy, Montanuniversitaet Leoben, 8700 Leoben, Austria
2 Department of Chemistry, Dnipro University of Technology, 49005 Dnipro, Ukraine
3 G.V. Kurdyumov Institute for Metal Physics of the National Academy of Sciences of Ukraine, 02000 Kyiv, Ukraine
4 Chair of Ceramics, Montanuniversitaet Leoben, 8700 Leoben, Austria
* Correspondence: lina.kieush@unileoben.ac.at

Abstract: The influence of different carbon sources, including anthracite, calcined petroleum coke, three samples of high-temperature coke, biochar, and a mixture of 50 wt.% biochar and 50 wt.% coke, on slag foaming behavior was studied. The slag's composition was set to FeO-CaO-Al_2O_3-MgO-SiO_2, and the temperature for slag foaming was 1600 °C. The effect of the carbon sources was evaluated using foaming characteristics (foam height, foam volume, relative foaming height, and gas fraction), X-ray diffraction (XRD), chemical analysis of the slag foams, Mossbauer spectroscopy, observation by scanning electron microscope (SEM), and energy-dispersive spectroscopy (EDS) mapping. Different foaming phenomena were found among conventional sources, biochar as a single source, and the mixture of coke and biochar. Biochar showed the most inferior foaming characteristics compared to the other studied carbon sources. Nevertheless, the slag foaming process was improved and showed slag foaming characteristics similar to results obtained using conventional carbon sources when the mixture of 50 wt.% coke and 50 wt.% biochar was used. The XRD analysis revealed a difference between the top and bottom of the slag foams. In almost all cases, a maghemite crystalline phase was detected at the top of the slag foams, indicating oxidation; metallic iron was found at the bottom. Furthermore, a difference in the slag foam (mixture of coke and biochar) was found in the presence of such crystalline phases as magnesium iron oxide (Fe_2MgO_4) and magnetite ($Mg_{0.4}Fe_{2.96}O_4$). Notwithstanding the carbon source applied, a layer between the foam slag and the crucible wall was found in many samples. Based on the SEM/EDS and XRD results, it was assumed this layer consists of gehlenite ($Ca_2(Al(AlSi)O_7)$) and two spinels: magnesium aluminate ($MgAl_2O_4$) and magnesium iron oxide (Fe_2MgO_4).

Keywords: electric arc furnace; slag foaming; carbon sources; biochar; steelmaking

1. Introduction

The practice of foaming slag is widely used in electric arc furnace (EAF) operations to achieve many benefits, including saving electricity, protecting the furnace lining, and reducing electrode consumption [1–8]. Slag foaming is also interesting for basic oxygen steelmaking [9–11] and ladle processing [12]. The foaming formation of the slag is influenced by the gas phase, which is determined by the amount of CO/CO_2 created, as well as by the slag phase, which determines the foamability [13–15].

In the interaction between carbon and slag in EAF, the phenomenon of slag foaming occurs, which involves the trapping of gas bubbles in the slag layer when a carbon-bearing source interacts with iron oxide in the slag. This gas foams the slag, increasing its volume depending on parameters such as the type of carbon source used, the FeO content of the slag, the slag's basicity, temperature, etc. Reduction reactions and slag foaming in EAF operations depend not only on the physicochemical properties of the slag [16] but also on the carbon source. Some technological limits regarding the use of carbon sources in EAF include low volatile matters (VM) of 2–7%, low moisture of 1–7%, a particle size of 0.5–5 mm, and low alkalis [17]. An important factor is the reactivity of the carbon source, which affects the chemical reactions that create the gas phase. On the other hand, the ash composition of the carbon-bearing source affects the slag phase and can change factors such as slag viscosity and surface tension. Changes in basicity and P_2O_5 content could be possible.

Several authors [12,18] have experimentally evaluated the phenomenon of slag foaming using conventional carbon sources. The effect of carbon-bearing particles (coke or coal) on the size and shape of slag foam has been studied. It has been established that carbon particles have a defoaming effect due to their non-wettability [19–22]. Therefore, the optimal amount of the FeO and C in the system is required, since excess carbon, instead of promoting gas-generating reactions, can reduce foam stability. Additionally, studies on the influence of the type of carbon-bearing source showed that the reduction in FeO and, therefore, the reaction kinetics depend on the type of carbon source.

Corbari et al. [23] used five different types of carbonaceous material—graphite, bituminous coal, bituminous char, anthracite coal, and anthracite char—to study the foaming process. The composition of the slag included CaO, SiO_2, MgO, and FeO within the range of 15 to 45 wt.%. This investigation showed that additional gas was released in the form of VM when using coal. However, VM is less effective than CO released from the slag foaming reaction. This is because the gas bubbles produced by the reaction were smaller, resulting in a more stable foam.

Ji et al. [24] used industrial slags of CaO–SiO_2–MgO–Al_2O_3–FeO with 30–40 wt.% FeO. As a carbon source, anthracite was used. According to the tests, the maximum carbon addition speed was 10 g of carbon per minute per kilo of slag. If the carbon amount was higher than this value, an excess of these particles was observed in the system. Additionally, it was noticed that the quantity of unreacted carbon particles acted to destabilize the formed foam.

Zhang and Fruehan [20] evaluated the influence of carbonaceous particles, namely, coke or coal char (with various sizes), in the foaming index when using forced argon injection at 1500 °C. The slag composition in this study used CaO–SiO_2–CaF_2–Al_2O_3. In some cases, the foam formation was suppressed due to the covering of the slag surface with carbon. Moreover, it was concluded that the bubbles generated by the reaction of FeO-C, which were small (from 1 to 3 mm), gave a more stable foam than bigger bubbles, consistent with the conclusions in the paper [23].

Aiming to reduce dependence on fossil fuels and decrease the contribution to greenhouse gas (GHG) emissions in EAF steelmaking, various authors have evaluated the slag foaming phenomenon and the impact on foaming behavior when using non-conventional carbon sources. Generally, studies focus on biochar use [25–27] and a mixture of biochar and coke [28]. Further, there are studies on waste plastics and rubber applications [29–31].

Maroufi et al. [32] studied the chemical interaction of EAF slag (Fe_2O_3-SiO_2-Al_2O_3-CaO-MgO-MnO) with three different carbon sources: coke, rubber-derived carbon, and a mixture of coke and rubber-derived carbon at atmospheric pressure at 1550 °C. The high reactivity of rubber-derived carbon led to a rapid interaction with molten slag. However, a low reduction rate of iron oxide from the slag was observed when using rubber-derived carbon, which was explained by the initial weak contact between rubber-derived carbon and molten slag. This led to the conclusion that the interaction between the carbonaceous material and the slag plays a significant role in the iron oxide reduction from slag.

Yunos et al. [33] investigated the possibility of using biomass in the EAF, namely, palm shell char, to partially replace coke in a laboratory-scale reactor at 1550 °C using the sessile-drop approach in an argon atmosphere. The slag composition in this study was set as Fe_2O_3–CaO–Al_2O_3–MgO–SiO_2–MnO. The test results showed an improved interaction with EAF slag compared to conventional coke.

Fidalgo et al. [34] studied two biochars obtained from agricultural residues, grape seed, and pumpkin seed chars for EAF steelmaking. Hard coal and three types of anthracite were also used to compare the results. The authors found that biochar could replace coal. Furthermore, it was noted that the highly volatile matters of biochar are an adequate stimulation for slag foaming.

Huang et al. [35] studied the interaction between synthetic slag with a composition of $CaO–Fe_2O_3–Al_2O_3–MgO–SiO_2–MnO$ and carbon materials obtained from various sources, including biochar, graphite, coke, and coal obtained from tire pyrolysis at a temperature of 1600 °C. Various interfacial phenomena were found between the slag and the investigated carbonaceous materials. The interaction between the biochar and slag was weak compared to other carbonaceous materials. The smooth surface of the biochar was claimed to decrease slag foaming.

The project entitled "Sustainable EAF steel production (GreenEAF)" [13] was devoted to studying the partial or full substitution of coal and natural gas by charcoal or syngas produced by the pyrolysis of biomass in EAF steel production. One of the interesting findings was that the tested charcoal gave only poor foaming results. The most obvious difference is the variable wettability, which, in the case of charcoal, prevents the reaction between the FeO in the slag and the carbon and, therefore, the formation of CO bubbles. The content of the VM in the carbon sources in these tests did not influence the foaming of the slag.

Echterhof [36] presented a review of the research related to the laboratory- and industrial-scale investigation of alternative carbon sources to replace the conventional carbon sources in EAF. The author also revealed that there is still much potential and a necessity for further research in this area.

This paper aims to analyze and compare slag foaming behavior using partially or entirely renewable carbon sources; to establish the effect of various carbon sources on foaming characteristics at 1600 °C as well as the composition of the slag foams; and to show the possibility of using renewable carbon sources for EAF applications to replace conventional carbon sources either partially or entirely.

2. Materials and Methods

For slag foaming tests, slag with the composition shown in Table 1 was prepared from a homogeneous mixture of FeO, MgO (≥98%), CaO (≥96%), SiO_2 (≥99%), and Al_2O_3 (≥99%) powders. The slag density and surface tension were calculated according to [37], and the foaming index was calculated according to [38]. Slag viscosity was determined via FactSage v.8.2.

Table 1. Target slag composition for slag foaming tests.

FeO, wt.%	SiO_2, wt.%	Al_2O_3, wt.%	MgO, wt.%	CaO, wt.%	B2	B3	ρ g/cm^3	η Pa·s	γ N·m^{-1}	Σ s
28.0	19.4	9.1	11.9	31.6	1.6	1.1	3.44	0.0382	0.568	0.863

Basicity (B2), CaO/SiO_2; basicity (B3), $CaO/SiO_2 + Al_2O_3$; ρ, density; η, viscosity; γ, surface tension; Σ, foaming index or average traveling time of the gas in the generated foam.

An alumina crucible (height 99 mm, inner diameter 63 mm, and wall thickness 4 mm) was filled with ultra-low carbon (ULC) steel of approximately 5 g (Table 2) and 100 g of slag.

Table 2. Chemical composition of ULC steel.

Elements	wt.%
C	0.007
Fe	99.6

The filled alumina crucible was placed in a graphite crucible (height 202 mm, inner diameter 70.3 mm, and wall thickness 20 mm) and flushed with nitrogen at a rate of 1000 L/h during each test.

The carbon content was calculated stoichiometrically for the required reduction according to the equation FeO (slag) + C = Fe (1) + CO [39,40]. The following carbon sources (Table 3) were used in this study: anthracite (Anth), calcined petroleum coke (CPC), high-temperature metallurgical coke samples (Coke1, Coke2, and Coke3), and wood biochar pellets (Bch), which were used with stable grain size from 0.5–1.0 mm to avoid the influence of particles on the size and behavior of slag foam.

Table 3. Characteristics of carbon sources used for slag foaming.

Characteristics	Anth	CPC	Coke1	Coke2	Coke3	Bch
Proximate analysis (wt.%)						
M	5.91	0.75	1.04	0.82	0.78	2.67
VM (db)	3.69	0.44	1.42	0.95	0.74	39.28
VM (daf)	4.12	0.45	1.57	1.03	0.80	41.04
Ash (db)	10.50	0.90	10.88	11.03	11.09	4.29
Elemental analysis (wt.%)						
S (db)	0.48	0.62	0.77	0.56	0.59	n.d.
C (db)	82.58	97.01	86.25	86.50	86.55	72.74
H (db)	2.19	0.26	0.29	0.23	0.21	4.62
N (db)	0.97	0.94	1.21	1.15	1.11	0.24
* Others, mainly O (db)	3.28	0.27	0.6	0.53	0.45	18.11
C_{fix} (db)	85.81	98.66	87.70	88.02	88.17	56.43
Characteristics of microstructural parameter (nm)						
L_c	2.08	1.58	1.86	1.56	1.32	0.69

M, moisture; C, carbon; H, hydrogen; N, nitrogen; S, sulfur; db, dry basis; daf, dry ash-free basis; VM(daf) = VM(db)·100/(100 − ash yield (db,%)); C_{fix}, wt.% = 100 − (wt.% VM (db) − wt.% Ash (db)); * calculated by difference; n.d., not detected; L_c, crystallite height.

All foaming tests were carried out using an induction high-temperature furnace MU-900 (Indutherm Erwärmungsanlagen GmbH) (Figure 1).

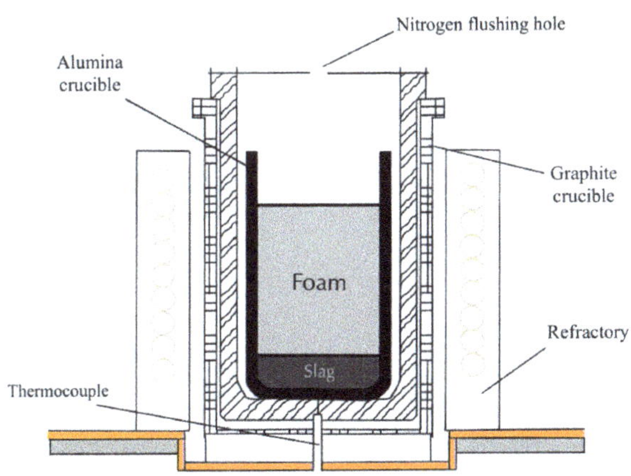

Figure 1. Experimental setup for the slag foaming tests.

When the temperature reached 1600 °C, a holding time of 15 min took place. The height of the molten slag was measured by dipping a molybdenum stick into the molten slag and then measuring the trace that remained on the stick with a ruler. Afterward, the carbon source was added, and the molten slag was mixed with a molybdenum stick, intensifying the carbon–slag reaction until the foam was stable. Due to the difference in density between the slag and the carbon, the carbon particles covered the surface of the slag and suppressed the foaming process. Stirring contributed to the penetration of carbon sources into the molten slag.

The standard time for foaming, regardless of the carbon source, was 6 min, which proved experimentally adequate for the carbon source to undergo a reaction. After the end of the test, the alumina crucible was quenched with liquid nitrogen to solidify the foam structure and prevent the foam from breaking. Each test was repeated four times for each carbon source.

A further alumina crucible with solidified foam was filled with the hardeners PUR 145 and SG 141/10 to prepare samples to be cut for examination. The samples were cut both horizontally and vertically and examined by microscopes.

Based on the foaming height, the volume of slag before (V_{slag}) and after foaming (V_{foam}) was calculated using Equations (1) and (2):

$$V_{slag} = \frac{d^2 \cdot \pi}{4} \cdot h_{slag}, \quad (1)$$

$$V_{foam} = \frac{d^2 \cdot \pi}{4} \cdot h_{foam}, \quad (2)$$

where d is the inner diameter of the crucible, cm; h_{slag} is slag height before foaming, cm; h_{foam} is slag height after foaming, cm.

Additionally, the slag weight (Table 1), the inner diameter of the alumina crucible, the slag density (Table 1), and Equation (1) were used to calculate the initial slag height without foaming.

The relative foaming height ($\frac{\Delta h}{h_0}$) was calculated by Equation (3):

$$\frac{\Delta h}{h_0} = \frac{h_{foam} - h_{slag}}{h_{slag}}, \quad (3)$$

The foaming height was estimated after the experiments, and the volumetric gas fraction contained in the foam (X_{gas}) was also calculated using Equation (4), according to [41,42].

$$X_{gas} = \frac{h_{foam} - h_{slag}}{h_{foam}}, \quad (4)$$

where h_{foam} is the slag height after foaming, cm; h_{slag} is the slag height before foaming, cm.

Additionally, the relative amount (ΔFeO) was calculated using Equation (5):

$$\Delta FeO = \frac{FeO_{slag} - FeO_{foam}}{FeO_{slag}} \times 100\%, \quad (5)$$

where FeO_{slag} is the amount of FeO in the initial slag composition, wt.%; FeO_{foam} is the amount of FeO in the slag composition after foaming, wt.%.

The chemical composition of the solidifying foam structure at the top and bottom of the crucible was determined by manganometrical titration according to Zimmermann-Reinhard (Fe_{tot} and Fe^{2+}), gravimetrical determination from lithium-tetraborate digestion (SiO_2), and inductive coupled plasma-optical emission spectroscopy from lithium tetraborate digestion (CaO, Al_2O_3, MgO).

X-ray diffraction (XRD) spectra of powdered slag foam samples were obtained using a Bruker AXS D8 advance diffractometer with a lynxeye detector and a Cu X-ray tube with Cu Kα radiation.

The magnetic microstructure of the samples and the state of their iron nuclei were studied using Mossbauer spectroscopy using the MS-1104Em spectrometer. The source of γ-quanta was the Co^{57} isotope in the chromium matrix. The natural width of the metal line α-Fe was 0.21 mm/s. The calibration of the isomeric shifts was carried out relative to α-Fe. Mossbauer spectra were interpreted using Univem MS analytical software.

Images of samples and mapping were obtained via a scanning electron microscope SEM FEI Quanta 200Mk2 equipped with an energy-dispersive (EDS) detector, back-scattered electron (BSE) detector, and a digital microscope Keyence VHX-E20.

3. Results and Discussion

Table 4 shows the calculated characteristics of the slag, namely, the height (h_{calc}) and volume (V_{calc}), which correspond to 0.93 cm and 29.1 cm^3.

Table 4. Calculated slag characteristics.

Calculated Slag Height h_{calc}, cm	Calculated Slag Volume V_{calc}, cm^3
0.93	29.1

Depending on the different carbon sources used, the foaming behavior of the slag was evaluated using the foaming characteristics shown in Table 5. It is noteworthy that with biochar, the height of the slag foam (SF) and the volume of the foam is lower compared with other carbon sources. Similar foaming behavior in the case of using biochar was noticed in [35] due to the low wettability of biochar by liquid slag, which prevents the reaction of carbon with FeO by the formation of a sufficient amount of CO bubbles [13]. Another reason is that due to the high reactivity of the biochar [43–45], FeO did not have time to react completely, and a smaller amount of CO was released, as evidenced by the value of the gas fraction (0.85). Among the carbon sources used, biochar is characterized by its low density, and therefore, when adding biochar particles, they were covered only on the surface of the slag foam. At the same time, stirring was difficult, because the biochar burned out on the slag foam surface very fast. As previously reported by [46], hydrogen in carbon sources can positively contribute to the FeO reduction reaction in the slag. However, based on the results presented in Table 5, when using biochar, which has the highest hydrogen value (4.62 wt.%), the lowest values in the height (6.5 cm) and volume of slag foam were obtained (201.9 cm^3).

It should be noted that when using anthracite with a hydrogen content of 2.19 wt.%, suitable slag foaming characteristics were obtained (8.7 cm and 272.1 cm^3 for foam height and foam volume, respectively). At the same time, the highest values were obtained when using coke2 (foam height of 8.9 cm and foam volume of 277.7 cm^3), as well as for coke1 (foam height of 8.3 cm and foam volume of 259.9 cm^3) and coke3 (foam height of 8.1 cm, and foam volume of 254.5 cm^3).

Using three samples of coke, anthracite, and calcined petroleum coke showed that the characteristics of the foams are similar.

Meanwhile, using 50 wt.% biochar in the mixture with coke 50 wt.% improved slag foaming characteristics (height, volume, relative foaming height, and gas fraction) compared to using biochar as a single carbon source for the foaming process. As noted during the tests, the improvement in these characteristics is due to the fact that 50 wt.% coke mitigated the drawbacks of using biochar and had a more significant effect on the foaming process than highly reactive biochar, particles of which burned very fast on the surface of the slag. Based on the results presented in Table 5, it can be concluded that the mixture of coke and biochar allows for obtaining foaming characteristics similar to using 100 wt.%

anthracite and at a level related to other coke samples. This can be explained by the fact that the change in physicochemical properties within the carbon source mixture, such as changes in surface area, density, and reactivity, exhibited a noticeable impact on the foaming mechanism, particularly when compared to individual carbon sources.

There is a statistically significant difference between the values of the height and the volume of slag foam samples as determined by one-way ANOVA (F value = 20.83, $p = 7.8 \cdot 10^{-8}$; and F value = 20.79, $p = 7.9 \cdot 10^{-8}$, respectively).

It is worth noting that the VM from the biochar (41.04 wt.%, daf) could also contribute to the foaming process, since the VM are released when the biochar is heated and could contribute to the foaming of the slag. However, the gas from the VM is less efficient than the gas from the chemical reaction of slag to foam, as previously shown in [12].

In the case of anthracite, this can be explained by the phenomenon of crushing coal particles in the process of releasing VM, which increased the surface area and, in turn, increased the rate of CO formation and, consequently, the height and volume of the slag foam. This phenomenon was also observed in [23]. Of note is that slag foam samples produced using anthracite as a carbon source show a vaster standard deviation for the height and volume characteristics of the slag foam compared to other carbon sources obtained by carbonization.

Notably, the slag is better wetted by a material with a similar chemical composition [35], and to achieve a good foaming process, it is required that carbon-bearing sources have relatively low ash content. However, biochar showed the most inferior foaming characteristics, although it is characterized by the lowest ash content (4.29 wt.%).

Typically, the gas fraction of slag foam is within the range of 0.80–0.95 [3]. The gas fractions vary between approximately 0.85–0.89, and the relative foaming heights between $\Delta h/h_0 = 5.8$–8.4. It has been observed that the lower the value of the gas fraction, the lower the relative foaming height (Figure 2).

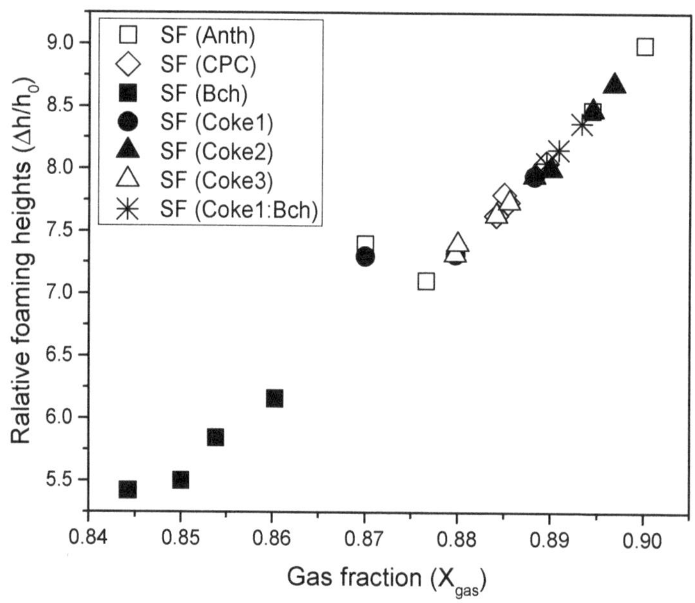

Figure 2. Relationship between gas fraction values and relative foaming heights, considering all values obtained after four tests.

Table 5. Summary results of slag foaming characteristics after four test sets.

Sample	Experimental Slag Height Mean h_{slag}, cm	std	Experimental Slag Volume Mean V_{slag}, cm^3	Std	Foam Height Mean h_{foam}, cm	std	Foam Volume Mean V_{foam}, cm^3	Std	Relative Foaming Height Mean $\Delta h/h_0$	Gas Fraction Mean X_{gas}
SF (Anth)					8.7	0.8	272.1	23.6	8.2	0.89
SF (CPC)					8.3	0.2	260.5	5.3	7.8	0.88
SF (Coke1)					8.3	0.3	259.9	9.2	7.7	0.89
SF (Coke2)	0.95	0.05	30	1.5	8.9	0.3	277.7	9.2	8.4	0.89
SF (Coke3)					8.1	0.2	254.5	5.8	7.6	0.88
SF (Bch)					6.5	0.3	201.9	9.0	5.8	0.85
SF (Coke1:Bch 50 wt.%:50 wt.%)					8.7	0.1	271.8	3.9	8.2	0.89

std, standard deviation.

Figure 3 shows the weight ratio of the carbon source to the slag and the effect on the foam height ratio. The amount of carbon material for tests was calculated stoichiometrically based on the amount of fixed carbon for each carbon source. Accordingly, the weight ratio of biochar to slag is the highest, and the foam height ratio is the lowest.

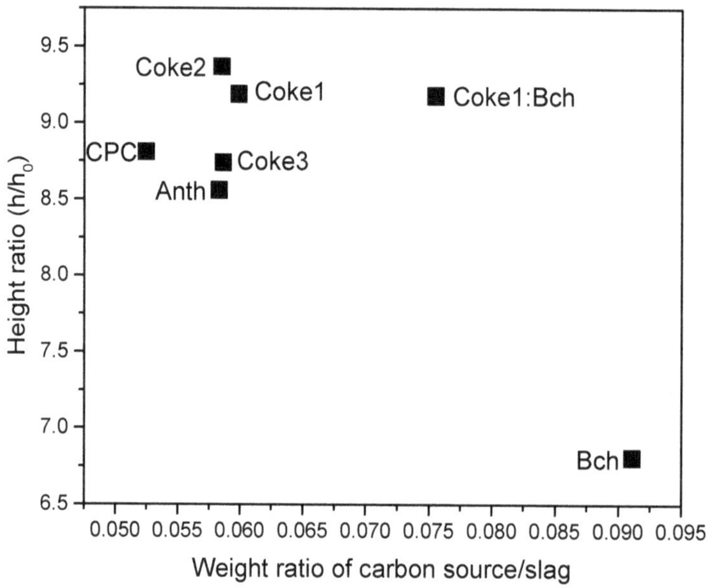

Figure 3. Weight ratio of carbon source/slag and ratio of slag heights.

However, the result of slag foaming changed with the partial use of biochar 50 wt.% with coke 50 wt.%, and the values became closer to the use of conventional carbon sources.

Selected examples of slag foams using different carbon sources are shown in Figure 4a–g. The results presented in Table 5 indicate that the VM and ash composition of carbon-bearing sources does not significantly contribute to slag foaming. Additionally, for VM, this can be explained by the large gas bubbles that formed, as shown in Figure 4a–g, which also complicated the process of creating stable foam.

Moreover, the results revealed the importance of the structural characteristics of the carbon sources used. A relationship was noticed between the XRD structural parameter crystallite height (L_c) of carbon sources and some slag foaming characteristics (Figure 5a). An increase in the structural ordering of the carbon material contributed to obtaining better slag foaming. Furthermore, Figure 5b shows the relationship between fixed carbon and slag foaming characteristics (slag foam height and slag foam volume), according to which the foaming process is better at higher values of fixed carbon in carbon sources, which confirms the conclusion that the higher fixed carbon, the better the slag foaming process [25].

Results of the ANOVA analysis showed the significance of the influence of fixed carbon content in carbon sources and the slag foaming characteristics of foam height and foam volume. Additionally, the influence of structural parameter crystallite height on foam height and foam volume was determined.

Figure 6a,b present the results of the XRD analysis of EAF slag foams obtained using different carbon sources. XRD analysis allowed for the identification of the following main crystalline phases at the top of the foams: gehlenite ($Ca_2(Al(AlSi)O_7)$), wuestite (FeO), bredigite ($Ca_{1.7}Mg_{0.3}SiO_4$), merwinite ($Ca_3Mg(SiO_4)_2$), and maghemite (Fe_2O_3).

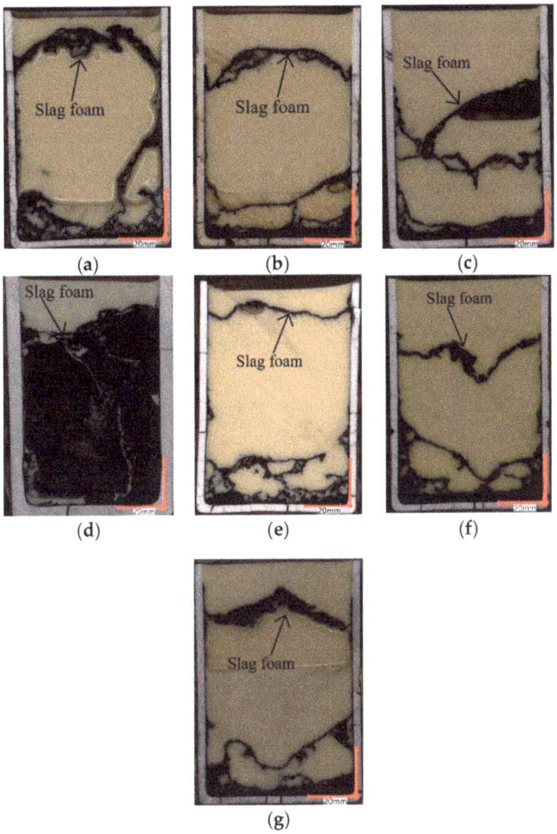

Figure 4. Examples of slag foams using different carbon materials: (**a**) slag foam obtained using anthracite; (**b**) calcined petroleum coke; (**c**) coke1; (**d**) biochar; (**e**) coke2; (**f**) coke3; and (**g**) 50 wt.% coke1: 50 wt.% biochar.

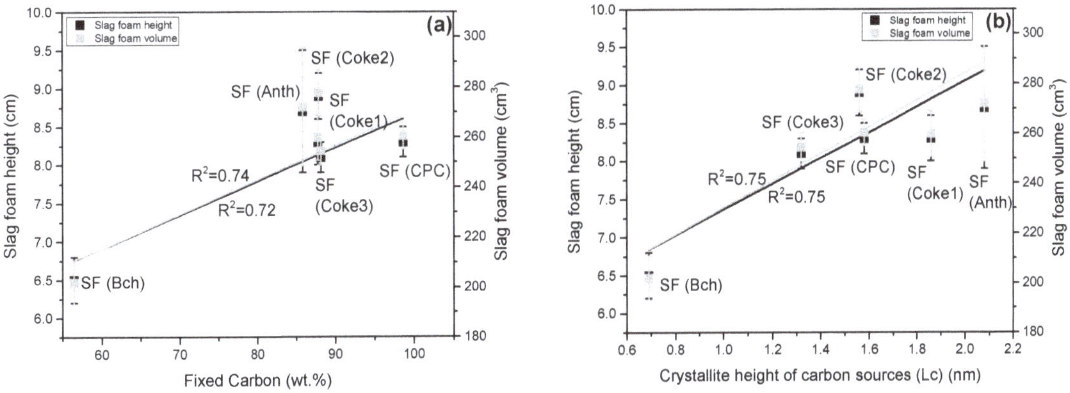

Figure 5. (**a**) Relationship between structural parameter crystallite height (L_C) for carbon source and slag foaming characteristics (foam height and foam volume); (**b**) relationship between amount of fixed carbon in carbon sources and slag foaming characteristics (foam height and foam volume).

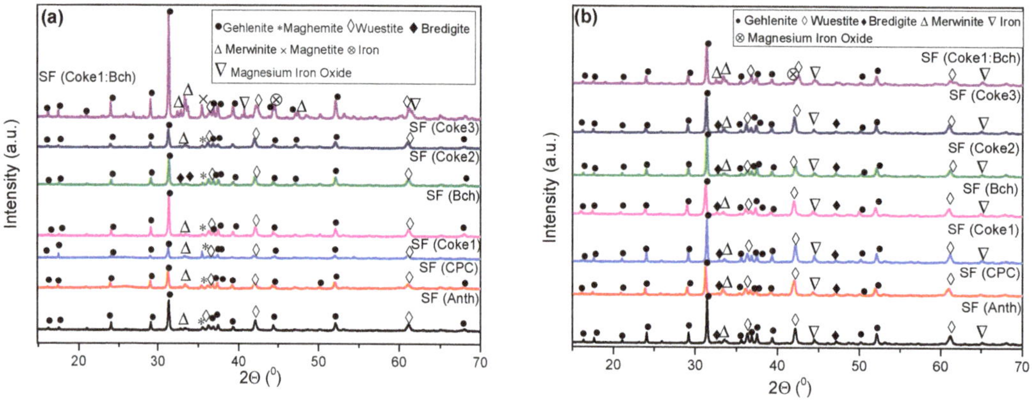

Figure 6. XRD patterns of slag foams obtained with different carbon sources: (**a**) at the top of the crucible; (**b**) and at the bottom of the crucible.

A small peak corresponding to the maghemite crystalline phase at the top of the slag foam can be explained as follows. Despite the fact that the crucible was flashed by nitrogen during the test, this did not contribute to a complete avoidance of oxidation. The maghemite crystalline phase was not detected for the lower part of the slag.

The slag foam at the bottom of the crucible indicated the following main crystalline phases: gehlenite, wuestite, merwinite, and metallic iron (Fe). Metallic iron was detected only in the samples from the bottom of the crucible. It should be noted that using non-conventional carbon sources did not significantly affect the crystalline phases compared to the presence of crystalline phases in slag foams obtained using conventional carbon sources.

In the case of slag foam, which was used as a carbon source, a mixture of coke1 and biochar XRD analysis confirmed the following crystalline phases at the top of the foams: gehlenite, wuestite, merwinite, and metallic iron (similar to other studied samples). The difference from the previous samples is the presence of such phases as magnesium iron oxide (Fe_2MgO_4) and magnetite ($Mg_{0.4}Fe_{2.96}O_4$). The slag foam at the bottom of the crucible contains gehlenite, wuestite, merwinite, magnesium iron oxide, and metallic iron. For this sample, no difference was observed between the crystalline phases for the top and bottom of the crucible.

The presence of magnetite in the top part of the slag foam, as in previous samples of slag foam where maghemite was found, is explained by the higher valence iron oxide, probably located in the slag foam region in contact with the air [47].

Due to the reduction in FeO in the slag during experiments, the slag foams' chemical composition was analyzed and is presented in Tables 6 and 7. The use of conventional or non-conventional carbon sources did not significantly affect the chemical composition of the slag selected at the top of the crucible after foaming. The range composition of the slag foams is 13.97–20.60 wt.% Fe_{tot}; 14.09–18.90 wt.% SiO_2; 29.72–33.03 wt.% CaO; 12.45–18.03 wt.% Al_2O_3; 5.51–9.80 wt.% MgO.

It should be noted that the high content of Al_2O_3 in the slag foam samples for both the top and bottom of the crucible can be explained by the presence of impurities from the alumina crucible.

Table 7 shows the chemical composition of the slag at the bottom of the crucible after foaming. The range composition of the slag foams is 17.69–34.10 wt.% Fe_{tot}; 12.13–17.68 wt.% SiO_2; 23.18–31.37 wt.% CaO; 14.16–17.00 wt.% Al_2O_3; 5.15–10.66 wt.% MgO.

Table 6. Chemical composition of the slag at the top of the crucible after foaming, wt.%.

Sample	Fe_{tot}	Fe^{2+}	SiO_2	CaO	Al_2O_3	MgO	B2
SF (Anth)	17.66	15.14	16.73	32.98	17.26	5.51	1.9
SF (CPC)	16.60	16.32	17.67	31.48	18.03	7.21	1.8
SF (Coke1)	16.71	16.37	16.88	30.79	16.95	7.13	1.8
SF (Coke2)	15.49	15.39	18.89	32.96	17.69	7.51	1.7
SF (Coke3)	19.21	18.99	17.05	29.72	17.55	6.90	1.7
SF (Bch)	13.97	11.57	18.90	33.03	17.49	6.88	1.7
SF (Coke1:Bch)	20.60	19.60	14.09	30.00	12.45	9.80	2.1

Table 7. Chemical composition of the slag at the bottom of the crucible after foaming, wt.%.

Sample	Fe_{tot}	Fe^{2+}	SiO_2	CaO	Al_2O_3	MgO	B2
SF (Anth)	23.72	23.67	15.77	28.89	15.10	6.97	1.8
SF (CPC)	18.92	16.06	15.69	30.90	15.96	6.73	1.9
SF (Coke1)	18.57	16.01	16.20	31.20	17.00	6.44	1.9
SF (Coke2)	17.69	17.14	17.68	31.37	16.99	7.39	1.8
SF (Coke3)	18.76	16.26	15.93	29.89	16.56	6.50	1.9
SF (Bch)	34.10	31.72	12.13	23.18	14.16	5.15	1.9
SF (Coke1:Bch)	21.12	20.00	12.36	28.07	14.64	10.66	2.3

The content of Si, Ca, and Mg oxides does not differ significantly between the samples from the top and bottom of the crucible and between different carbon sources. The reason for the decrease in the MgO content in all samples can be explained by the activity of MgO with some other slag components [48,49].

The decrease in the amount of Fe^{2+} after the foaming of the slag indicates a reduction reaction. Notably, high iron content was observed only for the slag foam SF (Bch) sample taken from the bottom of the crucible. As evidenced by XRD analysis (Figure 6b), metallic iron was detected at the bottom of each slag foam sample. Thus, the increase in iron content for SF (Bch) can be explained by the influence of oxidation of the slag foams from Fe to FeO during tests. When biochar was used, additional gas in the form of VM was released, contributing to oxidation and eventually decreasing foaming. This is confirmed by the slag foam height and volume results when using biochar that are given in Table 5.

Slag foaming characteristics were measured and evaluated in a comparison of changes in Fe (total) content in the slag after tests both for the top and bottom sides of crucibles, and the results are summarized and plotted in Figure 7a,b. Notably, when using biochar, the Fe (total) content is the lowest for the upper part of the crucibles, as are the slag foaming values. At the same time, the Fe (total) content is the highest in the lower part of the crucibles.

The relative amount of FeO is presented in Table 8. The decrease in FeO content in the slag foam samples both for the top and bottom of the crucible is irregular and indicates the presence of a reduction in FeO. Considering the results presented in Tables 5 and 8, it becomes evident that the incorporation of a coke1 and biochar mixture not only positively influenced the foaming process, but this also showed a more significant effect on the reduction in FeO compared to using individual carbon sources.

Figure 8a shows the SEM image of the interface between the foaming slag and the crucible. The EDS layered image (Figure 8b) shows that between the foamed slag and the crucible, there is a layer mainly composed of Mg, Al, Ca, Fe, and O. This phenomenon was observed for many samples under study, unrelated to the type of carbon source in the experiments.

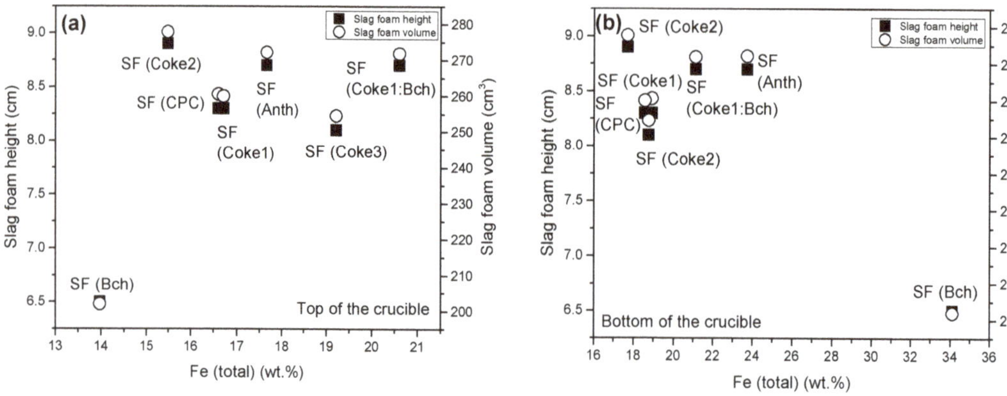

Figure 7. (**a**) Content of Fe (total) and slag foaming characteristics (height and volume) at the top of the crucibles; (**b**) content of Fe (total) and slag foaming characteristics (height and volume) at the bottom of the crucibles.

Table 8. Relative amount of FeO of the slag foam collected at the top and bottom of the crucible.

Sample	Relative amount ΔFeO_{top}, wt.%	Relative amount ΔFeO_{bottom}, wt.%
SF (Anth)	45.9	15.5
SF (CPC)	41.7	42.6
SF (Coke1)	41.5	42.8
SF (Coke2)	45.0	38.8
SF (Coke3)	32.2	41.9
SF (Bch)	58.7	n.a.
SF (Coke1:Bch)	30.0	28.6

n.a., not available.

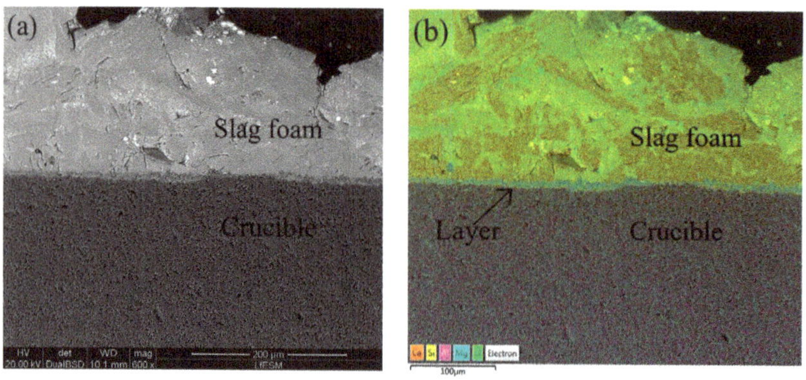

Figure 8. (**a**) SEM image of slag foam, layer between slag foam and wall of crucible and crucible; (**b**) EDS layered image of slag foam, layer between slag foam and wall of crucible and crucible.

EDS mapping of the spatial distribution of the elements of the slag foam, the layer between the slag foam and the wall of the crucible, and the crucible detected Al, O, Ca, Si, Mg, and Fe, as shown in Figure 9.

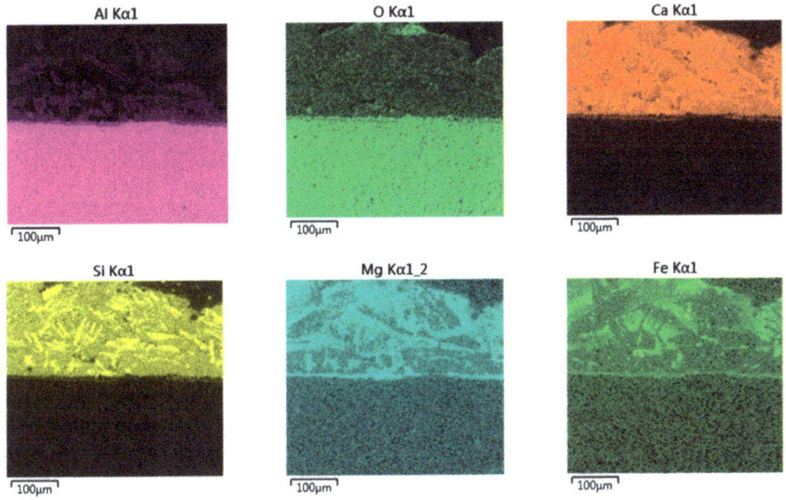

Figure 9. Energy-dispersive spectroscopy (EDS) mapping from scanning electron microscopy of elements of slag foam, layer between slag foam, and wall of crucible and crucible.

XRD analysis (Figure 10) of the slag foam and the layer between the slag foam and the wall of crucible identified the presence of the following phases: gehlenite, merwinite, corundum (Al_2O_3), and wuestite, as well as two spinels: magnesium aluminate ($MgAl_2O_4$) and magnesium iron oxide (Fe_2MgO_4). The presence of gehlenite, merwinite, and wuestite can be attributed to the slag foam, as XRD previously confirmed these phases (Figure 6a,b).

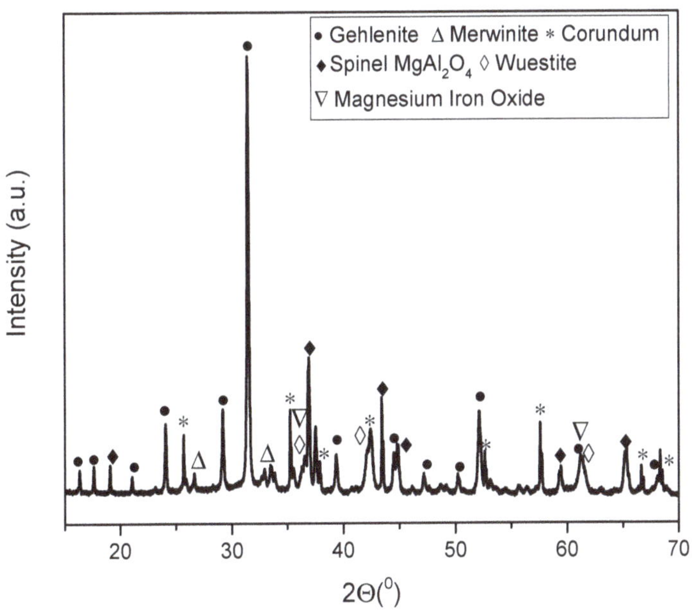

Figure 10. XRD analysis of slag foam, layer between slag foam and wall of crucible, and crucible.

The presence of peaks characterizing corundum is substantiated using alumina crucibles for this study. Based on the results of SEM/EDS, it can be assumed that the presence of phases such as gehlenite and two spinels refers to the layer between the slag foam and the crucible and can be explained by the activity of FeO and MgO, as well as by MgO and Al_2O_3 in liquid slag [50]. As reported in [19], slag with low basicity B3 and which has high amounts of FeO can be saturated by the spinel phases, including $MgAl_2O_4$ in its composition.

For all studied samples (Figure 11a–g), the Mossbauer spectra represent a doublet corresponding to the paramagnetic state of metallic iron Fe^{57} nuclei. The absence of sextuplet lines in the samples indicates that no magnetically ordered structures are formed with the participation of iron. The present paramagnetic component corresponds to iron nuclei in the valence state of Fe^{2+}.

Analysis of Mossbauer spectra revealed that two doublet lines in all cases approximate the paramagnetic component. The obtained values of the isomer shift (IS) and quadrupole splitting (QS) for the doublet components are in agreement with the parameters for crystalline wuestite FeO [51]. Both doublet components have close values of IS = 1.02 ± 0.05 mms^{-1} and different values of QS (Table 9).

The value of QS is determined by the symmetry of the close surroundings of the Fe^{57} nuclei and depends on the magnitude of the elastic deformation of the lattice; namely, the greater the deformation of the lattice, the higher the value of the QS. The presence of two doublet components with different quadrupole splitting values in the spectrum can be explained by the difference in the near environment of the Fe^{57} nuclei, which are in crystalline non-equivalent positions.

The non-equivalence of crystal positions of ferrous iron nuclei in the FeO structure can be related to the structural diversity of positions in the volume of the iron oxide particle and its surface layer. A higher QS value can reflect a more disordered nuclear environment of Fe^{57} than nuclei in the bulk of the particle, for which the near environment is more ordered [52].

In addition, the difference in the QS values of both components is possible due to different mechanisms of dense packing of anionic substructures, in particular, by packing octahedra ($Fe^{(II)}O_6$) connected to each other through faces or vertices [53]. These approaches correspond to the cubic and rhombohedral phases of FeO that were identified by XRD analysis (Figure 6a,b) and the presence of other reflexes, respectively, along with the main maxima for the rhombohedral phase [54].

Mossbauer spectroscopy allows for the detection of differences in Fe^{57} nuclei's positions for the cubic and rhombohedral phases by differentiating IS and QS of different doublet components. Therefore, the presence of both spectral doublets can be interpreted by different octahedron packings ($Fe^{(II)}O_6$) in the crystal structure of FeO of the studied slag foam samples.

Figure 12 shows the values of the quadrupole splitting of both doublets for each of the samples. A clear tendency of interdependence between the quadrupole splitting of both doublet components is observed: a decrease in the QS1 first doublet component of the sample is characterized by an increase in the QS2 second doublet component, respectively. Moreover, the value of QS1 represents a sequential order for all slag foam samples: Coke3 > Coke1:Bch > Anth > Bch > Coke2 > Coke1 > CPC. Similarly, the same consecutive order of values can be shown for QS2, which is increasing for slag foam samples: Coke3 < Coke1:Bch < Anth < Bch < Coke2 < Coke1 < CPC. Summarizing the dependence between QS1 and QS2, it can be concluded that the increase in the disorder of the environment of Fe^{57} nuclei in one of the non-equivalent positions leads to an ordering of the surroundings of the Fe^{57} nuclei in another position. Thus, there is a connection between the near-surface region and the particle volume: an increase in the ordering of the surface region corresponds to more ordering of the particle volume.

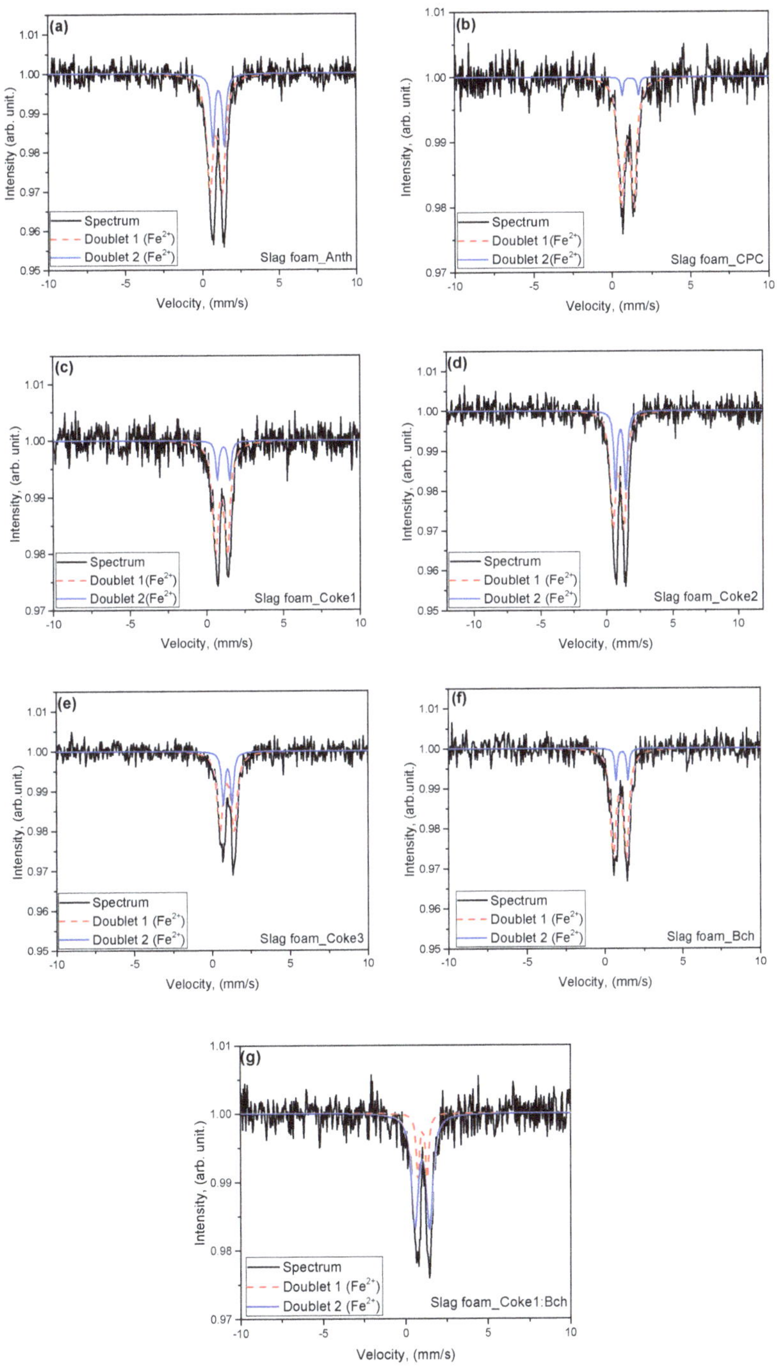

Figure 11. Mossbauer spectra of the slag foams using different carbon sources: (**a**) anthracite; (**b**) calcined petroleum coke (CPC); (**c**) coke1; (**d**) coke2; (**e**) coke3; and (**f**) biochar; and (**g**) 50 wt.% coke1:50 wt.% biochar.

Table 9. Mossbauer data on slag foam obtained using different carbon sources.

Sample	Component	IS, $mm \cdot s^{-1}$	QS, $mm \cdot s^{-1}$	H, Tesla	Relative intensity (%)	G, $mm \cdot s^{-1}$	Interpretation
SF (Anth)	Doublet1 (Fe^{2+})	0.99	0.87	0	76	0.57	$Fe^{(II)}O_6$
	Doublet2 (Fe^{2+})	1.07	0.71	0	24	0.25	$Fe^{(II)}O_6$
SF (CPC)	Doublet1 (Fe^{2+})	1.02	0.77	0	94	0.46	$Fe^{(II)}O_6$
	Doublet2 (Fe^{2+})	1.22	1.02	0	6	0.17	$Fe^{(II)}O_6$
SF (Coke1)	Doublet1 (Fe^{2+})	1.01	0.77	0	84	0.47	$Fe^{(II)}O_6$
	Doublet2 (Fe^{2+})	1.17	0.81	0	16	0.24	$Fe^{(II)}O_6$
SF (Coke2)	Doublet1 (Fe^{2+})	0.95	0.80	0	71	0.49	$Fe^{(II)}O_6$
	Doublet2 (Fe^{2+})	1.12	0.76	0	29	0.29	$Fe^{(II)}O_6$
SF (Coke3)	Doublet1 (Fe^{2+})	1.02	0.91	0	74	0.46	$Fe^{(II)}O_6$
	Doublet2 (Fe^{2+})	1.06	0.57	0	26	0.25	$Fe^{(II)}O_6$
SF (Bch)	Doublet1 (Fe^{2+})	1.00	0.81	0	89	0.42	$Fe^{(II)}O_6$
	Doublet2 (Fe^{2+})	1.15	0.74	0	11	0.17	$Fe^{(II)}O_6$
SF (Coke1:Bch)	Doublet1 (Fe^{2+})	1.03	0.89	0	78	0.49	$Fe^{(II)}O_6$
	Doublet2 (Fe^{2+})	1.06	0.57	0	22	0.24	$Fe^{(II)}O_6$

IS, isomer shift relative to α-Fe; QS, quadrupole splitting; H, hyperfine magnetic field; G, ground state.

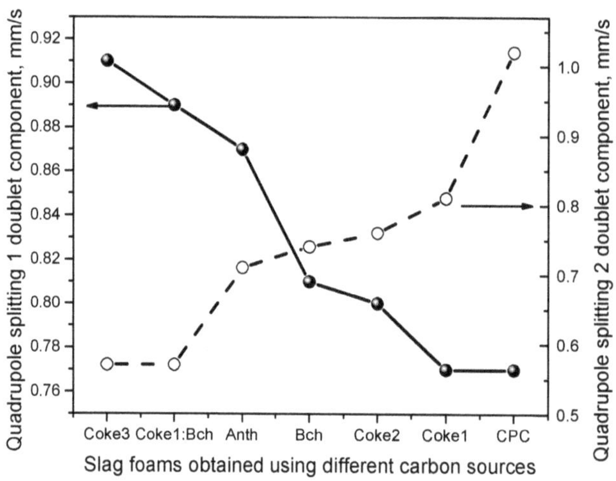

Figure 12. Quadrupole splitting of both doublets for each of the slag foam samples.

Figure 13 shows the contribution of the intensities of each doublet component to the total integrated intensity of the Mossbauer spectra of the samples.

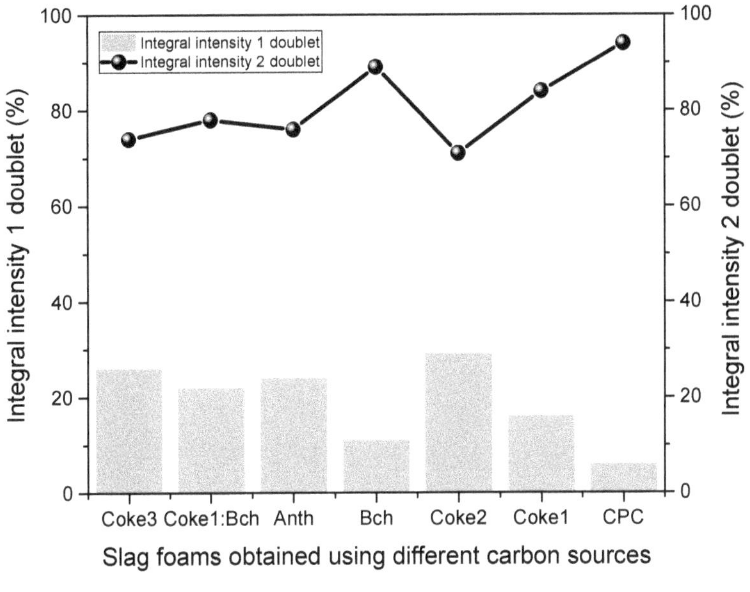

Figure 13. Integral intensity of each of the doublet components of the slag foam samples.

For all samples, the integrated intensity of the first doublet line is dominant and varies in the range of 72–94%. For this component, a higher value is observed in most samples of QS1, reflecting a more significant disorder in the environment around these Fe^{57} nuclei. Considering that the contribution of the doublet line intensity is proportional to the number of Fe^{57} nuclei and resonance absorption γ-quanta on which it is formed, it can be interpreted that exactly the first doublet line corresponds to iron nuclei located in the near-surface region of iron oxide particles. Moreover, they are characterized by rhombohedral packaging of packages ($Fe^{(II)}O_6$) with the possibility of including defects in the form of vacancies or implementation atoms. The second doublet line corresponds to Fe^{57} nuclei in the volume of iron oxide particles with cubic packing. Exceptions are slag foam (Coke1 and CPC) samples that are characterized by a more considerable QS value for the second doublet line.

Moreover, the contribution of the intensity of the second doublet line to the total integrated intensity is small for both samples and does not exceed 16%. The reason for this can be related to the low content of the wuestite phase and the small number of oxygen positions in these samples (Table 10). In this case, due to a lack of oxygen, the non-equivalent positions of the iron nuclei are not clearly expressed in the sense of forming a clear boundary near the surface area/volume of the particle. The formation of an oxide material that is homogeneous in terms of iron positions takes place, which includes regions with a more defective (disordered) environment, where there are inclusion regions with a more asymmetric defective environment of iron nuclei and, accordingly, a higher QS value.

Values of IS isomeric shifts of the doublet components of the Mossbauer spectra analyzed relative to the metallic α-Fe determine the chemical environment of Fe^{57} nuclei and their valence state. As mentioned, this made detecting the presence of Fe^{2+} ions and formation possible due to the synthesis of FeO oxide phases. The presence of two doublet lines with averaged values of isomeric shifts IS1 = 1.02 ± 0.05 mms^{-1} and IS2 = 1.11 ± 0.11 mms^{-1} indicates a partially different, although insignificant, chemical environment of the Fe^{57} nuclei, which arises as a result of their different surroundings in the volume of particles and their surface region. The dynamics of the change in the isomeric shift for each sample (Figure 14) indicate that for the case of the first doublet component, a relatively constant value of IS1 is observed, which, according to the interpretation, corresponds to Fe^{57} iron nuclei in the volume of iron oxide particles.

Table 10. SEM/EDS analysis results of chemical composition for slag foam obtained using different carbon sources, wt.%.

Sample	O	Mg	Al	Si	Ca	Fe
SF (Anth)	35.98	4.11	11.67	8.70	21.85	17.70
SF (CPC)	36.30	4.42	8.37	10.57	23.76	16.58
SF (Coke1)	38.80	4.76	10.94	10.71	23.57	11.22
SF (Coke2)	38.63	3.82	8.79	10.32	23.06	15.39
SF (Coke3)	37.38	4.54	9.73	10.10	23.96	14.29
SF (Bch)	37.21	3.17	10.26	9.93	24.12	15.31
SF (Coke1:Bch)	38.18	4.63	9.48	10.77	25.52	11.42

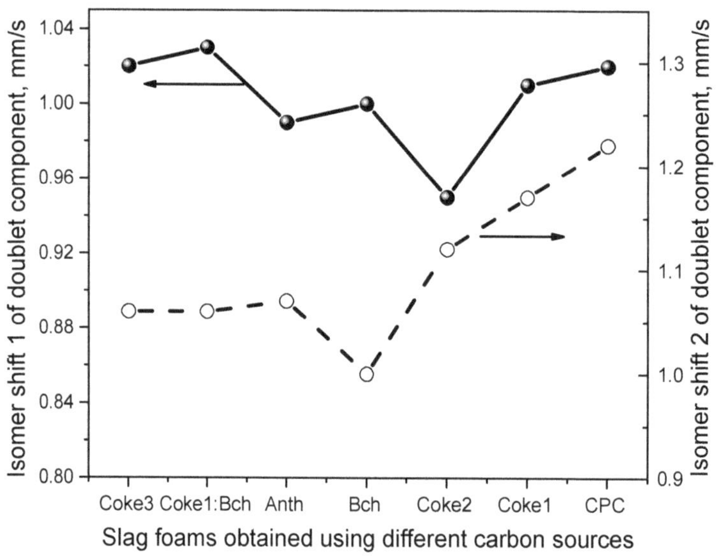

Figure 14. Magnitudes of isomeric shifts of doublet components of Mossbauer spectra.

The value of the isomeric shifts of the second doublet component IS2 for the samples forms a series of values that increase linearly for slag foam samples: Coke3 $\leq$ Coke1:Bch < Anth < Coke2 < Coke1 < CPC. The same trend was observed for the QS2 quadrupole moments, associated with increased disordering at the surface regions of iron oxide particles. Considering the values of the isomeric shifts, it can be assumed that an increase in disorder in the environment of Fe^{57} surface nuclei is associated with a growth in defectiveness of the structure. This causes a change in the chemical environment, particularly a possible change or delocalization of electrons, which in turn causes a difference in the covalence of chemical bonds. The exception is the sample SF (Bch), for which a lower value of the isomer shift IS2 = 1.00 mms^{-1} was noted. For this sample, the maximum value of the integrated intensity of the second component of the Mossbauer spectrum (89%) was observed, corresponding to the defective surface area of iron oxide particles.

Table 10 depicts the SEM/EDS analysis results of the chemical composition of slag foams. The elements Mg, Al, Si, Ca, and Fe were detected, previously determined by chemical analysis.

The investigated slag foam samples were characterized by scanning electron microscopy (SEM) and energy-dispersive spectroscopy (EDS) mapping from scanning electron microscopy of elements. The study by EDS mapping of the spatial distribution of the elements Ca, C, Al, O, Si, Fe, and Mg (Figures 15–21) clearly establishes correlations between elements, which provides information on the minerals that comprise the slag foams and confirmed the presence of all the elements identified by XRD (Figure 6a,b) and chemical analysis (Tables 6 and 7).

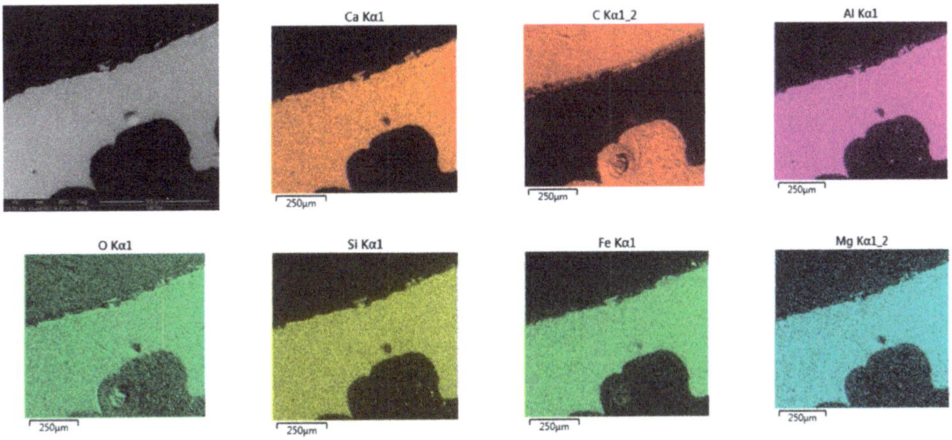

Figure 15. Energy-dispersive spectroscopy (EDS) mapping from scanning electron microscopy of elements in slag foam obtained using anthracite (Anth).

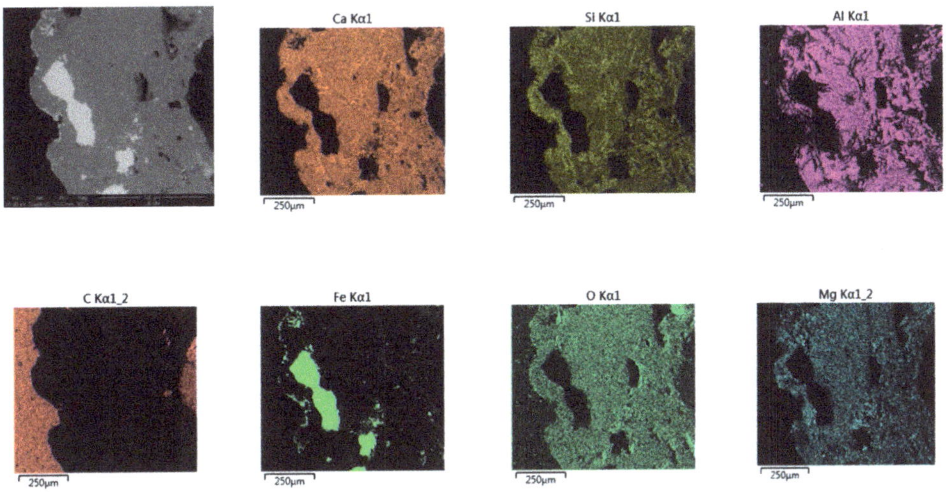

Figure 16. Energy-dispersive spectroscopy (EDS) mapping from scanning electron microscopy of elements in slag foam obtained using calcined petroleum coke (CPC).

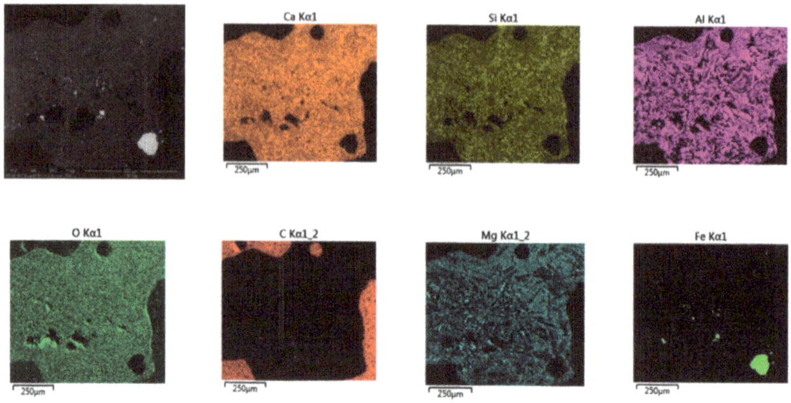

Figure 17. Energy-dispersive spectroscopy (EDS) mapping from scanning electron microscopy of elements in slag foam obtained using coke1.

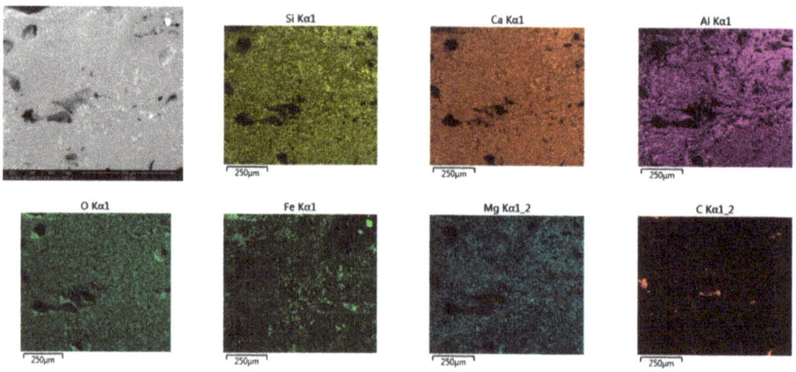

Figure 18. Energy-dispersive spectroscopy (EDS) mapping from scanning electron microscopy of elements in slag foam obtained using coke2.

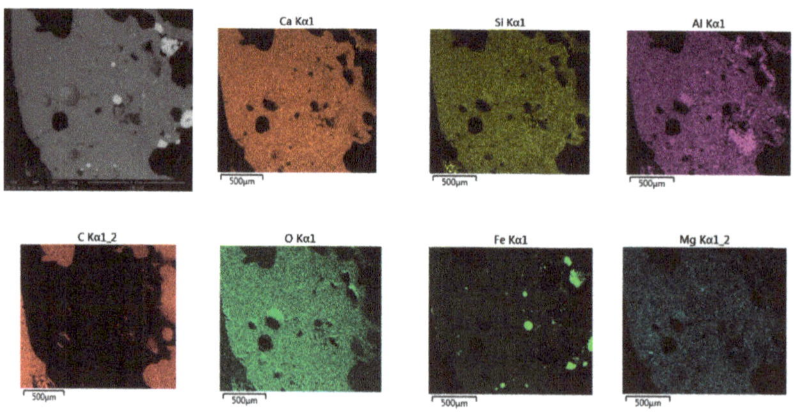

Figure 19. Energy-dispersive spectroscopy (EDS) mapping from scanning electron microscopy of elements in slag foam obtained using coke3.

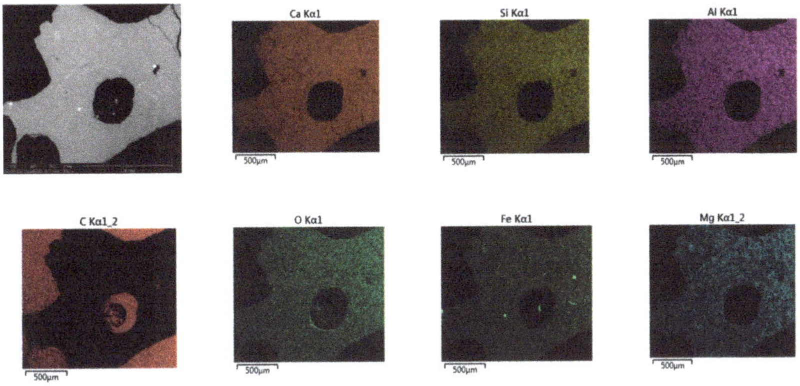

Figure 20. Energy-dispersive spectroscopy (EDS) mapping from scanning electron microscopy of elements in slag foam obtained using biochar (Bch).

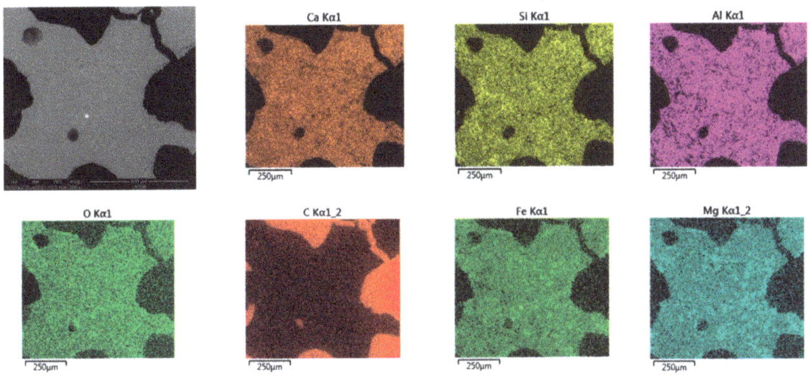

Figure 21. Energy-dispersive spectroscopy (EDS) mapping from scanning electron microscopy of elements in slag foam obtained using 50 wt.% coke:50 wt.% biochar.

The results of the slag foams' analysis by SEM/EDS analysis and EDS mapping in general show no adverse effect detectable on the chemical content of the slag foams when using biochar fully or partially.

4. Conclusions

The effect of six carbon sources on the foaming process was studied, of which three were conventional carbon-bearing sources: anthracite, calcined petroleum coke, and three samples of high-temperature cokes; and two were non-conventional carbon-bearing sources: biochar and a mixture of 50 wt.% biochar and 50 wt.% coke. Based on the results, the following conclusions can be drawn:

- Biochar showed the most inferior foaming characteristics due to the following. The low wettability of biochar by liquid slag prevents the reaction of carbon with FeO. The high reactivity of the biochar meant that FeO did not have time to react completely, and a smaller amount of CO was released, as evidenced by the value of the gas fraction (0.85). The low density meant that when adding biochar particles, they covered only the surface of slag foam and burned out on the slag foam surface very fast.
- Using the mixture of 50 wt.% coke and 50 wt.% biochar, the slag foaming process was improved and showed slag foaming characteristics similar to results obtained using conventional carbon sources. Based on the results obtained, it can be concluded that

biochar as a single carbon source cannot be recommended for use in slag foaming; however, when mixed with a conventional source, namely, coke, it showed promising results in EAF practice. Utilizing a biochar and coke1 mixture led to enhanced foaming compared to using biochar as the individual carbon source. Additionally, the observed foaming results obtained from the biochar and coke1 mixture were at a similar level as those achieved using other coke samples.

- A relationship was noticed between the XRD structural parameter crystallite height (L_c) of carbon sources and slag foaming characteristics (height and volume of slag foam). An increase in the structural ordering of the carbon material contributed to obtaining better slag foaming.
- The chemical composition of all slag foam samples showed no significant influence from the type of carbon source used. The decrease in the amount of Fe^{2+} after the foaming of the slag indicates a reduction reaction. The difference is between the top and bottom of the slag foam, as revealed by XRD. A maghemite (Fe_2O_3) and magnetite ($Mg_{0.4}Fe_{2.96}O_4$) crystalline phase in the top part of the slag foams is explained by the higher-valence iron oxide, probably located in the slag foam region in contact with air. At the same time, metallic iron was found at the bottom of all samples. A difference in the slag foam (mixture of coke and biochar) was found in the presence of such crystalline phases as magnesium iron oxide (Fe_2MgO_4) and magnetite ($Mg_{0.4}Fe_{2.96}O_4$).
- Based on SEM/EDS, it was noticed in many samples that there was a layer between the slag foam and the crucible wall, irrespective of the type of carbon source used. XRD analysis identified the following phases: gehlenite, merwinite, corundum (Al_2O_3), and wuestite, as well as two spinels: magnesium aluminate ($MgAl_2O_4$) and magnesium iron oxide (Fe_2MgO_4). It was assumed that the gehlenite and the two spinels could be attributed to the composition of the layer based on SEM/EDS.
- Mossbauer spectroscopy made it possible to obtain values of the isomer shift (IS) and quadrupole splitting (QS) for the doublet components that are in agreement with the parameters for crystalline wuestite (FeO). Through the analysis of interatomic magnetic interactions and the non-equivalent environment surrounding the nuclei, it was found that an increase in the disordering of nuclei in surface positions leads to a higher degree of order within the core of the particles. Additionally, a change in the chemical environment of the iron nuclei was observed, which was attributed to the defects present in the surface regions.

The findings of this study provide insight into the different impacts of various carbon sources, both conventional and non-conventional, on the foaming process. A noteworthy practical implication derived from the results is the viability of incorporating 50 wt.% biochar, a renewable carbon source, in combination with conventional carbon sources such as coke.

Author Contributions: Conceptualization, L.K. and J.S.; Investigation, L.K., A.K., A.H., H.H. and H.Z.; Methodology, L.K.; Supervision, J.S.; Writing—original draft, L.K.; Writing—review and editing, J.S., A.K., A.H. and H.H. All authors have read and agreed to the published version of the manuscript.

Funding: This research received no external funding.

Data Availability Statement: Not applicable.

Acknowledgments: Support by the scholarship program "Scholarship of the Scholarship Foundation of the Republic of Austria, Postdocs," (MPC-2022-02241), financed by the Federal Ministry of Education, Science, and Research of Austria, is gratefully acknowledged. Stahl- und Walzwerk Marienhütte GmbH, Graz, Austria, and voestalpine Stahl GmbH, Linz, Austria are also gratefully acknowledged for providing samples for research. The authors are also grateful to the reviewers for their insightful comments and efforts in improving the manuscript's text.

Conflicts of Interest: The authors declare no conflict of interest.

References

1. Agnihotri, A.; Singh, P.K.; Singh, D.; Gupta, M. Foamy Slag Practice to Enhance the Energy Efficiency of Electric Arc Furnace: An Industrial Scale Validation. *Mater. Today Proc.* **2021**, *46*, 1537–1542. [CrossRef]
2. Morales, R.D.; Rubén, L.G.; López, F.; Camacho, J.; Romero, J.A. The Slag Foaming Practice in EAF and Its Influence on the Steelmaking Shop Productivity. *ISIJ Int.* **1995**, *35*, 1054–1062. [CrossRef]
3. Matsuura, H.; Fruehan, R.J. Slag Foaming in an Electric Arc Furnace. *ISIJ Int.* **2009**, *49*, 1530–1535. [CrossRef]
4. Vieira, D.; de Almeida, R.A.M.; Bielefeldt, W.V.; Vilela, A.C.F. Slag Evaluation to Reduce Energy Consumption and EAF Electrical Instability. *Mat. Res.* **2016**, *19*, 1127–1131. [CrossRef]
5. Molloseau, C.L.; Fruehan, R.J. The Reaction Behavior of Fe-C-S Droplets in CaO-SiO_2-MgO-FeO Slags. *Met. Mater. Trans B* **2002**, *33*, 335–344. [CrossRef]
6. Kitamura, S.; Okohira, K. Influence of Slag Composition and Temperature on Slag Foaming. *ISIJ Int.* **1992**, *32*, 741–746. [CrossRef]
7. Zhu, T.X.; King, M.P.; Coley, K.S.; Irons, G.A. Erratum to: Progress in Slag Foaming in Metallurgical Processes. *Met. Mater. Trans. B* **2013**, *44*, 478. [CrossRef]
8. Vidacak, B.; Arvanitidis, I.; Jönsson, P.G.; Sjöberg, P. Observation on Foaming of EAF Slags in the Production of Stainless Steel: Observation on Foaming of EAF Slags. *Scand. J. Metall.* **2002**, *31*, 321–327. [CrossRef]
9. Jung, S.-M.; Fruehan, R.J. Foaming Characteristics of BOF Slags. *ISIJ Int.* **2000**, *40*, 348–355. [CrossRef]
10. Wang, R.; Zhang, B.; Liang, Y.; Liu, C.; Jiang, M. Micro Insight into Foaming Behavior of CaO–SiO_2–FeO–MgO-Based Slag Induced by Slag/Metal Reaction. *Ceram. Int.* **2023**, *49*, 17185–17193. [CrossRef]
11. Wang, R.; Zhang, B.; Hu, C.; Liu, C.; Jiang, M. Study on Splashing Mechanism in Basic Oxygen Furnace Based on Slag Foaming Modeling. *Steel Res. Int.* **2022**, *93*, 2100318. [CrossRef]
12. Zhang, Y.; Fruehan, R.J. Effect of the Bubble Size and Chemical Reactions on Slag Foaming. *MMTB* **1995**, *26*, 803–812. [CrossRef]
13. European Commission. *Directorate General for Research and Innovation. Sustainable EAF Steel Production (GREENEAF)*; Publications Office: Luxembourg, 2013.
14. Luz, A.P.; Tomba Martinez, A.G.; López, F.; Bonadia, P.; Pandolfelli, V.C. Slag Foaming Practice in the Steelmaking Process. *Ceram. Int.* **2018**, *44*, 8727–8741. [CrossRef]
15. Chychko, A.; Seetharaman, S. Foaming in Electric Arc Furnace Part I: Laboratory Studies of Enthalpy Changes of Carbonate Additions to Slag Melts. *Met. Mater. Trans. B* **2011**, *42*, 20–29. [CrossRef]
16. Kim, H.S.; Min, D.J.; Park, J.H. Foaming Behavior of CaO-SiO_2-FeO-MgOsatd-X (X=Al_2O_3,MnO, P_2O_5, and CaF_2) Slags at High Temperatures. *ISIJ Int.* **2001**, *41*, 317–324. [CrossRef]
17. Leonard, D.C.; Bonte, L.; Dufour, A.; Ferstl, A.; Raipala, K.; Scmole, P.; Schoone, P.; Verduras, J.L.; Willmers, R.R. Coke Quality Requirements of European Blast Furnace Engineers. In Proceedings of the Proceedings 3rd European Cokemaking Congress, Gent, Belgium, 16–18 September 1996.
18. Bhoi, B.; Jouhari, A.K.; Ray, H.S.; Misra, V.N. Smelting Reduction Reactions by Solid Carbon Using Induction Furnace: Foaming Behaviour and Kinetics of FeO Reduction in CaO–SiO_2–FeO Slag. *Ironmak. Steelmak.* **2006**, *33*, 245–252. [CrossRef]
19. Luz, A.P.; Avila, T.A.; Bonadia, P.; Pandolfelli, V.C. Slag Foaming: Fundamentals, Experimental Evaluation and Application in the Steelmaking Industry. *Refract. Worldforum* **2011**, *3*, 91–98.
20. Zhang, Y.; Fruehan, R.J. Effect of Carbonaceous Particles on Slag Foaming. *MMTB* **1995**, *26*, 813–819. [CrossRef]
21. Ogawa, Y.; Katayama, H.; Hirata, H.; Tokumitsu, N.; Yamauchi, M. Slag Foaming in Smelting Reduction and Its Control with Carbonaceous Materials. *ISIJ Int.* **1992**, *32*, 87–94. [CrossRef]
22. Wu, K.; Qian, W.; Chu, S.; Niu, Q.; Luo, H. Behavior of Slag Foaming Caused by Blowing Gas in Molten Slags. *ISIJ Int.* **2000**, *40*, 954–957. [CrossRef]
23. Corbari, R.; Matsuura, H.; Halder, S.; Walker, M.; Fruehan, R.J. Foaming and the Rate of the Carbon-Iron Oxide Reaction in Slag. *Met. Mater. Trans. B* **2009**, *40*, 940–948. [CrossRef]
24. Ji, F.-Z.; Barati, M.; Coley, K.; Irons, G.A. Kinetics of coal injection into iron oxide containing slags. *Can. Metall. Q.* **2005**, *44*, 85–94. [CrossRef]
25. Kieush, L.; Rieger, J.; Schenk, J.; Brondi, C.; Rovelli, D.; Echterhof, T.; Cirilli, F.; Thaler, C.; Jaeger, N.; Snaet, D.; et al. A Comprehensive Review of Secondary Carbon Bio-Carriers for Application in Metallurgical Processes: Utilization of Torrefied Biomass in Steel Production. *Metals* **2022**, *12*, 2005. [CrossRef]
26. Kieush, L.; Schenk, J.; Koveria, A.; Rantitsch, G.; Hrubiak, A.; Hopfinger, H. Utilization of Renewable Carbon in Electric Arc Furnace-Based Steel Production: Comparative Evaluation of Properties of Conventional and Non-Conventional Carbon-Bearing Sources. *Metals* **2023**, *13*, 722. [CrossRef]
27. DiGiovanni, C.; Li, D.; Ng, K.W.; Huang, X. Ranking of Injection Biochar for Slag Foaming Applications in Steelmaking. *Metals* **2023**, *13*, 1003. [CrossRef]
28. Ismail, A.N.; Yunos, N.F.D.M.; Rajeswari, M.S.R.; Idris, M.A. Production of Metallic Iron from Electric Arc Furnace (EAF) Slag Using Palm Shells Char/Coke Blends. In Proceedings of the AIP Conference Proceedings, Arau, Malaysia, 3 May 2021; p. 020202.
29. Sahajwalla, V. Recycling Waste Plastics in EAF Steelmaking: Carbon/Slag Interactions of HDPE-Coke Blends. *Steel Res. Int.* **2009**, *80*, 535–543. [CrossRef]
30. Kongkarat, S.; Khanna, R.; Koshy, P.; O'kane, P.; Sahajwalla, V. Recycling Waste Polymers in EAF Steelmaking: Influence of Polymer Composition on Carbon/Slag Interactions. *ISIJ Int.* **2012**, *52*, 385–393. [CrossRef]

31. Zaharia, M.; Sahajwalla, V.; Kim, B.-C.; Khanna, R.; Saha-Chaudhury, N.; O'Kane, P.; Dicker, J.; Skidmore, C.; Knights, D. Recycling of Rubber Tires in Electric Arc Furnace Steelmaking: Simultaneous Combustion of Metallurgical Coke and Rubber Tyres Blends. *Energy Fuels* **2009**, *23*, 2467–2474. [CrossRef]
32. Maroufi, S.; Mayyas, M.; Mansuri, I.; O'Kane, P.; Skidmore, C.; Jin, Z.; Fontana, A.; Sahajwalla, V. Study of Reaction Between Slag and Carbonaceous Materials. *Met. Mater. Trans. B* **2017**, *48*, 2316–2323. [CrossRef]
33. Yunos, N.F.M.; Zaharia, M.; Idris, M.A.; Nath, D.; Khanna, R.; Sahajwalla, V. Recycling Agricultural Waste from Palm Shells during Electric Arc Furnace Steelmaking. *Energy Fuels* **2012**, *26*, 278–286. [CrossRef]
34. Fidalgo, B.; Berrueco, C.; Millan, M. Chars from Agricultural Wastes as Greener Fuels for Electric Arc Furnaces. *J. Anal. Appl. Pyrolysis* **2015**, *113*, 274–280. [CrossRef]
35. Huang, X.-A.; Ng, K.W.; Giroux, L.; Duchesne, M. Carbonaceous Material Properties and Their Interactions with Slag During Electric Arc Furnace Steelmaking. *Met. Mater. Trans. B* **2019**, *50*, 1387–1398. [CrossRef]
36. Echterhof, T. Review on the Use of Alternative Carbon Sources in EAF Steelmaking. *Metals* **2021**, *11*, 222. [CrossRef]
37. Heo, J.H.; Park, J.H. Assessment of Physicochemical Properties of Electrical Arc Furnace Slag and Their Effects on Foamability. *Met. Mater Trans. B* **2019**, *50*, 2959–2968. [CrossRef]
38. Stadler, S.A.C.; Eksteen, J.J.; Aldrich, C. Physical Modelling of Slag Foaming in Two-Phase and Three-Phase Systems in the Churn-Flow Regime. *Miner. Eng.* **2006**, *19*, 237–245. [CrossRef]
39. Matsuura, H.; Manning, C.P.; Fortes, R.A.F.O.; Fruehan, R.J. Development of a Decarburization and Slag Formation Model for the Electric Arc Furnace. *ISIJ Int.* **2008**, *48*, 1197–1205. [CrossRef]
40. Fruehan, R.J.; Goldstein, D.; Sarma, B.; Story, S.R.; Glaws, P.C.; Pasewicz, H.U. Recent Advances in the Fundamentals of the Kinetics of Steelmaking Reactions. *Met. Mater Trans. B* **2000**, *31*, 891–898. [CrossRef]
41. Hong, L.; Hirasawa, M.; Sano, M. Behavior of Slag Foaming with Reduction of Iron Oxide in Molten Slags by Graphite. *ISIJ Int.* **1998**, *38*, 1339–1345. [CrossRef]
42. Xiang, J.; Wang, X.; Yang, M.; Wang, J.; Shan, C.; Fan, G.; Qiu, G.; Lv, X. Slag-Foaming Phenomenon Originating from Reaction of Titanium-Bearing Blast Furnace Slag: Continuous Monitoring of Foaming Height and Calibration. *J. Mater. Res. Technol.* **2021**, *11*, 1184–1192. [CrossRef]
43. Kieush, L.; Schenk, J.; Pfeiffer, A.; Koveria, A.; Rantitsch, G.; Hopfinger, H. Investigation on the Influence of Wood Pellets on the Reactivity of Coke with CO_2 and Its Microstructure Properties. *Fuel* **2022**, *309*, 122151. [CrossRef]
44. Bazaluk, O.; Kieush, L.; Koveria, A.; Schenk, J.; Pfeiffer, A.; Zheng, H.; Lozynskyi, V. Metallurgical Coke Production with Biomass Additives: Study of Biocoke Properties for Blast Furnace and Submerged Arc Furnace Purposes. *Materials* **2022**, *15*, 1147. [CrossRef]
45. Kieush, L.; Boyko, M.; Koveria, A.; Yaholnyk, M.; Poliakova, N. Manganese Sinter Production with Wood Biomass Application. *KEM* **2020**, *844*, 124–134. [CrossRef]
46. Sahajwalla, V.; Zaharia, M.; Kongkarat, S.; Khanna, R.; Rahman, M.; Saha-Chaudhury, N.; O'Kane, P.; Dicker, J.; Skidmore, C.; Knights, D. Recycling End-of-Life Polymers in an Electric Arc Furnace Steelmaking Process: Fundamentals of Polymer Reactions with Slag and Metal. *Energy Fuels* **2012**, *26*, 58–66. [CrossRef]
47. Almeida, R.A.M.D.; Vieira, D.; Bielefeldt, W.V.; Vilela, A.C.F. MgO Saturation Analisys of $CaO\text{-}SiO_2\text{-}FeO\text{-}MgO\text{-}Al_2O_3$ Slag System. *Mat. Res.* **2017**, *21*, 1–8. [CrossRef]
48. Luz, A.P.; Vivaldini, D.O.; López, F.; Brant, P.O.R.C.; Pandolfelli, V.C. Recycling MgO–C Refractories and Dolomite Fines as Slag Foaming Conditioners: Experimental and Thermodynamic Evaluations. *Ceram. Int.* **2013**, *39*, 8079–8085. [CrossRef]
49. Chang, Y.-E.; Lin, C.-M.; Shen, J.-M.; Chang, W.-T.; Wu, W. Effect of MgO Content on the Viscosity, Foaming Life, and Bonding in Liquid and Liquid/Solid $CaO\text{-}SiO_2\text{-}MgO\text{-}5Al_2O_3\text{-}30FeO$ Slags. *Metals* **2021**, *11*, 249. [CrossRef]
50. Pretorius, E.; Carlisle, R.C. Foamy Slag Fundamentals and Their Practical Application to Electric Furnace Steelmaking. *Iron Steelmak.* **1999**, *26*, 79–88.
51. Aldon, L.; Jumas, J.-C. Lithium-Induced Conversion Reaction in Wüstite $Fe_{1-x}O$ Studied by 57Fe Mössbauer Spectroscopy. *Solid State Sci.* **2012**, *14*, 354–361. [CrossRef]
52. Bazaluk, O.; Hrubiak, A.; Moklyak, V.; Moklyak, M.; Kieush, L.; Rachiy, B.; Gasyuk, I.; Yavorskyi, Y.; Koveria, A.; Lozynskyi, V.; et al. Structurally Dependent Electrochemical Properties of Ultrafine Superparamagnetic 'Core/Shell' $\gamma\text{-}Fe_2O_3$/Defective $\alpha\text{-}Fe_2O_3$ Composites in Hybrid Supercapacitors. *Materials* **2021**, *14*, 6977. [CrossRef]
53. Cornell, R.M.; Schwertmann, U. *The Iron Oxides: Structure, Properties, Reactions, Occurences and Uses*, 1st ed.; Wiley: Hoboken, NJ, USA, 2003; ISBN 978-3-527-30274-1.
54. Ajinkya, N.; Yu, X.; Kaithal, P.; Luo, H.; Somani, P.; Ramakrishna, S. Magnetic Iron Oxide Nanoparticle (IONP) Synthesis to Applications: Present and Future. *Materials* **2020**, *13*, 4644. [CrossRef]

energies

MDPI

Article

Sewage Sludge Gasification Process Optimization for Combined Heat and Power Generation

Alberto Carotenuto [1], Simona Di Fraia [1], Nicola Massarotti [1], Szymon Sobek [2], M. Rakib Uddin [1,*], Laura Vanoli [1] and Sebastian Werle [3,*]

1 Dipartimento di Ingegneria, Università degli Studi di Napoli "Parthenope", 80143 Napoli, Italy
2 Department of Heating, Ventilation and Dust Removal Technology, Silesian University of Technology, Stanisława Konarskiego St. 20, 44100 Gliwice, Poland
3 Department of Thermal Technology, Silesian University of Technology, Konarskiego 22, 44100 Gliwice, Poland
* Correspondence: mohammadrakib.uddin001@studenti.uniparthenope.it (M.R.U.); sebastian.werle@polsl.pl (S.W.)

Abstract: This work aims to assess the effect of the operating parameters of the gasifying agent preheating temperature and equivalence ratio (ER) on the conversion of sewage sludge (SS) to syngas through gasification and combined heat and power (CHP) generation. A novel gasification model was simulated in Aspen Plus to represent a fixed-bed updraft gasifier to generate syngas from SS through an equilibrium approach restricted by temperature. The novelty of this work is that the model was developed by applying the gasifying agent preheating temperature as an operating variable instead of the gasification temperature. It was calibrated by using a set of experimental values and then validated by comparing the numerical results with the experimental outcomes related to nine different operating conditions of air preheating temperatures and ER. A good agreement between the simulation and experimental results was observed. The optimum gasification process parameters of the air preheating temperature and ER were predicted to be 150 °C and 0.2, respectively. The CHP generation potentiality of SS was assessed to be 2.54 kW/kg SS as dry solids (DS), of which 0.81 kW was electrical and the remainder was thermal power. The conversion of SS to CHP through the proposed treatment can reduce 0.59 kg CO_2/kg SS as DS emissions compared with that of natural gas combustion to generate a similar quantity of energy.

Keywords: sewage sludge; gasification; air preheating temperature; equivalence ratio; syngas; combined heat and power; Aspen Plus

1. Introduction

Global primary energy demand increased at an average rate of 2.2% per year from 1990 (9128 Mtoe/yr) to 2018 (14,421 Mtoe/yr) due to the rapid growth in population, urbanization, and industrial productivity. The major contributor to meeting the global primary energy demand are fossil sources, accounting for around 78–82%, and the remainder comes from nuclear power and renewable sources, such as solar, wind, geothermal, biomass, hydropower, tide, and ocean energy [1,2]. The energy demand is expected to reach around 17,100 Mtoe/yr by 2030 based on the current energy consumption and population growth rate [3]. The ease of extraction and cost-effective processing make fossil fuels cheaper compared with renewable energy and are responsible for their choice as the prime source to meet the global primary energy demand [4]. However, excessive dependency on fossil fuels exerts tremendous pressure on finite reserves, with consequent rapid depletion [5]. Thus, a severe global energy crisis will be observed within the next few decades [6]. Furthermore, greenhouse gases (GHGs), including CO_2, CO, N_2O, NOx, SOx, hydrofluorocarbons, and sulfur hexafluoride, are generated continuously and emitted into the atmosphere during energy utilization from fossil sources in the residential, commercial, transportation, and industrial sectors [4,7]. The CO_2 emissions rate due to primary energy consumption reached 33.5 Gt/yr in 2018 from 20.5 Gt/yr in 1990 [8].

Among the potential sources of GHGs, waste management, including sewage sludge (SS) plays a significant role. Currently, the most common practices for SS management are use in agriculture as fertilizer (either preparing compost or directly), landfilling, and incineration [9]. SS managed through agricultural land spreading [10,11] and landfilling is decomposed continuously by the microbial activities of aerobic and anaerobic bacteria on the carbonaceous matter and nutrients present in SS. As a result of decomposition, CO, CO_2, N_2O, CH_4, NO, SO_2, and acetate are emitted to the atmosphere [12]. CO_2, CO, HCl, SOx, NOx, dioxins, furan, polychlorinated biphenyls, and polycyclic aromatic hydrocarbons are formed and emitted into the environment during the incineration of SS due to excessive air use [12]. CO_2 is recognized to be the highest contributor (74.4%) to the global GHGs, followed by CH_4 (17.3%) [4]. The rise in GHG levels has caused a global average temperature increase of 0.62 °C above the reference value between 1990 and 2020 [13].

Research on the identification of potential sources for renewable energy generation and the development of cost-effective conversion techniques is an attractive field to reduce the pressure on finite fossil fuel reserves and GHG emissions. For these aims, in 2021 the European Union (EU) announced a new green deal that set different targets for the member states to be achieved by 2030, such as 40% of primary energy required to be covered by renewable sources and the reduction of GHG emissions by 55% compared with the 1990 level [14]. In this framework, energy recovery from waste containing SS offers two benefits for reducing GHG emissions: a reduction in waste transportation and disposal and an increase in renewable energy production.

The application of stringent wastewater treatment directives drives the improvement of existing wastewater treatment facilities, causing a global increase in the SS generation rate. Moreover, the rate of increase in the population connected to sewage systems is increasing worldwide to fulfill the wastewater treatment directives. For instance, a continuous increase in the SS generation rate has been observed from 6.10 to 8.50 Mt as dry solids (DS) per year in the period of 2010 to 2018 in 27 EU countries (EU–27) [15]. By considering the most recently updated value of the per capita SS generation rate by Eurostat 2022 [15,16] and that, in the future, the sewage system will be improved to collect and treat all the generated wastewater, the SS generation rate is expected to reach about 10 Mt DS/yr by 2030.

The characteristics of generated SS strongly depend on the living standards of the population of a country, which control the quality of wastewater, and the treatment techniques (physical, chemical, or biological) utilized in wastewater treatment plants (WWTPs). The energy content present in SS depends on the composition of organic and inorganic constituents present in it and fluctuates in terms of a lower heating value (LHV) in the range of 11 to 20 MJ/kg SS as DS [9,17,18].

As mentioned above, producing adsorbent materials or energy from SS could reduce the problems associated with the current SS management strategies [19–21]. As an example, Miricioiu et al. (2021) [19] estimated a CO_2 adsorption capacity of 11.87 cm^3/g under operating conditions of a −196 °C temperature and 1 bar pressure for the char derived from SS through pyrolysis treatment. Regarding energy recovery from SS, thermal treatment techniques are more advantageous due to their contribution to global primary energy demand, as well as their higher conversion efficiency (more than 80%) [22,23] and lower processing time (from 30 to 70 min) compared with biological methods (from 30 to 60% and between 7 and 105 days, respectively) [24].

Gasification is considered a promising thermal treatment for energy recovery from SS over other thermal routes due to its capability to produce products with higher energy content. Additionally, gasification offers higher conversion efficiency (carbon conversion efficiency (CCE) and cold gas efficiency (CGE)), and mild operating parameters, leading to lower operating costs and GHG emissions [9,25]. Gasification is the thermal decomposition of SS by supplying limited oxygen through a gasifying agent to ensure incomplete combustion to form carbon monoxide, hydrogen, carbon dioxide, methane, and other lighter hydrocarbons. The mixture of gaseous components that exits the gasifier during the gasification of biomass or SS is commonly known as syngas. The LHV of syngas obtained through gasification of SS or biomass varies from 3 to 9.73 MJ/Nm^3 of raw syngas, based on the quality of the feed materials and operating conditions [26,27]. The oxygen required to complete the gasification process is commonly supplied through air due to its low cost and better availability. Depending on the final syngas specifications, other potential candidates as gasifying agents are pure O_2, steam, CO_2, and mixtures of air–steam, O_2–steam, and CO_2–steam [20].

The performance of the gasification process using air as a gasifying agent depends on the operating variables of the temperature and equivalence ratio (ER) [28,29]. The weight ratio of the actual air fed into the reactor to the stoichiometric amount of air essential for biomass combustion is the ER [30]. The gasification process performance may be estimated by assessing the syngas LHV, CCE, CGE, and net power $(\dot{P}_{net})$ obtained from the transformation process. Several authors identified optimum gasification operating parameters of temperature in the range of 770–850 °C and ER between 0.15 and 0.21 for energy recovery from SS. To complete the optimization, researchers have to conduct numerous experimental tests, in the range of 4 to 23 [30–35]. Carrying out these experimental campaigns makes the optimization process time-consuming and costly. In this context, numerical modeling may significantly reduce the time and cost of biomass gasification process optimization, including SS, as the experimental outcomes under two different conditions are sufficient to develop a model in chemical process simulation software, such as Aspen Plus. One set of gasification experimental conditions (temperature and ER) with the generated syngas composition is used for model calibration, whereas another set is applied for the validation [9].

However, gasification has some drawbacks of tar formation during syngas generation from biomass, including SS, and requires their complete removal before further application [36]. The operating conditions influence tar formation. As an example, syngas generated from pine sawdust at 900 °C temperature and with an ER of 0.21 contains 1.19 g/Nm^3 of tar, whereas the tar content raises to 1.89 g/Nm^3 for the gasification conditions of 800 °C and ER of 0.39 [37].

The syngas produced from biomass, including SS, can be employed to power internal combustion engines (ICEs) to produce combined heat and power (CHP) or be used as feed materials for the generation of chemicals (methanol, acetic acid, formaldehyde, etc.), hydrogen, and alternative fuel (gasoline, kerosene, and diesel through the Fischer–Tropsch reaction) [38,39]. CHP generation from SS through the thermal treatment of gasification offers two advantages:

- It reduces the pressure on finite fossil fuel reserves to generate electricity to run WWTPs, as the electrical energy can support a fraction of the demand to complete the wastewater treatment;
- Thermal power may be used to dry mechanically dewatered SS to a moisture content lower than 10 wt.%, as suggested for carrying out gasification in the pertinent literature [9,40,41].

The prediction of CHP generation from biomass through gasification integrated with an ICE by numerical process simulation is limited. Electrical and thermal efficiencies between 19.9–27% and 17.8–39%, respectively, have been estimated for a cogeneration system powered by syngas generated through wood gasification [36,42], whereas a thermal efficiency of 33.5% has been predicted for the use of olive kernels [43]. Concerning SS, the electrical and thermal efficiencies have been estimated in the ranges of 19.3–29.2% and 44.93–48.7% respectively [9,20,44]. Finally, considering waste paper and deinking sludge blends, electrical, thermal, and system efficiencies of 18.32, 46.86, and 72.87%, respectively, have been assessed [27].

In the current research, the conversion of SS to syngas by simulating a fixed-bed updraft air-gasifier considering an autothermal mode of operation is simulated. The variation in syngas quality and gasification process performances with differences in the operating conditions of air preheating temperature and ER are examined. The gasification process is numerically analyzed in the Aspen Plus V8.8 software (Bedford, MA, USA) by developing a model based on experimental outcomes. Considering the available literature, gasification models are mainly employed to assess the effect of temperature and ER as operating variables on syngas generation by analyzing different feedstocks, such as kitchen waste [45], sawdust [46], Napier grass [47], and SS [9,20,48]. Differently, this paper focuses on the effect of air preheating temperature on the autothermal gasification process. To the best of the authors' knowledge, this analysis is carried out here for the first time. During gasification model calibration, a temperature-restricted chemical equilibrium approach is applied, as proposed in some studies available in the literature [9,28,29,49]. The calibrated model is validated against nine distinct sets of experimental outcomes based on distinct ER and air preheating temperatures. Differently, in the literature, a maximum of five distinct gasification temperatures at unique ER have been used [20]. The optimum operating points are estimated through the developed model by analyzing the variations in syngas composition, LHV, CGE, CCE, and $\dot{P}_{net}$ with variations in the gasifying agent preheating temperature and ER. Finally, the use of the produced syngas to power an ICE system is considered to estimate the:

- Energy recovery potential of SS as CHP;
- CO_2 emission reduction over a gas turbine system;
- Emission profile to assess the pollutant concentration.

2. Experimental Analysis: Sewage Sludge to Syngas

2.1. Sewage Sludge Collection and Characterization

A description of the experimental campaign of SS gasification used to develop the numerical model presented in this study is available in other publications [50,51].

Dry SS was obtained from a WWTP in Poland where wastewater is treated by applying a combination of mechanical and biological processes with stabilization by anaerobic digestion. Anaerobically digested SS is dewatered through a mechanical belt press and drying is completed in a convective dryer by applying hot air with a temperature of 260 °C in the WWTP. The thermally dried SS presented a granulated form due to the application of air with high temperatures as a drying agent. A picture of the analyzed SS sample is presented in Figure 1.

Figure 1. Picture of the sewage sludge used in the current research work.

Standard methods of PN-EN 14774-3:2010, PN-EN 15402:2011, and PN-EN 15403:2011 were applied to quantify the moisture, volatile matter, and ash content, respectively, available in the thermally dried SS used in the current analysis. The major chemical constituents present in SS were measured by an infrared analyzer. Proximate and ultimate elemental analysis of the SS was completed on a dry basis (db). A description of the analysis techniques employed to characterize the SS used in the present study can be found in a previous publication [52]. The characteristics of the SS are illustrated in Table 1.

Table 1. Quality of SS with energy content.

Proximate Analysis (wt.%, db)		Ultimate Analysis (wt.%, db)		LHV (MJ/kg, db)	LHV (kWh/kg, db)
Moisture	5.30 ± 0.06	C	31.79 ± 0.2	12.96 ± 0.4	3.60 ± 0.1
		H_2	4.36 ± 0.05		
Volatile matter	51.00 ± 0.96	N_2	4.88 ± 0.07		
		S	1.67 ± 0.01		
Fixed carbon	7.20 ± 0.06	F_2	0.013 ± 0.001		
		Cl_2	0.22 ± 0.001		
Ash	36.50 ± 0.1	O_2	20.57		

2.2. Energy Recovery from Sewage Sludge

The energy content of the SS was transformed into syngas through a gasification experimental campaign conducted in a fixed-bed updraft reactor made of stainless steel with dimensions of a 150 mm internal diameter and 300 mm height. The gasifier had a maximum capacity of 1.39 g/s and was insulated properly to avoid energy loss as heat to the atmosphere. The gasifier had 6 thermocouples combined with an Agilent recording system placed every 50 mm from the bottom to the top, with the initial one placed 10 mm above the grate, to record the temperature in different zones inside the reactor. The SS was directed downward from the top and the gasifying agent (air) passed in the opposite direction of the feed flow from the bottom. Four consecutive distinct zones of drying (for moisture removal), pyrolysis (volatile matter, char, and ash formation through the decomposition of SS), reduction (volatile matter and carbon transformed to syngas), and combustion zones (ignition of char to supply energy to the above three regions to ensure autothermal gasification) inside the reactor from top to bottom. Air passed through a heater with a control system to ensure a specific air preheating temperature. A schematic view of the system considered for the gasification of the SS with a temperature profile at an ER of 0.16 is shown in Figure 2. Every experiment was conducted three times to check the reproducibility of the results. Average results with uncertainties for the gasification of SS under certain operating conditions of air preheating temperature and ER are illustrated in Table 2.

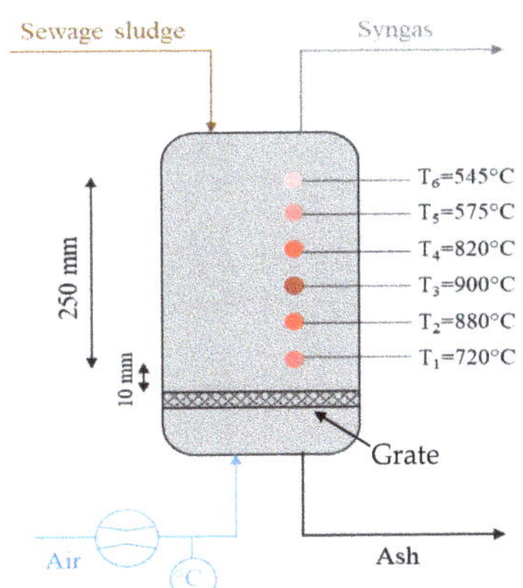

Figure 2. Representation of fixed-bed updraft gasifier with temperature profile for the SS gasification at an ER of 0.16 and air preheating temperature of 25 °C.

Table 2. Operating conditions applied in the gasification campaign conducted in the current research and corresponding syngas compositions with LHV.

Test Condition	ER (-)	SS Flow Rate (g/s)	Air Flow Rate (g/s)	* APT (°C)	Dry Syngas Composition (vol.%)				LHV (MJ/Nm^3)
					H_2	CO	CO_2	CH_4	
I	0.16	1.0	0.78	25	5.80 ± 0.1	23.90 ± 0.6	15.90 ± 0.4	0.85 ± 0.04	3.75 ± 0.6
II				100	5.90 ± 0.1	24.00 ± 0.7	15.40 ± 0.5	0.85 ± 0.05	3.77 ± 0.7
III				150	6.00 ± 0.2	24.20 ± 0.7	15.35 ± 0.4	0.87 ± 0.04	3.81 ± 0.6
IV				200	6.20 ± 0.1	24.50 ± 0.6	15.00 ± 0.5	0.89 ± 0.03	3.87 ± 0.5
V				250	6.80 ± 0.1	24.90 ± 0.7	14.80 ± 0.6	0.96 ± 0.04	3.99 ± 0.4
VI	0.27		1.32	25	7.90 ± 0.2	19.50 ± 0.8	15.90 ± 0.5	1.00 ± 0.05	3.44 ± 0.5
VII				100	7.92 ± 0.1	19.70 ± 0.6	15.70 ± 0.4	1.05 ± 0.06	3.47 ± 0.4
VIII				150	8.00 ± 0.2	20.00 ± 0.8	15.65 ± 0.6	1.06 ± 0.05	3.52 ± 0.5
IX				200	8.02 ± 0.1	20.50 ± 0.7	14.60 ± 0.7	1.10 ± 0.07	3.59 ± 0.6
X				250	8.05 ± 0.2	20.80 ± 0.6	14.80 ± 0.7	1.21 ± 0.06	3.64 ± 0.5

* APT = Air preheating temperature.

Fixed-bed gasifiers have low operating costs and generate syngas with a lower quantity of tar, which significantly reduces the cleaning costs before being used in further processes [53,54].

The experimental campaigns were completed by varying the incoming gasifying agent preheating temperatures (25, 100, 150, 200, and 250 °C) at two different ERs (0.16 and 0.27). The ER values were chosen based on previous experimental analysis to test the alteration in the syngas composition and LHV with operating parameters. The generated syngas was collected in a Tedlar gas bag for offline measurement. The syngas composition was measured using a calibrated gas chromatograph Agilent 6980N series. The experimental campaign was performed by maintaining a SS flow rate of 1.11 g/s. The tar content of the produced syngas was not analyzed due to the limitations of the analyzer.

3. Aspen plus Simulation on the Transformation of Sewage Sludge Energy Content to Combined Heat and Power

The simulation of the SS conversion to CHP was completed by integrating gasification in a fixed-bed updraft reactor with an ICE system in Aspen Plus V8.8 software. The model was developed through the sequential connection of different unit operation blocks for specific tasks, as unique blocks for either gasification or ICE were not available in the Aspen Plus library. The block diagram related to the cogeneration of electrical and thermal energy from SS through the proposed route is illustrated in Figure 3.

The raw material used to generate electrical and thermal energy in the current analysis was SS, which, in Aspen Plus, was treated as a non-conventional component. SS was decomposed into the conventional components of C, H_2, N_2, H_2O, S, and O_2, and the non-conventional component of ash in the pyrolysis step. A Boston–Matthias-modified Peng–Robinson (PR-BM) equation of state was chosen to assess the thermodynamic properties of all conventional components involved in the simulation [43], whereas the enthalpy and density of non-conventional components (SS and Ash) were estimated through some of the models available in Aspen Plus (HCOALGEN and DCOALIGT, respectively).

In the following text, the simulated system is described by referring to the blocks and streams illustrated in Figure 3.

The drying, pyrolysis, combustion, and reduction of SS to generate syngas in the gasification process were simulated by integrating different blocks. The drying and pyrolysis of SS were completed in an RYield reactor (DECOMP) at a temperature of 545 °C, which was measured in the experimental campaign (presented in Figure 2). The separation of the decomposed SS stream exiting the RYield reactor was completed in a separator block (SEPR) to three streams: CHAR (the fraction of carbon that was combusted to supply thermal energy to complete the gasification process in autothermal mode), GASFEED (the remaining fraction of carbon with other conventional components), and ASH. The simulation of the combustion of char with ambient air (CHRBAIR stream) to ensure the autothermal gasification mode was completed in an RGibbs reactor (CHRBRN). Gasification reactions were simulated by minimizing the Gibbs's free energy and were completed in another RGibbs reactor (GASIFIER). The preheating of the incoming gasifying agent to the reactor based on ER was completed in a heat exchanger (AIRHTR) by supplying thermal energy from an external source (QAIR stream). Preheated air was mixed with the conventional components coming from the SEPR block and the product exiting the char combustion reactor (CHRBP stream) in a MIXER prior to the entrance of the gasifier. The thermal energy of the char combustion and pyrolysis reactor was integrated with the gasifier, similar to the experimental campaign. The difference between the numerical results evaluated from the simulated model and the experimental campaign was reduced to an acceptable limit of less than ±20% through the restriction of the equilibrium position of individual gasification reactions to a specific temperature [9,55]. The product exiting the gasifier (RAWSYNG stream) was mixed with the ash (HTASH stream), whose temperature was fictitiously increased to the gasification temperature by passing through a heater (ASHTR). Ash and dust from the syngas were separated by passing through an SSplit (CYCLONE). The syngas exiting the gasifier was cooled down to a temperature of 30 °C, suitable for ICE operation, by passing through a heat exchanger (COOLER). The thermal energy could be recovered for other processes, e.g., air preheating in the gasification or other purposes based on the production capacity [36].

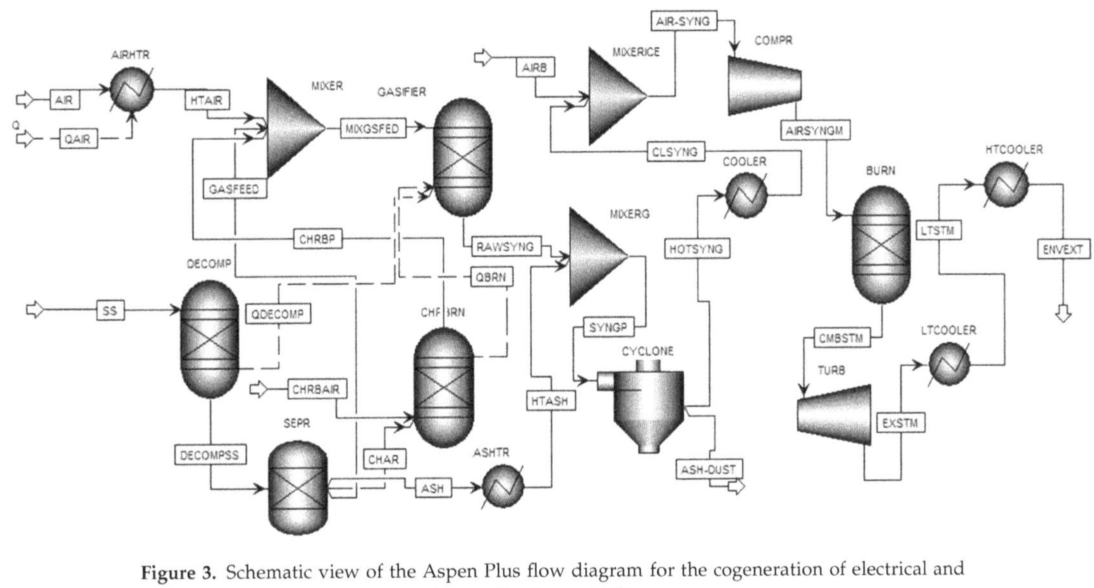

Figure 3. Schematic view of the Aspen Plus flow diagram for the cogeneration of electrical and thermal energy from SS by integrating gasification and ICE systems.

The ICE system was simulated in Aspen Plus by considering three consecutive thermodynamic processes of isentropic compression, constant volume combustion, and isentropic expansion to generate mechanical energy from the potential energy of syngas [20,36]. The potential energy of the syngas and air mixture was raised through isentropic compression, whereas combustion at a constant volume generated thermal energy through the ignition of syngas. Finally, isentropic expansion produced mechanical energy from the thermal energy of the air–syngas combustion streams to drive a generator and produce electrical energy [36,43,56–59].

Thermal energy recovery from the ICE exhaust was simulated by considering two steps of cooling in two heat exchangers. The first stage (LTCOOLER) simulated the recovery of low-temperature (LT) thermal energy available from the engine coolant stream that was cooled down from 90 °C to 80 °C [36]. The exhaust stream from ICE passed through a heat exchanger (HTCOOLER) to extract the high-temperature (HT) thermal energy. Both the LT and HT thermal energies could be used to reduce the moisture content of mechanically dewatered SS to a gasification standard of <10 wt.% or for district heating based on the production capacity of the plant [20].

3.1. Gasification Modeling

Biomass gasification simulation by using the chemical equilibrium method had a better agreement with the experimental outcomes compared with the kinetic model [59]. The following assumptions were taken into account to simplify the gasification model's development [9,59–61].

- The model was zero-dimensional;
- The stream fed to the pyrolysis zone was instantaneously disintegrated;
- There was no temperature gradient inside the gasifier in the radial direction, which ensured isothermal conditions;
- The hydrodynamic features of the gasifier were neglected;
- Equilibrium conditions were reached by all the considered gasification reactions;
- Intermediate product formation pathways during the gasification process were not considered;
- Char was full of carbon and combusted to generate thermal energy;
- Ideal behavior was shown by all the gaseous components;
- Gasification was completed under atmospheric pressure;
- Tar formation was not assessed as assumed by other researchers during SS gasification model development [9,28,29,44,49]. The LHV of syngas generated from biomass, including SS deposition, changed with the tar content deposition due to the alteration in the operating conditions. The tar content was deposited at higher temperatures and lower ER. However, the effect on syngas LHV alteration was less than 5% [37,62]. Therefore, neglecting tar formation did not significantly affect the potentiality of electrical and thermal energy cogeneration, which was the main aim of the current research.

Seven reactions were considered to simulate the SS gasification process through the Rgibbs reactor (GASIFIER block of Figure 3) and are listed in Table 3 with the heat of the reaction (ΔH) [63,64].

Table 3. Counted chemical reactions with ΔH during SS gasification process simulation.

Reaction No.	Reaction Scheme	Reaction Name	ΔH, (kJ/mol)
R1	$C + H_2O \rightarrow H_2 + CO$	Water gas	+131.0
R2	$C + O_2 \rightarrow CO_2$	Carbon combustion	−393.0
R3	$C + 2H_2 \rightarrow CH_4$	Methanation	−74.0
R4	$CO + H_2O \rightarrow H_2 + CO_2$	Water gas shift	−41.0
R5	$2C_2H_6 + 7O_2 \rightarrow 6H_2O + 4CO_2$	Ethane combustion	−1559.8
R6	$C_3H_8 + 5O_2 \rightarrow 4H_2O + 3CO_2$	Propane combustion	−2220.0
R7	$2H_2 + O_2 \rightarrow 2H_2O$	Hydrogen combustion	−242.0

The Boudouard reaction $(CO_2 + C \rightarrow 2CO)$ is not able to achieve the equilibrium point due to the shorter holding time of the reactants in the Rgibbs reactor of block GASIFIER [65]. For this reason, such a reaction was not taken into account in the formulation of the gasification model in the present analysis.

Calibration and Validation of Gasification Model

The reactions mentioned in Table 3, considered for the simulation of SS gasification, did not reach equilibrium for a unique temperature (gasification temperature) due to the distinct value of the equilibrium constants for each reaction. The movement of the equilibrium point of a chemical reaction in the forward or backward directions highly depends on temperature [66]. For this reason, the results related to syngas quality (composition), energy content (LHV), and conversion efficiencies (CCE and CGE), estimated through the equilibrium approach, significantly deviated from the experimental data [28,67]. However, the maximum deviation suggested for a model to argue for a reliable representation of experimental outcomes is ±20% [9,28,64]. As mentioned above, this condition was fulfilled by controlling the equilibrium position of individual gasification reactions to a particular temperature. The new equilibrium temperature assigned for each reaction completed in the block GASIFIER of the proposed conversion route illustrated in Figure 3 was calculated through a restricted chemical equilibrium approach according to Equation (1):

$$T_{Eqlm} = T_{Gasf} + \Delta T_{Appr} \quad (1)$$

where T_{Eqlm} stands for the equilibrium temperature in °C, T_{Gasf} denotes gasification the temperature in °C, and ΔT_{Appr} represents an approximate value of temperature in °C to inhibit the equilibrium position of each gasification reaction.

The deviation of the numerical results from experimental outcomes was calculated according to Equation (2):

$$Deviation\ (\%) = \frac{Simulation\ result - Experimental\ result}{Experimental\ result} \cdot 100 \quad (2)$$

As the syngas was composed of different constituents, calculating its average deviation may have been of interest and was calculated as reported in Equation (3):

$$Average\ Deviation\ (\%) = \frac{1}{N}\sum_{i=1}^{N} |Deviation| \quad (3)$$

where N is the number of the considered syngas component.

The gasification model was calibrated by considering the operating points and syngas properties under condition I depicted in Table 2, whereas conditions II to X were employed for the validation of the model.

The numerical value of ΔT_{Appr} (see Equation (1)) for each gasification reaction was identified by allocating a 5% standard deviation for each syngas component compared with the experimental campaign. The measured gasification (reduction zone) temperature for test condition I was 900 °C. The gasification temperature was used to identify the fraction of carbon converted to char that was burnt in the combustion zone to ensure autothermal operation. This was implemented in the model through an external Fortran subroutine.

Pearson's correlation coefficient was evaluated according to Equation (4) to examine the strength of the linear relationship between the air preheating temperature (x) and gasification process response (y) of the syngas composition and LHV [68].

$$r_{xy} = \frac{n\sum xy - \sum x \sum y}{\sqrt{\left[n\sum x^2 - \left(\sum x\right)^2\right]\left[n\sum y^2 - \left(\sum y\right)^2\right]}} \quad (4)$$

where r_{xy} is the correlation coefficient, x is the air preheating temperature, and y is the composition of the considered syngas components and the LHV of syngas.

3.2. Cogeneration Modeling

CHP cogeneration was simulated by counting the following assumptions [69]:

- Steady-state condition;
- Potential and kinetic energy remains unchanged, and pressure drops in the different blocks connected to complete the ICE system are neglected

3.3. Process Performance Evaluation

3.3.1. Gasification System

The parameters of the syngas *LHV*, *CGE*, *CCE*, and $\dot{P}_{net}$ gained from the products generated through *SS* gasification were considered to predict the gasification process performance.

The *LHV* of syngas was assessed from the composition corresponding to Equation (5) [70]:

$$LHV_{syng}(\mathrm{MJ/Nm^3}) = 0.108y_{\mathrm{H_2}} + 0.126y_{\mathrm{CO}} + 0.358y_{\mathrm{CH_4}} \tag{5}$$

where $y_{\mathrm{H_2}}$, y_{CO}, and $y_{\mathrm{CH_4}}$ are the fractions of H_2, CO, and CH_4 by volume available in the syngas, respectively.

The ratio of the available energy exiting from the gasification reactor as syngas to the investment as SS was designated by CGE and was calculated based on Equation (6) [71]:

$$CGE\ (\%) = \frac{LHV_{syng} \cdot \dot{V}_{syng}}{LHV_{SS} \cdot \dot{m}_{SS}} \cdot 100 \tag{6}$$

where LHV_{syng} and LHV_{SS} correspond to the *LHV* of the syngas in MJ/Nm^3 and SS in MJ/kg, respectively, whereas $\dot{V}_{syng}$ and $\dot{m}_{SS}$ are the syngas volumetric flow rate in Nm^3/h and mass flow rate of *SS* in kg/h.

The fractional transformation by weight of the carbon content from the feed materials of SS to the product of syngas during the gasification process is CCE, and is expressed by Equation (7) [23]:

$$CCE(\%) = \frac{12}{22.4} \cdot \frac{\dot{V}_{syng}}{\dot{m}_{SS} \cdot C\% \cdot \sum_{i=1}^{5} n_i \cdot y_i} \cdot 100 \tag{7}$$

where i represents the syngas carbon-containing component (CO, CO_2, and CH_4), $C\%$ represents the *SS* carbon portion by weight, n_i represents the carbon number, and y_i is the volume fraction of compound i that exists in the syngas.

The discrepancy between the summation of primary power available in syngas and offered during syngas cooling and investment to complete air preheating to reach a specific temperature is $\dot{P}_{net}$, and is described in Equation (8):

$$\dot{P}_{net} = \dot{P}_{syng} + \dot{P}_{syngcl} - \dot{P}_{prht} \tag{8}$$

where $\dot{P}_{syng}$ and $\dot{P}_{syngcl}$ are the primary power in kW offered by syngas and syngas cooling to the useable temperature for ICE, respectively, and $\dot{P}_{prht}$ represents the thermal power demand in kW for air preheating.

3.3.2. Cogeneration System

The electrical (η_{el}), thermal (η_{th}), and system (η_{sys}) efficiencies of the CHP system were determined according to Equations from (9) to (11), respectively.

$$\eta_{el}(\%) = \frac{\dot{P}_{Eff\_el}}{LHV_{Syng} \cdot \dot{V}_{syng}} \cdot 100 \tag{9}$$

$$\eta_{th}(\%) = \frac{\dot{P}_{th}}{LHV_{Syng} \cdot \dot{V}_{syng}} \cdot 100 \tag{10}$$

$$\eta_{sys}(\%) = \frac{\dot{P}_{Eff\_el} + \dot{P}_{syngcl} + \dot{P}_{th}}{LHV_{SS} \cdot \dot{m}_{SS} + \dot{P}_{prht}} \cdot 100 \quad (11)$$

where $\dot{P}_{Eff\_el}$ represents the effective rate of electric power obtained from the ICE in kW, LHV_{Syng} is the syngas LHV in kWh/Nm^3, $\dot{P}_{th}$ is the rate of thermal power available (summation of both LT and HT thermal energy) in kW, $\dot{P}_{syngcl}$ denotes the thermal power generation from syngas cooling in kW, LHV_{SS} is the LHV of SS in kWh/kg, and $\dot{P}_{prht}$ is the rate of power invested to complete air preheating in kW.

3.4. Cogeneration Process Simulation: Operating Parameters

The simulation of the CHP system fueled by syngas was completed by collecting the operating conditions of a real biomass gasification-based cogeneration plant, i.e., CMD–ECO20x where the gasification process was integrated with an ICE with a capacity of 20.0 kW_{el}. These operating data are presented in Table 4 [36].

Table 4. Operating parameters involved in the simulation of the ICE system.

Operating Parameters, Unit		Value
Incoming syngas temperature to the ICE combustion chamber (°C)		30.0
Incoming air temperature to the ICE (°C)		20.0
Equivalence ratio used for syngas combustion (-)		1.11
Pressure of ICE system for compression and combustion (bar)		9.45
Energy losses during syngas combustion (% of syngas energy)		10.0
Isentropic efficiency (%)	Compression process	85.0
	Expansion process	87.0
Mechanical efficiency (%)	Compression process	99.0
	Expansion process	
Exhaust stream pressure from the ICE (bar)		1.0
Exhaust stream Temperature from the ICE (°C)		338.0
Exhaust fume utilization temperature (°C)		80.0
Cooling water temperature to the ICE radiator	Inlet (°C)	80.0
	Exit (°C)	90.0

4. Results and Discussion

4.1. Development of Gasification Model

The gasification model was calibrated to identify the ΔT_{Appr} for each gasification reaction of test condition I by using built-in Aspen Plus regression tools, and the outcomes are shown in Table 5.

Table 5. Limiting temperature to restrict each gasification reaction equilibrium and the fraction of carbon that moved to char formation.

Reaction No.	ΔT_{Appr} (°C)
R1	−292.7
R2	51.1
R3	−493.2
R4	500
R5	−320.8
R6	252.2
R7	−469.7
Fraction of carbon moving to char formation	0.0879

Considering the model calibration findings shown in Table 5, four reactions of those considered in the current simulations (R1, R3, R5, and R7) were stopped below the equilibrium temperature, whereas the other three (R2, R4, and R6) moved to above the equilibrium point. Finally, 8.79% of the carbon from SS moved to the combustion zone as char to ensure autothermal gasification.

The difference between the predicted syngas composition with the corresponding energy content (LHV) and experimental results is presented in Figure 4 for model calibration and in Figure 5 for validation.

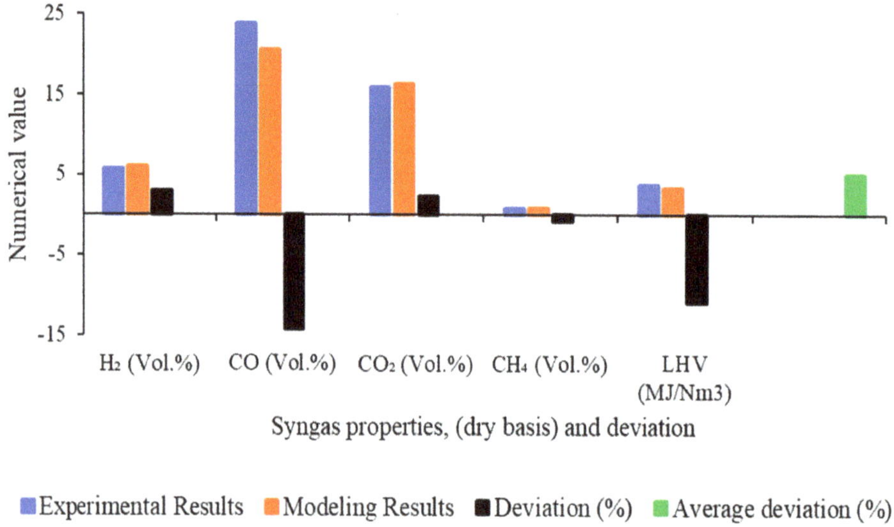

Figure 4. Syngas composition with LHV: deviation of numerical results from experimental outcomes during model calibration (test condition I).

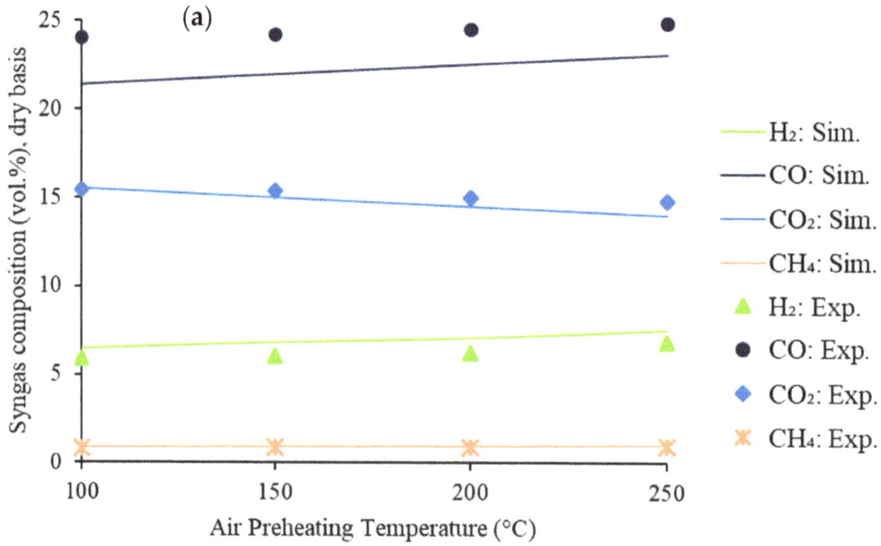

Figure 5. *Cont.*

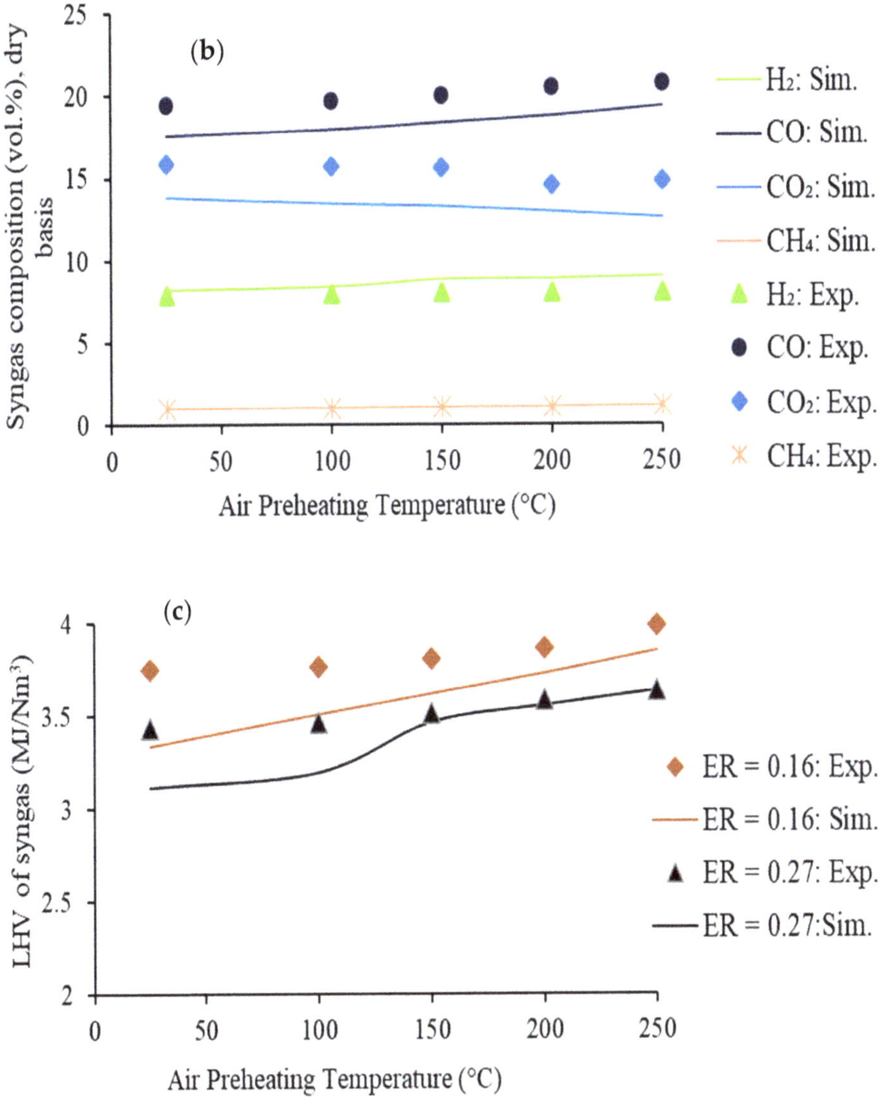

Figure 5. Syngas composition with LHV: deviation of numerical results from experimental outcomes during model validation ((**a**) conditions II to V, (**b**) conditions VI to X, and (**c**) LHV of syngas for the gasification at ER of 0.16 and 0.27).

The Pearson's correlation between the air preheating temperature and composition of the syngas components for gasification at ER of 0.16 and 0.27 with syngas LHV is presented in Table 6.

Considering the results depicted in Table 6, the variables of the gasifying agent preheating temperature, the considered syngas components, and LHV had a strong correlation, as the correlation coefficient was higher than 0.8 for all cases [68,72].

The predicted ethane and propane are not presented in Figures 4 and 5, as these two components were not identified during the experimental campaign.

Table 6. Pearson's correlation coefficient between gasifying agent preheating temperature and process response.

		ER = 0.16			
Syngas component	H_2	CO	CO_2	CH_4	LHV
Experimental	0.93	0.99	−0.97	0.95	0.93
Simulation	0.999	0.999	−0.999	0.998	0.999
		ER = 0.27			
Experimental	0.97	0.98	−0.88	0.93	0.98
Simulation	0.96	0.93	−0.96	0.997	0.97

As clearly shown from the comparative results predicted from the simulated model and experimental campaign presented in Figures 4 and 5a–c, the numerical model represented the experimental outcomes with good agreement, as the deviation for the concentration of individual syngas components (varied in the range of −0.77% to 14.97%), as well as LHV (between 5.46% and 11.23%) of syngas, was lower than ±15% during model calibration and validation, satisfying the tolerable limit mentioned in the available literature of less than ±20% [9,64]. Hence the developed gasification model represented the experimental campaign accurately and efficiently. The syngas LHV was underpredicted due to the underestimation of the concentration of the syngas components of CO and CH_4 from the experimental campaign outcomes, as these two constituents were the major contributors to the syngas LHV [70,73]. The higher ash content present in the examined SS sample was the main reason for the highest deviation in terms of CO and CO_2 prediction through the formulated gasification model from the experimental campaign results [74]. However, the deviation of the individual syngas components with LHV from the experimental data was in the range of −0.77% to 14.97%, which was substantially lower compared with the other relevant literature, between 5.94 to 31.69% for the gasification model development related to SS, oil sludge, and Almond Shell [9,28,55,75,76].

4.2. Sensitivity Analysis

The alteration in syngas quality in terms of the composition, LHV, and gasification process performances of CGE, CCE, and $\dot{P}_{net}$ with air preheating temperature and ER were analyzed to determine the optimum parameters for the examined SS sample. A mass flow rate of 27.78 g/s was considered in the analysis.

4.2.1. Effect of Air Preheating Temperature

The trend of syngas quality (composition and LHV) and gasification process performance ($\dot{P}_{net}$, CCE, and CGE) with the air preheating temperature from room temperature (25 °C) to 500 °C was examined at a constant ER of 0.16. The results are illustrated in Figure 6.

As air preheating temperature increases:

- the concentration of two syngas components (CO and H_2) in syngas increases continuously
- the concentration of other two components (CO_2 and C_3H_8) shows a decreasing trend
- the concentration of CH_4 shows a slowly increasing trend.

The fluctuation of syngas composition with incoming air temperature is explained by the change in exothermic and endothermic reaction rates [9,55,76,77]. The uninterrupted reduction of CO_2 and C_3H_8 composition is explained by the backward movement of exothermic reactions (propane formation and combustion reactions) with air preheating temperature. This occurs due to a fixed quantity of char (8.79% of carbon present in SS) combusted to ensure the autothermal mode of gasification which is independent of operating parameters. On the other hand, the endothermic reactions (water gas and water gas shift) rate raises with air preheating temperature, causing the increase of CO and H_2 composition in syngas [33,78].

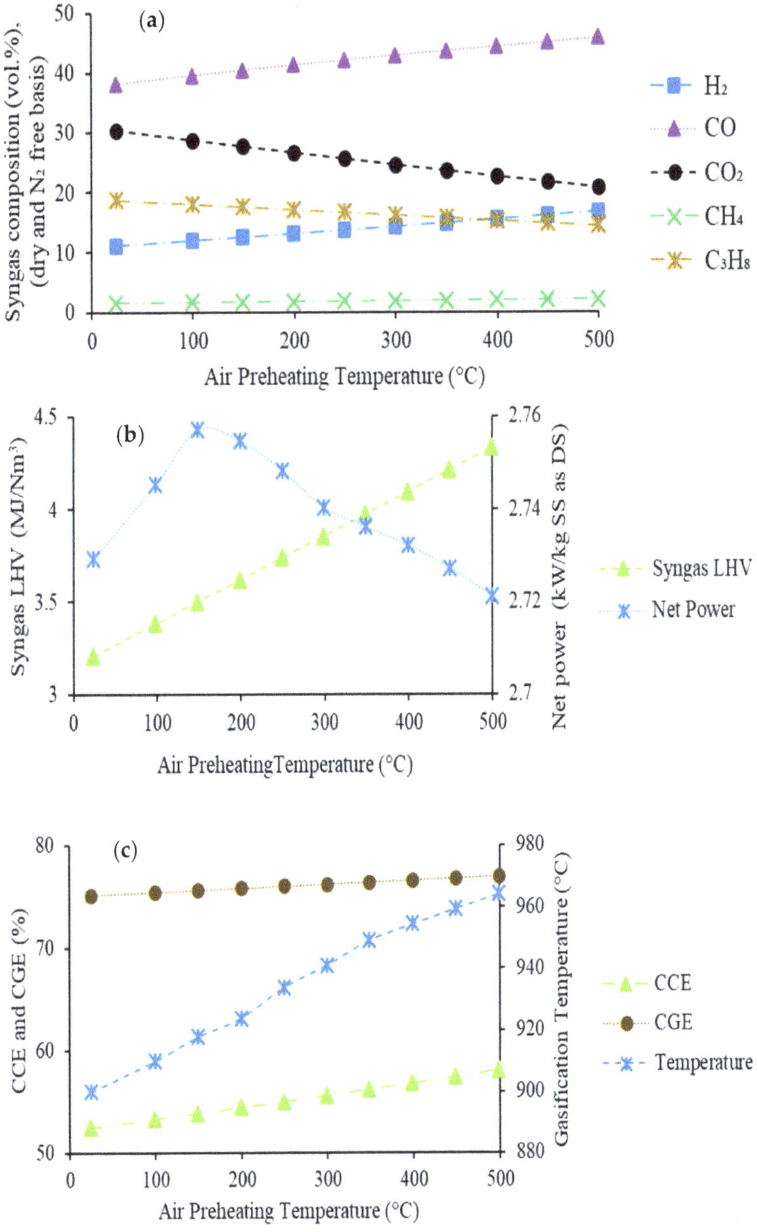

Figure 6. Fluctuation of (**a**) syngas composition, (**b**) syngas LHV and $\dot{P}_{net}$ from gasification product, and (**c**) CCE, CGE, and gasification temperature with air preheating temperature.

Syngas LHV continuously increases with air preheating temperature as the concentration of H_2, CO, and CH_4 raises as shown in Figure 6a [70,73]. Consequently, the energy content of syngas increases as well as gain from syngas cooling to ambient temperature (30 °C), and the thermal power needed for air preheating also increases. The trend of $\dot{P}_{net}$ reflects these three aspects, thus it increases up to an air preheating temperature of 150 °C and decreases further. Indeed, the increase in energy gain for air preheating temperatures from 150 to 200 °C does not compensate for the increase in thermal power demand.

CGE increased with air preheating temperature due to the increase in syngas energy content [23], as presented in Figure 6b. The cumulative increment in the carbon fractions of the syngas through the formation of CO and CH_4 with air preheating temperature was greater compared with the decrement by CO_2 and C_3H_8 concentration decrement, as illustrated in Figure 6a. Subsequently, a continuous increment in CCE was observed with air preheating temperature. The gasification temperature increased continuously with the air preheating temperature due to the increase in the endothermic reaction rate [55,76,77]. This was due to the fixed quantity of char being combusted to ensure the autothermal mode of gasification.

Based on the simulation results presented in Figure 6, the predicted optimum air preheating temperature was 150 °C because a further increase caused a reduction in the available $\dot{P}_{net}$ during energy recovery from SS through gasification treatment in the autothermal operation mode.

4.2.2. Influence of the Equivalence Ratio

The effect of varying ER between 0.1 and 0.4 of the individual syngas components' concentration, LHV, CCE, CGE, $\dot{P}_{net}$, and gasification temperature was examined at an estimated ideal air preheating temperature of 150 °C. The outcomes are depicted in Figure 7.

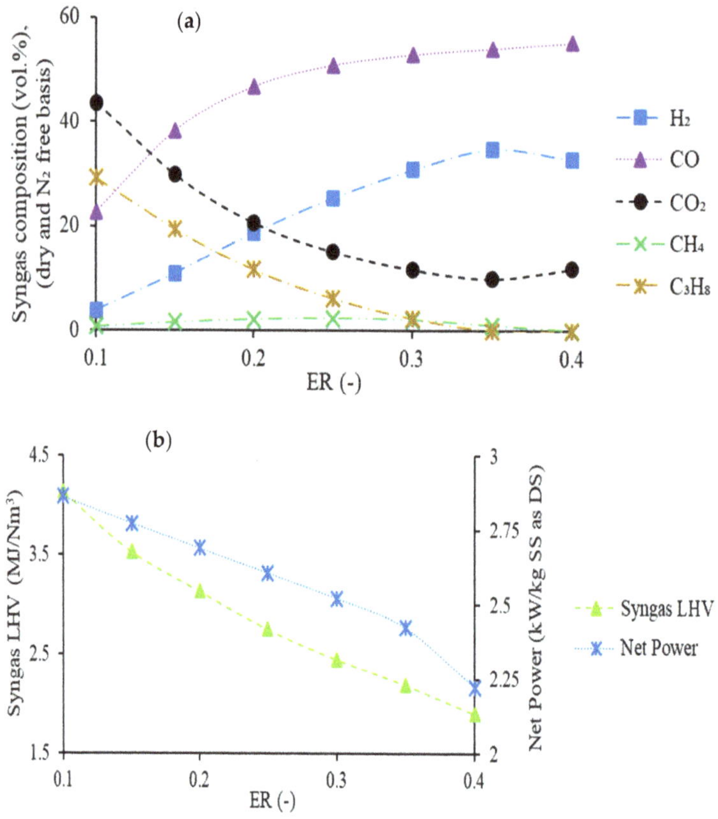

Figure 7. *Cont.*

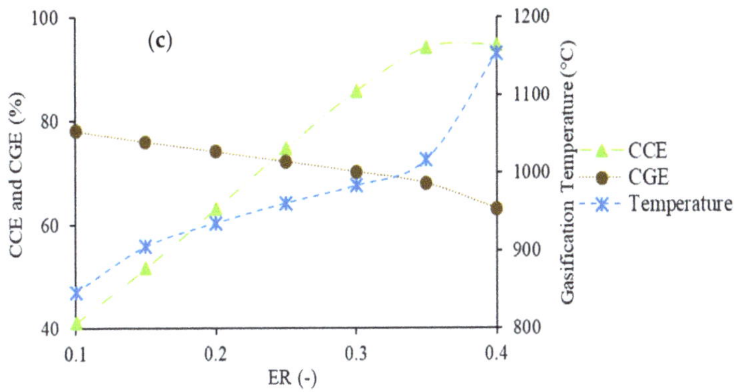

Figure 7. Variation in the (**a**) composition of individual syngas components, (**b**) syngas LHV and $\dot{P}_{net}$ obtained from SS gasification, and (**c**) CCE, CGE, and gasification temperature with ER.

Increasing the ER caused a continuous increase in the oxygen concentration inside the gasifier, which drives drove the onward movement of the combustion reaction to lessen the impact, which was fully described by Le Chatelier's principle [60]. Consequently, the product concentration involved in the combustion reactions (CO_2 and H_2O) grew continuously, whereas a reverse pattern was visible for the reactant concentrations of C_3H_8, H_2, and CH_4. The reaction rate of the water gas, water gas shift, and steam methane reforming reactions increased with an increase in the CO_2 and H_2O concentrations, which drove the increase in the concentrations of the syngas components of CO, H_2, CH_4, [33,65,78]. A reverse trend was observed for CO_2.

The syngas LHV decreased steadily with ER as N_2 entrance to the gasifier increased, which created a dilution effect and was responsible for the continuous decrease in the syngas energy content [60,79,80]. On the other hand, $\dot{P}_{net}$ decreased continuously with ER due to the decrease in the available primary power (LHV) obtained from the syngas. At the same time, the thermal power needed to complete air preheating increased due to the increase in the incoming air volume. Although the incoming air volume increased, the power obtained from syngas cooling increased very slowly with ER. The gasification efficiency in terms of CGE decreased with ER due to the decrease in the syngas energy content [23]. The combustion reaction rate increased with ER; consequently, more carbon was transformed from the feed to the products, causing an increase in CCE. The increase in ER during SS gasification in the autothermal mode was responsible for the increase in the gasification temperature due to the forward movement of oxidation reactions, which were exothermic and generated more heat inside the gasifier [55,76,77].

A minimum value of 0.2 was predicted as an ideal ER to generate syngas from SS via gasification, although the syngas LHV, $\dot{P}_{net}$, and CGE were higher at an ER of 0.15. However, the conversion of SS to syngas at an ER of <0.20 favored the pyrolysis reaction, instead of gasification, and was liable for the formation of more tar, ash, and other impurities [9,31,76]. The complete removal of these pollutants from syngas, which is mandatory before further utilization in ICEs or chemical synthesis, would be more costly compared with the net power increase for the gasification of SS at ER of 0.15, instead of 0.20 [38].

The effect of air preheating temperature and ER on syngas quality in terms of composition, LHV, and process performances (CCE and CGE) predicted in the present analysis was in accordance with the available literature related to syngas generation from SS or waste biomass [9,28–31,43,55,60,75,76,79–83].

4.3. Cogeneration Process Performances

The evaluated electrical and thermal power generation potentiality of the SS used in the current study were 0.81 kW_{el}/kg SS as DS and 1.73 kW_{th}/kg SS as DS (of which 0.39 kW was HT and 1.34 kW was LT thermal power), respectively. The comparison between CGE, η_{el}, η_{th}, and η_{sys} predicted from the present analysis with the available literature on similar simulation studies is presented in Table 7.

Table 7. Comparison of CGE, η_{el}, η_{th}, and η_{sys} predicted in the present study with the available literature.

Biomass	CGE (%)	η_{el} (%)	η_{th} (%)	η_{sys} (%)	Reference
SS	74.1	25.7	51.5	74.9	Present study
	74.6	25.6	47.8	76.7	[20]
	71.9	24.6	45.9	76.8	
	72.3	24.8	46.3	67.4	[48]
	n.r.	32.0	n.r.	n.r.	[84]
	69.0	20.0	n.r.	n.r.	[85]
Wood	n.r.	27.0	40.0	67.0	[58]
MSW	59.0	19.1	20.0	40.1	[59]

n.r. = mot reported, MSW = municipal solid waste.

The predicted emission profile from ICE contained pollutants, as follows: CO_2 = 1.36 kg/kW$_{el}$; SO_2 = 0.039 kg/kW$_{el}$, and NOx = 0.015 kg/kW$_{el}$.

The conversion of SS to electrical and thermal energy through the proposed system could reduce the use of fossil fuels to generate CHP for the wastewater treatment sector, which is highly energy-consuming, especially in terms of electricity.

According to the International Energy Agency data, the national electrical energy demand in Poland in 2019 was equal to 166 TWh. WWTPs consume around 1% of the national electricity demand in EU countries [86]. The transformation of the energy content present in all of the produced SS in Poland in 2019 (575 kt as DS, according to the Eurostat data) to CHP through the proposed treatment scheme would allow supplying 30% of the electrical energy consumption of wastewater treatment and 54% of the thermal energy required for the drying of mechanically dewatered SS [9]. Thus, in terms of environmental impact, 1.97 Mt of equivalent CO_2/yr would be saved over a similar amount of CHP generation using natural gas.

Such an aspect appears to be particularly important in Poland, where the national energy demand is mainly supplied by coal and the share of renewables in the gross final energy consumption is only 16.1% against 22.1% of EU–27.

The main challenge of the energy generation from SS through the proposed system is decreasing organic (tar) and inorganic (hydrogen sulfide, hydrogen chloride, and alkali metals) pollutants formation or their removal from the generated syngas through a gasification process. The tar, HCl, H_2S, and alkali metal contents in the syngas produced in the gasification of SS could be further reduced by selecting a suitable catalyst, optimizing the operating parameters and reactor configuration, or the filtration of the syngas [87–89]

Life-cycle assessment and the economic analysis of the proposed system were not within the scope of the present work; however, these aspects will be investigated in future analysis to evaluate the environmental and economic benefits of the proposed plant for the valorization of SS to energy products. Moreover, the gasification simulation model will be extended to account for the tar content.

5. Conclusions

An experimental campaign was conducted in a fixed-bed updraft gasifier to complete the energy recovery from SS as syngas through air gasification in the autothermal mode. SS gasification was carried out at two different ERs of 0.16 and 0.27 and five distinct air preheating temperatures of 25, 100, 150, 200, and 250 °C. A computer-aided numerical model in Aspen Plus V8.8 was calibrated based on the experimental data obtained at an air preheating temperature and ER of 25 °C and 0.16, respectively, and validated at nine different operating conditions. The impact of air the preheating temperature and ER on the syngas composition and energy content, as well as on the process performances of CCE, CGE, and $\dot{P}_{net}$ obtained from the gasification treatment, was analyzed to identify the optimum operating parameters, which were found to be 150 °C and at least 0.2, respectively. Further, an ICE fueled by the produced syngas was modeled to assess the CHP generation potential, finding 0.81 kW_{el}/kg SS as DS and 1.73 kW_{th}/kg SS as DS of electrical and thermal power, respectively. The ICE emission profiles were predicted to estimate the concentration of pollutants.

The work carried out allows assessing syngas production from SS that, as a renewable source, may contribute to decreasing fossil fuel dependency. Considering Poland as an example, if all the produced SS in 2019 was used for energy recovery, it would cover 30% of the electricity demand of the wastewater treatment sector and 54% of the thermal energy required to decrease the moisture content of mechanically dewatered SS to the gasification standard (from 70 wt.% to 10 wt.%), saving 1.97 Mt of equivalent CO_2/yr over the natural gas power CHP-generation technique needed to cover the same energy loads.

Author Contributions: Conceptualization, A.C., S.D.F., N.M., S.S., M.R.U. and L.V.; methodology, S.D.F. and M.R.U.; validation, S.D.F. and M.R.U.; formal analysis, S.D.F. and M.R.U.; investigation, S.D.F., N.M. and M.R.U.; resources, S.D.F. and M.R.U.; data curation, S.D.F. and M.R.U.; writing—original draft preparation, M.R.U.; writing—review and editing, A.C., S.D.F., N.M., S.S., L.V. and S.W.; visualization, S.D.F. and M.R.U.; supervision, S.D.F. and N.M. All authors have read and agreed to the published version of the manuscript.

Funding: This research received no external funding.

Conflicts of Interest: The authors declare no conflict of interest.

References

1. IEA Key World Energy Statistics 2021, IEA, Paris. Available online: https://www.iea.org/reports/key-world-energy-statistics-2021 (accessed on 12 December 2022).
2. Ritchie, H.; Roser, M. Energy. Available online: https://ourworldindata.org/energy (accessed on 10 January 2022).
3. Roser, M. Future Population Growth. Available online: https://ourworldindata.org/future-population-growth (accessed on 13 December 2022).
4. Londoño-Pulgarin, D.; Cardona-Montoya, G.; Restrepo, J.C.; Muñoz-Leiva, F. Fossil or bioenergy? Global fuel market trends. *Renew. Sustain. Energy Rev.* **2021**, *143*, 110905. [CrossRef]
5. Sharma, P.; Sivaramakrishnaiah, M.; Deepanraj, B.; Saravanan, R.; Reddy, M.V. A novel optimization approach for biohydrogen production using algal biomass. *Int. J. Hydrogen Energy* 2022, *in press*. [CrossRef]
6. Hossain, M.S.; ul Karim, T.; Onik, M.H.; Kumar, D.; Rahman, M.A.; Yousuf, A.; Uddin, M.R. Impact of temperature, inoculum flow pattern, inoculum type, and their ratio on dry anaerobic digestion for biogas production. *Sci. Rep.* **2022**, *12*, 6162. [CrossRef]
7. Uddin, M.N.; Salam, M.A.; Uddin, M.R.; Khan, M.R. Biodiesel from Nahor Seed Oil: Synthesis, Evaluation, and Compatibility with Petrodiesel. *Pet. Chem.* **2022**, *62*, 1211–1222. [CrossRef]
8. IEA Global Energy Review: CO2 Emissions in 2020, IEA, Paris. Available online: https://www.iea.org/articles/global-energy-review-co2-emissions-in-2020 (accessed on 12 December 2022).
9. Di Fraia, S.; Massarotti, N.; Uddin, M.R.; Vanoli, L. Conversion of Sewage Sludge to combined heat and power: Modeling and optimization. *Smart Energy* **2021**, *5*, 100061. [CrossRef]
10. European Commission Environmental, Economic and Social Impacts of the Use of Sewage Sludge on Land. Consult. Rep. Options Impacts, Rep. by RPA. Milieu Ltd. WRc Eur. Comm. DG Environ. Under Study Contract DG ENV.G.4/ ETU/2008/0076r. 2009, pp. 1–20. Available online: https://www.efar.be/wp-content/uploads/Draft-Impact-Assessment-Sludge-Directive.pdf (accessed on 10 December 2022).
11. Mininni, G.; Mauro, E.; Piccioli, B.; Colarullo, G.; Brandolini, F.; Giacomelli, P. Production and characteristics of sewage sludge in Italy. *Water Sci. Technol.* **2019**, *79*, 619–626. [CrossRef] [PubMed]
12. Monte, M.C.; Fuente, E.; Blanco, A.; Negro, C. Waste management from pulp and paper production in the European Union. *Waste Manag.* **2009**, *29*, 293–308. [CrossRef]
13. Lenssen, N.J.L.; Schmidt, G.A.; Hansen, J.E.; Menne, M.J.; Persin, A.; Ruedy, R.; Zyss, D. Improvements in the GISTEMP Uncertainty Model. *J. Geophys. Res. Atmos.* **2019**, *124*, 6307–6326. [CrossRef]

14. European Commission. European Green Deal : Commission Proposes Transformation of EU Economy and Society to Meet Climate Ambitions. 2021. Available online: https://commission.europa.eu/strategy-and-policy/priorities-2019-2024/european-green-deal/delivering-european-green-deal_en (accessed on 15 December 2022).
15. Eurostat Sewage Sludge Generation and Management. Available online: https://appsso.eurostat.ec.europa.eu/nui/submitViewTableAction.do (accessed on 30 January 2022).
16. Eurostat Population Projection. Available online: https://ec.europa.eu/eurostat/statistics-explained/index.php?oldid=497115#Population_projections (accessed on 27 July 2022).
17. Liu, X.; Chang, F.; Wang, C.; Jin, Z.; Wu, J.; Zuo, J.; Wang, K. Pyrolysis and subsequent direct combustion of pyrolytic gases for sewage sludge treatment in China. *Appl. Therm. Eng.* **2018**, *128*, 464–470. [CrossRef]
18. Zaker, A.; Chen, Z.; Zaheer-Uddin, M. Catalytic pyrolysis of sewage sludge with HZSM5 and sludge-derived activated char: A comparative study using TGA-MS and artificial neural networks. *J. Environ. Chem. Eng.* **2021**, *9*, 105891. [CrossRef]
19. Miricioiu, M.G.; Zaharioiu, A.; Oancea, S.; Bucura, F.; Raboaca, M.S.; Filote, C.; Ionete, R.E.; Niculescu, V.C.; Constantinescu, M. Sewage sludge derived materials for CO_2 adsorption. *Appl. Sci.* **2021**, *11*, 7139. [CrossRef]
20. Carotenuto, A.; Di Fraia, S.; Massarotti, N.; Sobek, S.; Uddin, M.R.; Vanoli, L.; Werle, S. Predictive modeling for energy recovery from sewage sludge gasification. *Energy* **2023**, *263*, 125838. [CrossRef]
21. Vishwajeet; Pawlak-Kruczek, H.; Baranowski, M.; Czerep, M.; Chorążyczewski, A.; Krochmalny, K.; Ostrycharczyk, M.; Ziółkowski, P.; Madejski, P.; Mączka, T.; et al. Entrained Flow Plasma Gasification of Sewage Sludge–Proof-of-Concept and Fate of Inorganics. *Energies* **2022**, *15*, 1948. [CrossRef]
22. Seggiani, M.; Vitolo, S.; Puccini, M.; Bellini, A. Cogasification of sewage sludge in an updraft gasifier. *Fuel* **2012**, *93*, 486–491. [CrossRef]
23. Jeong, Y.S.; Choi, Y.K.; Park, K.B.; Kim, J.S. Air co-gasification of coal and dried sewage sludge in a two-stage gasifier: Effect of blending ratio on the producer gas composition and tar removal. *Energy* **2019**, *185*, 708–716. [CrossRef]
24. Martínez, E.J.; Sotres, A.; Arenas, C.B.; Blanco, D.; Martínez, O.; Gómez, X. Improving Anaerobic Digestion of Sewage Sludge by Hydrogen Addition: Analysis of Microbial Populations and Process Performance. *Energies* **2019**, *12*, 1228. [CrossRef]
25. Costa, M.; Buono, A.; Caputo, C.; Carotenuto, A.; Cirillo, D.; Costagliola, M.A.; Di Blasio, G.; La Villetta, M.; Macaluso, A.; Martoriello, G.; et al. The "INNOVARE" project: Innovative plants for distributed poly-generation by residual biomass. *Energies* **2020**, *13*, 4020. [CrossRef]
26. Werle, S.; Wilk, R.K. A review of methods for the thermal utilization of sewage sludge: The Polish perspective. *Renew. Energy* **2010**, *35*, 1914–1919. [CrossRef]
27. Di Fraia, S.; Uddin, M.R. Energy Recovery from Waste Paper and Deinking Sludge to Support the Demand of the Paper Industry: A Numerical Analysis. *Sustainability* **2022**, *14*, 4669. [CrossRef]
28. Abdelrahim, A.; Brachi, P.; Ruoppolo, G.; Di Fraia, S.; Vanoli, L. Experimental and Numerical Investigation of Biosolid Gasification: Equilibrium-Based Modeling with Emphasis on the Effects of Different Pretreatment Methods. *Ind. Eng. Chem. Res.* **2020**, *59*, 299–307. [CrossRef]
29. de Andrés, J.M.; Vedrenne, M.; Brambilla, M.; Rodríguez, E. Modeling and model performance evaluation of sewage sludge gasification in fluidized-bed gasifiers using Aspen Plus. *J. Air Waste Manag. Assoc.* **2019**, *69*, 23–33. [CrossRef]
30. De Andrés, J.M.; Narros, A.; Rodríguez, M.E. Air-steam gasification of sewage sludge in a bubbling bed reactor: Effect of alumina as a primary catalyst. *Fuel Process. Technol.* **2011**, *92*, 433–440. [CrossRef]
31. Freda, C.; Cornacchia, G.; Romanelli, A.; Valerio, V.; Grieco, M. Sewage sludge gasification in a bench scale rotary kiln. *Fuel* **2018**, *212*, 88–94. [CrossRef]
32. Mun, T.Y.; Kim, J.W.; Kim, J.S. Air gasification of dried sewage sludge in a two-stage gasifier: Part 1. the effects and reusability of additives on the removal of tar and hydrogen production. *Int. J. Hydrogen Energy* **2013**, *38*, 5226–5234. [CrossRef]
33. Mun, T.Y.; Kim, J.S. Air gasification of dried sewage sludge in a two-stage gasifier. Part 2: Calcined dolomite as a bed material and effect of moisture content of dried sewage sludge for the hydrogen production and tar removal. *Int. J. Hydrogen Energy* **2013**, *38*, 5235–5242. [CrossRef]
34. Mun, T.Y.; Cho, M.H.; Kim, J.S. Air gasification of dried sewage sludge in a two-stage gasifier. Part 3: Application of olivine as a bed material and nickel coated distributor for the production of a clean hydrogen-rich producer gas. *Int. J. Hydrogen Energy* **2014**, *39*, 5634–5643. [CrossRef]
35. Chen, Y.H.; Lan Thao Ngo, T.N.; Chiang, K.Y. Enhanced hydrogen production in co-gasification of sewage sludge and industrial wastewater sludge by a pilot-scale fluidized bed gasifier. *Int. J. Hydrogen Energy* **2021**, *46*, 14083–14095. [CrossRef]
36. Cirillo, D.; Di Palma, M.; La Villetta, M.; Macaluso, A.; Mauro, A.; Vanoli, L. A novel biomass gasification micro-cogeneration plant: Experimental and numerical analysis. *Energy Convers. Manag.* **2021**, *243*, 114349. [CrossRef]
37. Tian, Y.; He, D.; Zeng, Y.; Hu, L.; Du, J.; Luo, Z.; Ma, W.; Zhang, Z. Experimental research on hydrogen-rich syngas yield by catalytic biomass air-gasification over Ni/olivine as in-situ tar destruction catalyst. *J. Energy Inst.* **2023**, *108*, 101263. [CrossRef]
38. Chae, H.J.; Kim, J.; Lee, S.C.; Kim, H.; Jo, S.B.; Ryu, J.; Kim, T.Y.; Lee, C.H.; Kim, S.J.; Kang, S. Catalytic Technologies for CO Hydrogenation for the Production of Light Hydrocarbons and Middle Distillates. *Catalysts* **2020**, *10*, 99. [CrossRef]
39. Wibowo, H.; Susanto, H.; Grisdanurak, N.; Hantoko, D.; Yoshikawa, K.; Qun, H.; Yan, M. Recent developments of deep eutectic solvent as absorbent for CO_2 removal from syngas produced from gasification: Current status, challenges, and further research. *J. Environ. Chem. Eng.* **2021**, *9*, 105439. [CrossRef]

40. Di Fraia, S.; Massarotti, N.; Vanoli, L.; Costa, M. Thermo-economic analysis of a novel cogeneration system for sewage sludge treatment. *Energy* **2016**, *115*, 1560–1571. [CrossRef]
41. Singh, V.; Phuleria, H.C.; Chandel, M.K. Estimation of energy recovery potential of sewage sludge in India : Waste to watt approach. *J. Clean. Prod.* **2020**, *276*, 122538. [CrossRef]
42. Francois, J.; Abdelouahed, L.; Mauviel, G.; Patisson, F.; Mirgaux, O.; Rogaume, C.; Rogaume, Y.; Feidt, M.; Dufour, A. Detailed process modeling of a wood gasification combined heat and power plant. *Biomass Bioenergy* **2013**, *51*, 68–82. [CrossRef]
43. Damartzis, T.; Michailos, S.; Zabaniotou, A. Energetic assessment of a combined heat and power integrated biomass gasification-internal combustion engine system by using Aspen Plus®. *Fuel Process. Technol.* **2012**, *95*, 37–44. [CrossRef]
44. Brachi, P.; Di Fraia, S.; Massarotti, N.; Vanoli, L. Combined heat and power production based on sewage sludge gasification: An energy-efficient solution for wastewater treatment plants. *Energy Convers. Manag. X* **2022**, *13*, 100171. [CrossRef]
45. Fatema, J.; Ahmed, T.; Islam, M.M.; Sakib, M.N.; Chowdhury, A.M.S.; Haque, P. Gasification of kitchen wastes in an updraft fluidized bed gasifier and simulation of the process with Aspen Plus. *J. Clean. Prod.* **2022**, *371*, 133670. [CrossRef]
46. Singh, M.; Salaudeen, S.A.; Gilroyed, B.H.; Dutta, A. Simulation of biomass-plastic co-gasification in a fluidized bed reactor using Aspen plus. *Fuel* **2022**, *319*, 123708. [CrossRef]
47. Hoo, K.K.; Md Said, M.S. Simulation of air gasification of Napier grass using Aspen Plus. *Sustain. Energy Technol. Assess.* **2022**, *50*, 101837. [CrossRef]
48. Carotenuto, A.; Di Fraia, S.; Massarotti, N.; Uddin, M.R.; Vanoli, L. Combined Heat and Power Generation from Mechanically Dewatered Sewage Sludge: Numerical Modelling. *Chem. Eng. Trans.* **2022**, *92*, 283–288. [CrossRef]
49. Migliaccio, R.; Brachi, P.; Montagnaro, F.; Papa, S.; Tavano, A.; Montesarchio, P.; Ruoppolo, G.; Urciuolo, M. Sewage Sludge Gasification in a Fluidized Bed: Experimental Investigation and Modeling. *Ind. Eng. Chem. Res.* **2021**, *60*, 5034–5047. [CrossRef]
50. Werle, S.; Dudziak, M. Gaseous fuels production from dried sewage sludge via air gasification. *Waste Manag. Res.* **2015**, *32*, 601–607. [CrossRef] [PubMed]
51. Werle, S. Impact of feedstock properties and operating conditions on sewage sludge gasification in a fixed bed gasifier. *Waste Manag. Res.* **2014**, *32*, 954–960. [CrossRef] [PubMed]
52. Werle, S. Gasification of a dried sewage sludge in a laboratory scale fixed bed reactor. *Energies* **2015**, *8*, 8562–8572. [CrossRef]
53. Giltrap, D.L.; McKibbin, R.; Barnes, G.R.G. A steady state model of gas-char reactions in a downdraft biomass gasifier. *Sol. Energy* **2003**, *74*, 85–91. [CrossRef]
54. Zainal, Z.A.; Ali, R.; Lean, C.H.; Seetharamu, K.N. Prediction of performance of a downdraft gasifier using equilibrium modeling for different biomass materials. *Energy Convers. Manag.* **2001**, *42*, 1499–1515. [CrossRef]
55. Acar, M.C.; Böke, Y.E. Simulation of biomass gasification in a BFBG using chemical equilibrium model and restricted chemical equilibrium method. *Biomass Bioenergy* **2019**, *125*, 131–138. [CrossRef]
56. Villarini, M.; Marcantonio, V.; Colantoni, A. Sensitivity Analysis of Different Parameters on the Performance of a CHP Internal Combustion Engine System Fed by a Biomass Waste Gasifier. *Energies* **2019**, *12*, 688. [CrossRef]
57. Lan, W.; Chen, G.; Zhu, X.; Wang, X.; Liu, C.; Xu, B. Biomass gasification-gas turbine combustion for power generation system model based on ASPEN PLUS. *Sci. Total Environ.* **2018**, *628–629*, 1278–1286. [CrossRef]
58. François, J.; Abdelouahed, L.; Mauviel, G.; Feidt, M.; Rogaume, C.; Mirgaux, O.; Patisson, F.; Dufour, A. Estimation of the energy efficiency of a wood gasification CHP plant using Aspen Plus. *Chem. Eng. Trans.* **2012**, *29*, 769–774. [CrossRef]
59. Násner, A.M.L.; Lora, E.E.S.; Palacio, J.C.E.; Rocha, M.H.; Restrepo, J.C.; Venturini, O.J.; Ratner, A. Refuse Derived Fuel (RDF) production and gasification in a pilot plant integrated with an Otto cycle ICE through Aspen plus™ modelling: Thermodynamic and economic viability. *Waste Manag.* **2017**, *69*, 187–201. [CrossRef]
60. Niu, M.; Huang, Y.; Jin, B.; Wang, X. Simulation of syngas production from municipal solid waste gasification in a bubbling fluidized bed using aspen plus. *Ind. Eng. Chem. Res.* **2013**, *52*, 14768–14775. [CrossRef]
61. Chen, C.; Jin, Y.Q.; Yan, J.H.; Chi, Y. Simulation of municipal solid waste gasification in two different types of fixed bed reactors. *Fuel* **2013**, *103*, 58–63. [CrossRef]
62. Roche, E.; De Andrés, J.M.; Narros, A.; Rodríguez, M.E. Air and air-steam gasification of sewage sludge. The influence of dolomite and throughput in tar production and composition. *Fuel* **2014**, *115*, 54–61. [CrossRef]
63. Sreejith, C.C.; Muraleedharan, C.; Arun, P. Performance prediction of steam gasification of wood using an ASPEN PLUS thermodynamic equilibrium model. *Int. J. Sustain. Energy* **2014**, *33*, 416–434. [CrossRef]
64. Pala, L.P.R.; Wang, Q.; Kolb, G.; Hessel, V. Steam gasification of biomass with subsequent syngas adjustment using shift reaction for syngas production: An Aspen Plus model. *Renew. Energy* **2017**, *101*, 484–492. [CrossRef]
65. Franco, C.; Pinto, F.; Gulyurtlu, I.; Cabrita, I. The study of reactions influencing the biomass steam gasification process. *Fuel* **2003**, *82*, 835–842. [CrossRef]
66. Barman, N.S.; Ghosh, S.; De, S. Gasification of biomass in a fixed bed downdraft gasifier—A realistic model including tar. *Bioresour. Technol.* **2012**, *107*, 505–511. [CrossRef] [PubMed]
67. Demirbaş, A. Carbonization ranking of selected biomass for charcoal, liquid and gaseous products. *Energy Convers. Manag.* **2001**, *42*, 1229–1238. [CrossRef]
68. Morris, R.D. Chapter Four. In *Composition with Pitch-Classes*; Yale University Press: London, UK, 2018; pp. 123–183.
69. Ruya, P.M.; Purwadi, R.; Lim, S.S. Supercritical water gasification of sewage sludge for power generation– thermodynamic study on auto-thermal operation using Aspen Plus. *Energy Convers. Manag.* **2020**, *206*, 112458. [CrossRef]

70. Zheng, X.; Chen, W.; Ying, Z.; Huang, J.; Ji, S.; Wang, B. Thermodynamic investigation on gasification performance of sewage sludge-derived hydrochar: Effect of hydrothermal carbonization. *Int. J. Hydrogen Energy* **2019**, *44*, 10374–10383. [CrossRef]
71. Hossain, M.S.; Kumar, D.; Paul, N.; Rahman, S.J.; Rahman, M.A.; Uddin, M.R.; Pirozzi, D.; Uddin, M.N.; Yousuf, A. Design and Development of Bubbling Fluidized Bed Gasifier for Non-Woody Biomass Gasification. *Chem. Eng. Trans.* **2022**, *92*, 289–294. [CrossRef]
72. Turkcan, A. The effects of different types of biodiesels and biodiesel-bioethanol-diesel blends on the cyclic variations and correlation coefficient. *Fuel* **2020**, *261*, 116453. [CrossRef]
73. Carotenuto, A.; Di Fraia, S.; Uddin, M.R.; Vanoli, L. Comparison of Combustion and Gasification for Energy Recovery from Residual Woody Biomass. *Int. J. Heat Technol.* **2022**, *40*, 888–894. [CrossRef]
74. Moretti, L.; Arpino, F.; Cortellessa, G.; Di Fraia, S.; Di Palma, M.; Vanoli, L. Reliability of equilibrium gasification models for selected biomass types and compositions: An overview. *Energies* **2022**, *15*, 61. [CrossRef]
75. Martínez González, A.; Silva Lora, E.E.; Escobar Palacio, J.C. Syngas production from oil sludge gasification and its potential use in power generation systems: An energy and exergy analysis. *Energy* **2019**, *169*, 1175–1190. [CrossRef]
76. Castillo Santiago, Y.; Martínez González, A.; Venturini, O.J.; Yepes Maya, D.M. Assessment of the energy recovery potential of oil sludge through gasification aiming electricity generation. *Energy* **2021**, *215*, 119210. [CrossRef]
77. Cavalaglio, G.; Coccia, V.; Cotana, F.; Gelosia, M.; Nicolini, A.; Petrozzi, A. Energy from poultry waste: An Aspen Plus-based approach to the thermo-chemical processes. *Waste Manag.* **2018**, *73*, 496–503. [CrossRef]
78. Sheth, P.N.; Babu, B.V. Production of hydrogen energy through biomass (waste wood) gasification. *Int. J. Hydrogen Energy* **2010**, *35*, 10803–10810. [CrossRef]
79. Kaewluan, S.; Pipatmanomai, S. Potential of synthesis gas production from rubber wood chip gasification in a bubbling fluidised bed gasifier. *Energy Convers. Manag.* **2011**, *52*, 75–84. [CrossRef]
80. Skoulou, V.; Koufodimos, G.; Samaras, Z.; Zabaniotou, A. Low temperature gasification of olive kernels in a 5-kW fluidized bed reactor for H2-rich producer gas. *Int. J. Hydrogen Energy* **2008**, *33*, 6515–6524. [CrossRef]
81. De Andrés, J.M.; Narros, A.; Rodríguez, M.E. Behaviour of dolomite, olivine and alumina as primary catalysts in air-steam gasification of sewage sludge. *Fuel* **2011**, *90*, 521–527. [CrossRef]
82. Ahmed, R.; Chandra, M.S. Equilibrium modeling in updraft gasifier for refinery sludge gasification using ASPEN PLUS simulator. *Appl. Mech. Mater.* **2013**, *393*, 729–734. [CrossRef]
83. Zhang, J.; Huo, X.; Dong, C.; Yang, Y. High-temperature oxygen gasification of sludge. In Proceedings of the 1st International Conference on Sustainable Power Generation and Supply, SUPERGEN '09, Nanjing, China, 6–7 April 2009; IEEE: Piscataway, NJ, USA, 2009; pp. 1–5.
84. Groß, B.; Eder, C.; Grziwa, P.; Horst, J.; Kimmerle, K. Energy recovery from sewage sludge by means of fluidised bed gasification. *Waste Manag.* **2008**, *28*, 1819–1826. [CrossRef]
85. Elsner, W.; Wysocki, M.; Niegodajew, P.; Borecki, R. Experimental and economic study of small-scale CHP installation equipped with downdraft gasifier and internal combustion engine. *Appl. Energy* **2017**, *202*, 213–227. [CrossRef]
86. Longo, S.; d'Antoni, B.M.; Bongards, M.; Chaparro, A.; Cronrath, A.; Fatone, F.; Lema, J.M.; Mauricio-Iglesias, M.; Soares, A.; Hospido, A. Monitoring and diagnosis of energy consumption in wastewater treatment plants. A state of the art and proposals for improvement. *Appl. Energy* **2016**, *179*, 1251–1268. [CrossRef]
87. Faizan, M.; Song, H. Critical review on catalytic biomass gasification: State-of-Art progress, technical challenges, and perspectives in future development. *J. Clean. Prod.* **2023**, *408*, 137224. [CrossRef]
88. Bocci, E.; Marcantonio, V.; Di Carlo, A. Tar and inorganic contaminant removal from syngas: Modeling and simulation. In *Advances in Synthesis Gas: Methods, Technologies and Applications*; Elsevier: San Diego, CA, USA, 2023; pp. 241–273.
89. Peck, D.; Holmes, W.; Zappi, M.; Hernandez, R.; Gang, D.; Buchireddy, P. Novel nickel ceramic filter for hot gas cleanup of tars from syngas. *Fuel Process. Technol.* **2023**, *244*, 107708. [CrossRef]

energies

MDPI

Article

Investigation of the Impact of Biochar Application on Foaming Slags with Varied Compositions in Electric Arc Furnace-Based Steel Production

Lina Kieush * and Johannes Schenk

Chair of Ferrous Metallurgy, Montanuniversitaet Leoben, 8700 Leoben, Austria
* Correspondence: lina.kieush@unileoben.ac.at

Abstract: This paper investigates the influence of biochar, either as an individual component or in combination with high-temperature coke, on the slag foaming behavior. High-temperature coke serves as a reference. Three scenarios were considered to study the slag foaming behavior, each characterized by different slag chemical compositions. The results indicate that biochar can promote steady foaming for specific slags when the basicity (CaO/SiO_2) falls within a range of 1.2 to 3.4. Experimental findings also reveal that stable foaming can be achieved when a mixture containing biochar and coke with a ratio of 1:1 is employed, with a minimum slag basicity of 1.0 and FeO content of 25 wt.%. The foaming range obtained using different FeO contents (15 wt.% to 40 wt.%) in the mixture surpasses the range observed with the individual application of coke or biochar. The X-ray diffraction (XRD) analysis showed that unrelated to the carbon source applied, the general pattern was that the phases larnite (Ca_2SiO_4) or dicalcium silicate were detected for slag foams with high basicity. Monticellite ($CaMgSiO_4$) and magnesium iron oxide (Fe_2MgO_4) were predominant in slag foam samples, with the highest MgO content. The presence of monticellite and merwinite ($Ca_3MgSi_2O_8$) occurred in samples with the lowest basicity. Eventually, the application of the mixture of coke and biochar showed the potential to obtain stable foaming across a wide range of slag compositions.

Keywords: biochar; coke; slag foaming; electric arc furnace; X-ray diffraction analysis

1. Introduction

For the electric arc furnace (EAF) process, the phenomenon of slag foaming is important because it not only improves reaction kinetics by increasing the effective surface area in electric steelmaking but also impedes the transfer of heat from the electric arc to the refractory lining [1–6]. In the slag foaming process, a carbon source is injected into the slag and reacts with FeO in the slag to form carbon monoxide (CO). In cases where CO gas bubbles cannot exit through the slag, the slag is considered to be undergoing a foaming [5]. At the same time, some iron in the slag will be reduced [2,7]. Fossil fuel carbon sources [8], namely, coal or cokes (metallurgical or petroleum), are commonly used as injected and charged carbon in the EAF process [9,10]. Applying conventional carbon sources has been extensively studied [10–13], and the main results have been analyzed in previous studies [14].

The practice of replacing fossil fuels with biomass-derived biochar by EAF has the potential to enhance environmental performance. This is because it avoids introducing additional carbon dioxide (CO_2) emissions into the atmosphere [15,16]. With the increasing emphasis on sustainable practices and the utilization of alternative carbon sources [17–19], it has turned to studying the feasibility of biochar as a potential substitute for conventional carbon sources in slag foaming applications. Biochar is a carbon-rich material produced by torrefaction [20,21] or pyrolysis [22,23] of biomass, such as wood chips, agricultural residues, or organic waste [24]. In the past, studies have been conducted to investigate using non-conventional carbon sources for slag foaming. The purpose of these studies was to measure and evaluate the impact of these sources, both in quantitative and qualitative terms. Additionally, these studies aimed to develop strategies and approaches for effectively utilizing these non-conventional carbon sources [25–27]. Moreover, multiple investigations have been undertaken to study the phenomenon of slag foaming by utilizing non-conventional carbon sources either independently or in combination with conventional sources. Herein, a summary overview of the main outcomes is presented.

Robinson et al. [28] investigated two variants of biochar derived from woody biomass, along with synthetic graphite and anthracite coal. The study findings indicated that substituting a portion of conventional carbon sources with biochar from woody biomass did not lead to any deviations from the operational conditions observed when using traditional carbon sources in the EAF. Moreover, the researchers emphasized that biochar derived from biomass should possess certain characteristics, such as undergoing torrefaction and pyrolysis, to meet the specific requirements for its application in EAF.

Demus et al. [29] investigated residuals from a wood gasification plant and a product from pyrolysis of agricultural residuals as substitutes for fossil coal used in EAF steelmaking. Melting tests in the pilot EAF showed that the agglomerated biochar reacted similarly to the reference anthracite coal. Briquetting results in reduced reactivity and slower burnout compared to fine biochar.

Cirilli et al. [16] carried out long-term industrial trials employing two types of biochar, namely, biochar derived from TORR COAL (obtained in the process of torrefaction at a temperature of 300–350 °C) and biochar obtained from thermal pyrolysis (800 °C). Anthracite was used as a reference carbon source. FeO content in the slag was set to 30.5 wt.% (standard) and 29.1 wt.% (biochar). The outcomes of these trials demonstrated the practicability of incorporating biochar as a carbon source, with no significant alterations observed in the analysis of steel and slag properties.

Yunos et al. [30] investigated the interactions of palm shells with EAF slag (34.9 wt.% Fe_2O_3, 30.4 wt.% CaO, 6.9 wt.% Al_2O_3, 10.9 wt.% MgO, 11.6 wt.% SiO_2, and 5.0 wt.% MnO), quantitatively estimating changes occurring in the slag volume as a function of time. In the case of palm char application, an increase in the slag volume and fluctuations has been observed. However, coke showed a low degree of fluctuations and poor volume ratios.

Hoikkaniemi et al. [31] studied slag foaming applied as carbon sources of coke dust (reference) and biochar, namely, spruce wood chips, after pyrolysis at 800 °C. The chemical composition of slag comprised 33.34 wt.% CaO, 16.59 wt.% SiO_2, 7.06 wt.% Al_2O_3, 7.72 wt.% MgO, 0.85 wt.% MnO, 1.08 wt.% TiO_2, 1.15 wt.% V_2O_5, and 29.02 wt.% FeO. According to the authors, biochar behaved comparably to coke dust as a foaming agent, suggesting its potential as a replacement for foaming agents derived from fossil fuels.

Contrariwise, several studies have reported that using biochar as an individual carbon source does not significantly impact slag foaming. DiGiovanni et al. [32] studied the feasibility of using biochar as an injection carbon material and establishing a methodology for ranking different injection carbon materials based on increased slag foam height. The composition of the slag was set to 23.0 wt.% FeO, 39.0 wt.% CaO, 14.0 wt.% SiO_2, 10.0 wt.% MgO, and 8.0 wt.% Al_2O_3. This study evaluated the potential of various carbon sources, namely, industrial injection coal, loose biochar, and bio-briquettes for slag foaming. Initially, loose biochar individually did not show significant slag foaming. Therefore, it was mixed with nut coke at a ratio of 70% nut coke and 30% loose biochar, and the height of the resulting slag was measured. The results indicated that the mixture of 70% nut coke and 30% loose biochar showed a noticeable increase in slag height, suggesting effective slag foaming.

Conversely, studying the effect of using biogenic material (torrefied wood pellets) on the composition of steel and slag (30.1 wt.% CaO, 33.9 wt.% Fe_2O_3, 6.1 wt.% Al_2O_3, 10.7 wt.% MgO, 13.0 wt.% SiO_2, and 6.2 wt.% MnO) using an induction melting furnace, Huang et al. [27] concluded that the potential increase in foaming when using biochar mixed with coke might be limited compared to the foaming achieved with coke individually.

Based on the analysis, it can be deduced that achieving a stable slag foam proves to be challenging when biochar is employed as the individual carbon source. Biochar shows limitations in its application compared to conventional carbon sources. Nevertheless, previous research results [14,33,34] have shown favorable effects of using a mixture of coke and biochar in slag foaming. Adding a mixture of coke and biochar to the slag has enhanced foam stability and increased slag foaming characteristics compared to applying biochar as an individual source.

The objective of this study is to quantify the foaming behavior based on the foaming characteristics of slag by analyzing foam height, foam volume, relative foam volume, and volumetric gas fraction using the slag samples prepared with different chemical compositions and carbon sources, including 100 wt.% high-temperature coke, 100 wt.% biochar, or a mixture of coke and biochar with a ratio of 1:1. Additionally, the influence of utilizing different carbon sources on the composition of the resulting slag foams is evaluated qualitatively by estimating the general chemical and phase composition of foam samples. This enables a practical explanation of the way the biochar application, either as an individual carbon source or in combination with conventional carbon sources, results in comparable outcomes to those achieved based on traditional carbon sources.

2. Materials and Methods

The slags used in this study were produced from powders consisting of specific chemical components, including FeO, CaO (at least 96% purity), SiO_2 (at least 99% purity), MgO (at least 98% purity), and Al_2O_3 (at least 99% purity). Since the main chemical composites of EAF slags can vary and generally consist of FeO (10–40 wt.%), CaO (22–60 wt.%), SiO_2 (6–34 wt.%), Al_2O_3 (3–14 wt.%), and MgO (3–13 wt.%) [35], this study of the application of different carbon sources considered the following three scenarios with compositions:

(1) In the first scenario, the slags had a constant FeO content of 25 wt.%, and the basicity (B2) was varied within the range of 1.0 to 3.4. Basicity was defined as the ratio of CaO to SiO_2. Additionally, another basicity (B3) was defined as the ratio of CaO to the sum of SiO_2 and Al_2O_3, and it varied between 0.8 and 2.0. The MgO content ranged from 15.0 to 6.0 wt.%, as shown in Table 3;

(2) In the second scenario, the slags had a constant basicity B2 of 1.4 and basicity B3 of 1.0, as well as the content of Al_2O_3 8 wt.%. Nevertheless, the FeO content varied between 24.0 and 30.7 wt.%, and the MgO content varied between 9.34 and 16.03 wt.%, as shown in Table 2;

(3) In the third scenario, the slags have a constant basicity B2 of 3.8 and a basicity B3 of 2.0. However, the FeO content varied between 10 and 50 wt.%, and the MgO content varied between 5.72 and 11.16 wt.%, as shown in Table 1.

Table 1. Slag compositions with stable basicity.

Sample	FeO, wt.%	CaO, wt.%	SiO_2, wt.%	MgO, wt.%	Al_2O_3, wt.%	Total	B2	B3	μ, Pa·s	ρ, g/cm^3	σ, N/m	Σ, s
S11	10.0	53.1	14.0	10.9	12.0	100.0	3.8	2.0	0.0466	3.20	0.58	3.94
S12	15.0	53.1	14.0	6.1	12.0	100.0	3.8	2.0	0.0420	3.24	0.58	3.52
S13	23.0	43.6	11.6	11.2	10.6	100.0	3.8	2.0	0.0348	3.43	0.59	2.81
S14	25.0	43.3	11.4	10.3	10.0	100.0	3.8	2.0	0.0328	3.45	0.59	2.64
S15	27.0	43.2	11.4	8.4	10.0	100.0	3.8	2.0	0.0314	3.47	0.60	2.51
S16	30.0	43.1	11.2	5.7	10.0	100.0	3.8	2.0	0.0291	3.51	0.60	2.31
S17	40.0	33.1	8.8	10.1	8.0	100.0	3.8	2.0	0.0023	3.72	0.61	0.18
S18	50.0	28.1	7.4	7.5	7.0	100.0	3.8	2.0	0.0018	3.89	0.62	0.13

μ, viscosity; ρ, density; σ, surface tension; Σ, foaming index.

Table 2. Slag compositions with stable basicity and Al_2O_3.

Sample	FeO, wt.%	CaO, wt.%	SiO_2, wt.%	MgO, wt.%	Al_2O_3, wt.%	Total	B2	B3	μ, Pa·s	ρ, g/cm³	σ, N/m	Σ, s
S8	24.0	30.0	22.0	16.0	8.0	100.0	1.4	1.0	0.044	4.41	0.73	2.83
S9	26.5	30.0	22.0	13.5	8.0	100.0	1.4	1.0	0.041	3.40	0.56	3.42
S10	30.7	30.0	22.0	9.3	8.0	100.0	1.4	1.0	0.038	3.44	0.56	3.15

μ, viscosity; ρ, density; σ, surface tension; Σ, foaming index.

Table 3. Slag compositions with stable FeO content.

Sample	FeO, wt.%	CaO, wt.%	SiO_2, wt.%	MgO, wt.%	Al_2O_3, wt.%	Total	B2	B3	μ, Pa·s	ρ, g/cm³	σ, N/m	Σ, s
S1	25.0	27.2	27.2	15.0	5.5	100.0	1.0	0.8	0.0475	3.33	0.54	4.07
S2	25.0	28.3	23.2	14.2	9.3	100.0	1.2	0.9	0.0465	3.38	0.54	3.95
S3	25.0	35.6	22.9	9.7	6.8	100.0	1.6	1.2	0.0412	3.34	0.56	3.48
S4	25.0	38.5	19.0	8.0	9.4	100.0	2.0	1.4	0.0391	3.37	0.57	3.25
S5	25.0	41.5	15.5	7.5	10.5	100.0	2.7	1.6	0.0363	3.40	0.58	2.97
S6	25.0	44.5	14.3	6.4	9.8	100.0	3.1	1.8	0.0341	3.40	0.59	2.78
S7	25.0	45.7	13.5	6.0	9.8	100.0	3.4	2.0	0.0333	3.40	0.59	2.71

μ, viscosity; ρ, density; σ, surface tension; Σ, foaming index.

The slag density and surface tension have been calculated according to [36]. Slag viscosity has been determined via FactSage v.8.2. Since the foaming index is supposed to be a function of the physical properties of the slag, it was calculated according to [37,38].

At the same time, two carbon sources (Table 4) were used in this study: high-temperature metallurgical coke (as a reference); wood biochar (Bch) obtained after pyrolysis at 600 °C [34]; and the mixture of coke and biochar with a ratio of 1:1 based on fixed carbon (C_{fix}) with stable grain size from 0.5 to 1.0 mm.

Table 4. Characters of carbon sources used for slag foaming, wt.%.

Characters	Coke	Bch
	Proximate analysis	
M	1.04	2.67
VM (daf)	1.57	41.04
Ash (db)	10.88	4.29
	Elemental analysis	
S (db)	0.77	n.d.
C (db)	86.25	72.74
H (db)	0.29	4.62
N (db)	1.21	0.24
* Others, mainly O (db)	0.60	18.11
C_{fix} (db)	87.70	56.43

M, moisture; C, carbon; H, hydrogen; N, nitrogen; S, sulfur; db, dry basis; daf, dry ash-free basis; VM(daf) = VM(db)·100/(100 − ash yield (db,%)); C_{fix}, wt.% = 100 − (wt.% VM (db) − wt.% Ash (db)); * Calculated by difference; n.d., not detected.

The Fourier transform infrared (FTIR) spectroscopy equipped with attenuated total reflectance (ATR) for two samples of carbon sources was performed by an Agilent Cary

630 FTIR spectrometer manufactured by Agilent Technologies, USA. The biochar or coke samples were scanned 32 times in the wavenumber range of 4000–650 cm^{-1} with a spectral resolution of 2 cm^{-1} and wavenumber accuracy of 0.05 cm^{-1}. OriginPro 2016 (9.3) software was used to evaluate the peaks of the spectra.

Proton nuclear magnetic resonance (1H NMR) spectroscopy for coke and biochar samples was performed via Bruker Ascend 400 MHz with Automation. The 1H 90° hard pulse was optimized for each sample. The duration of the low-power pulse for mixing was 400 ms in every case. Both NMR experiments were performed at 25 °C.

For slag foaming experiments, the carbon content was determined using stoichiometric calculations based on the required reduction described by the following equation FeO(slag) + C = Fe(1) + CO [12] and is presented in Table 5.

Table 5. Amount of carbon source for required reduction, g.

Sample	Coke	Biochar	Coke + Biochar
S1–S7	4.75	7.38	2.38 + 3.69
S8	4.56	7.09	2.28 + 3.55
S9	5.04	7.83	2.52 + 3.92
S10	5.83	9.07	2.92 + 4.54
S11	1.90	2.95	0.95 + 1.48
S12	2.85	4.43	1.43 + 2.22
S13	4.37	6.79	2.19 + 3.40
S14	4.75	7.38	2.38 + 3.69
S15	5.13	7.97	2.57 + 3.99
S16	5.70	8.86	2.85 + 4.43
S17	7.60	11.81	3.80 + 5.90
S18	9.50	14.77	4.75 + 7.40

All foaming experiments were conducted using an induction high-temperature furnace MU-900 (Indutherm Erwärmungsanlagen GmbH, Walzbachtal, Germany). Prior to experiments, approximately 5 g of ultra-low carbon (ULC) steel (containing 0.007 wt.% carbon) was placed at the bottom of the crucible of alumina (with an inner diameter of 63 mm, wall thickness of 4 mm, and height of 99 mm). Subsequently, 100 g of slag powder was added to the crucible. The filled alumina crucible was placed inside a graphite crucible (with an inner diameter of 70.3 mm, wall thickness of 20 mm, and height of 202 mm). Throughout each test, the system was continuously flushed with nitrogen at a rate of 1000 L/h. Once the temperature reached 1600 °C, it was held constant for a duration of 30 min. The height of the molten slag was measured by inserting a molybdenum stick into the slag and observing the resulting mark, which was then measured using a ruler. Afterward, a carbon source was added to the molten slag to initiate the foaming process. The carbon-slag reaction was enhanced by mixing the molten slag and carbon source using a molybdenum stick at the beginning of the foaming process. The foaming process was over after reaching a steady foam height and no subsequent changes in slag foam height, alongside a complete carbon material reaction. Following the conclusion of the test, the alumina crucible was extracted from the graphite crucible and subjected to cooling using liquid nitrogen. This cooling process served to solidify the slag foam, facilitating further investigation. The foaming continued depending on the carbon source and slag composition for 2 to 4 min in the case of biochar and 4 to 9 min for coke or a mixture of coke and biochar.

The height of the slag foam was determined by calculating the difference between the upper surface of the foam and the non-foamed layer located at the bottom of the crucible

(Figure 1). This measurement was repeated at three different positions within the crucible, and the resulting values were averaged to obtain the mean height of the slag foam.

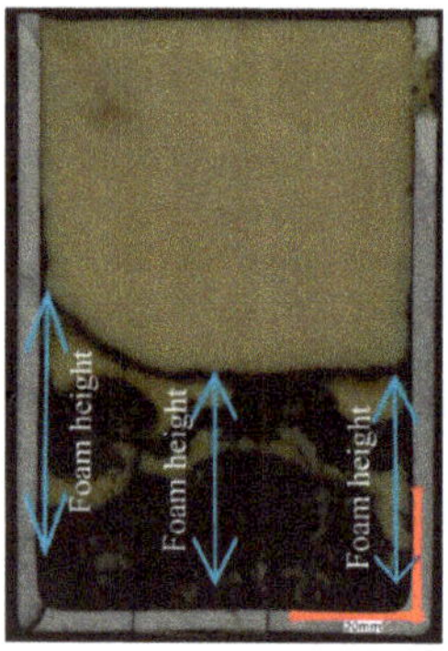

Figure 1. Example of foam height measurement.

Based on the foam height, the volume of slag before (V_{slag}) and after foaming (V_{foam}) was calculated using Equations (1) and (2):

$$V_{slag} = \frac{d^2 \cdot \pi}{4} \cdot h_{slag}, \quad (1)$$

$$V_{foam} = \frac{d^2 \cdot \pi}{4} \cdot h_{foam}, \quad (2)$$

where d is the inner diameter of the crucible, cm; h_{slag} is slag height before foaming, cm; h_{foam} is slag height after foaming, cm.

Additionally, the slag weight (Table 3), the inner diameter of the alumina crucible, slag density (Table 3), and Equation (1) were used to calculate the initial slag height without foaming.

The relative foaming volume $\left(\frac{\Delta V}{V_0}\right)$ was calculated by Equation (3):

$$\frac{\Delta V}{V_0} = \frac{V_{foam} - V_{slag}}{V_{slag}}, \quad (3)$$

The foaming height was estimated after the experiments, and the volumetric gas fraction contained in the foam (X_{gas}) was also calculated using Equation (4), according to [39,40].

$$X_{gas} = \frac{h_{foam} - h_{slag}}{h_{foam}}, \quad (4)$$

where h_{foam} is slag height after foaming, cm; h_{slag} is slag height before foaming, cm.

The powdered slag foam samples were subjected to X-ray diffraction (XRD) analysis using a Bruker AXS D8 advance diffractometer. The XRD spectra were recorded using a lynxeye detector and a Cu X-ray tube emitting Cu Kα radiation.

General chemical compositions of powdered slag foam samples were measured using EDS.

3. Results and Discussion

3.1. Evaluation of Carbon Sources via Fourier Transform Infrared Spectroscopy and Proton Nuclear Magnetic Resonance Spectroscopy

The two carbon sources, namely, coke and biochar, are investigated using an FTIR spectroscopy, and the spectra are presented in Figure 2.

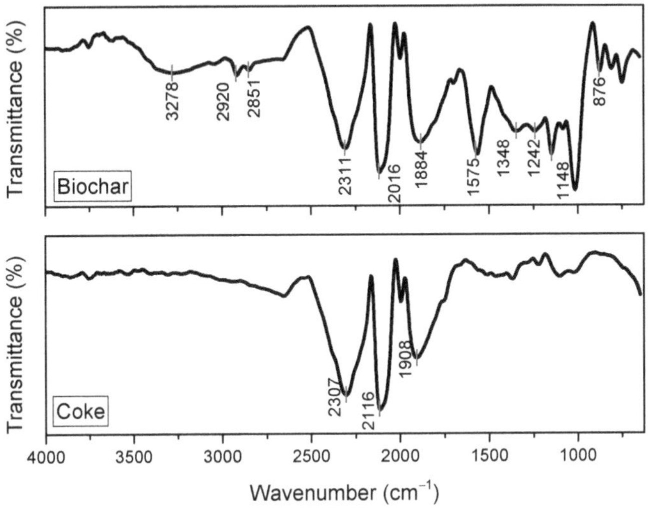

Figure 2. FTIR spectra of two carbon sources. Assignments of peaks are given within the text.

Figure 2 evidences the chemical differences between biochar and coke. It is seen that various functional groups in the FTIR spectra of coke and biochar are represented by the peaks located at 3520–3200 cm^{-1} -OH stretching (alcohols), 2915–2842 cm^{-1} C-H stretching (aliphatic), 1870–1540 cm^{-1} (oxygen-containing functional group), and 900–700 cm^{-1} C-H_{ar} out-of-plane deformation (aromatic group). In the coke samples, the stretching of –OH and other volatile groups was reduced compared to their presence in the corresponding biochar (3278 cm^{-1}). This decrease in -OH stretching in coke is likely due to the loss of -OH during carbonization. Additionally, the C-H functional groups (2920 and 2851 cm^{-1}) were found in biochar but were lacking in coke, as they were released as gaseous products (CO, CO_2, and CH_4) during carbonization. These agree with the conclusion reported by [41,42]. Nevertheless, the most pronounced peaks in coke samples were identified at 2307 cm^{-1} as an O=C=O stretching (carbon dioxide), at 2115 cm^{-1} C triple-bond (alkyne), and 1908 cm^{-1} C=C=C stretching (allene). In the case of biochar, approximately at position 2311 cm^{-1}, O=C=O stretching (carbon dioxide) was found; at 2016 cm^{-1}, C=C=N stretching (ketenimine) was identified, and at 1884 cm^{-1}, C-H bending (aromatic compound) was determined.

Furthermore, stretching vibration peaks in the range of 1500 to 600 cm^{-1} were more evident in biochar compared to coke. The breakage of C-H and C-O bonds, corresponding to the formation of general gases, occurs during carbonization. The peaks related to oxygen-containing functional groups are attributed to the presence of an aromatic C double-bond and possibly other O-containing functional groups. The intensity of these peaks in coke is reduced due to the breaking of the C double-bond compared to the biochar sample (1575 cm^{-1}) or the loss of O-containing functional groups caused by the higher carbonization temperature used in coke production. Additionally, the out-of-plane deformation of C-H_{ar} (aromatic group) stretching in biochar appeared more intense compared to coke.

Figure 3a,b present the proton nuclear magnetic resonance spectra for biochar and coke. It can be observed that nearly all peaks (A–E) are presented as singlets (s) both for biochar and coke. Only one peak F as multiplet (m) was detected for the biochar sample. Peaks (A–F) with chemical shifts in the region from 0.5 ppm to 3.0 ppm are evident, with the most common aliphatic proton [43]. Protons at the position around 1.0 ppm could be associated with alkyl groups, such as methyl (CH_3) or methylene (CH_2) groups, in organic compounds. The chemical shift range from 3.98 to 4.64 ppm generally corresponds to the region associated with proton signals near the region of aliphatic (saturated) hydrogens (biochar). Notably, most aliphatic groups disappeared after carbonization, as shown in the coke sample spectrum and as observed in FTIR results (Figure 2).

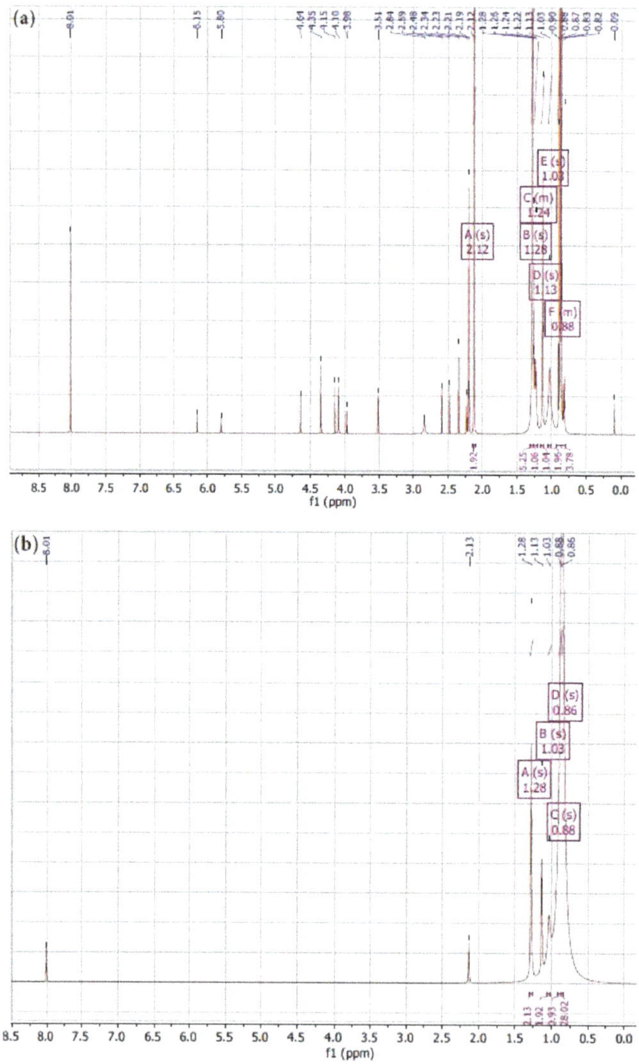

Figure 3. 1H NMR spectra chemical shifts (f1, ppm) of two carbon sources: (**a**) biochar; (**b**) coke. Assignments of peaks A–F are given within the text.

Protons attached to aromatic rings usually appear in the region from 6.0 ppm to 8.0 ppm. A pronounced peak at 8.0 ppm in the biochar sample suggests the presence of protons in an aromatic environment. The decrease in this peak can be observed for the coke sample.

The presence of more oxygen-containing functional groups and aliphatic groups indicates that biochar is a more reactive carbon source compared to coke, which may have an impact on the different foaming behavior of the slag.

3.2. Evaluation of Slag Foaming Behavior: The Case of 100 wt.% Coke as a Carbon Source

Before assessing the foaming of the slag, the initial height of the molten slag was estimated through calculations and measured during experiments. The slag weight, the inner diameter of the alumina crucible, and slag density were used to calculate the initial slag height without foaming. The slag volume was estimated using the calculated and measured heights obtained from the experiments. Figure 4a,b shows the combined results of comparing the difference between the calculated and measured values for slag height and the calculated and measured values for slag volume.

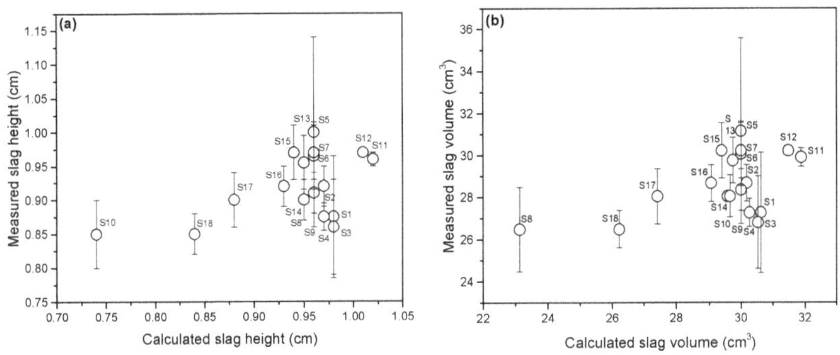

Figure 4. (**a**) Comparison between calculated slag height and measured slag height. (**b**) Comparison between calculated slag volume and measured slag volume.

As has been reported in some papers [44,45], generally, foam height should decrease with the increase in the foaming index of slag and an increase in viscosity μ, a decrease in density ρ, and surface tension σ. Table 6 presents the results of evaluating the foaming characters of slags with a stable FeO content of 25 wt.%. These experiments utilized coke as the carbon source. The findings indicate that as the basicity increases, the viscosity decreases while the density and surface tension increase. Consequently, the foam height and volume of the slags also increase, ranging from 3.5 cm to 11.0 cm for foam height and from 109.05 cm^3 to 342.72 cm^3 for foam volume. However, for slags S6 and S7, further increases in basicity 3.1 and 3.4 resulted in a slight decrease in both the height and volume of the slag foam, measuring 8.5 cm and 8.0 cm for foam height and 264.83 cm^3 and 249.25 cm^3 for foam volume, respectively. Notably, the slags with basicity 1.0 had unstable foaming, with several collapses oscillating up and down. In Table 6, the value for slag with basicity 1.0 represents not so much a foam height as the highest point of oscillation after adding coke.

Table 6. Slag foaming characters (coke as a carbon source; stable content of FeO 25 wt.%).

Sample	Foam Height h_{foam}, cm	Foam Volume V_{foam}, cm^3	Relative Foaming Volume, $\Delta V/V_0$	Volumetric Gas Fraction, X_{gas}
S1	3.5 *	109.05	3.32	0.77
S2	7.0	218.10	6.45	0.87
S3	9.6	299.10	10.85	0.92
S4	10.6	330.26	11.33	0.92
S5	11.0	342.72	11.22	0.92
S6	8.5	264.83	8.04	0.89
S7	8.0	249.25	7.60	0.88

* height of oscillation.

Since the characters of the relative volume of foaming and the volumetric gas fraction are dependent parameters on the height and volume of foam of the slag, a similar tendency was noticed for them: an increase in values and reaching the maximum value for S5 slag with basicity 2.7 and then a decrease.

All slag foam samples underwent qualitative assessment via XRD analysis, revealing different phases in each sample. Mainly, it detected such phases as silicates, namely, larnite Ca_2SiO_4 (00-033-0302) or dicalcium silicate (00-002-0843), gehlenite $Ca_2(Al(AlSi)O_7)$ (01-079-2421), merwinite $Ca_3MgSi_2O_8$ (01-074-0382), and monticellite $CaMgSiO_4$ (01-084-1320); ferrite, namely, brownmillerite $Ca_2(Al,Fe)_2O_5$ (00-030-0226); metal oxide, namely, wuestite FeO (00-046-1312); spinels, namely, magnesium iron oxide Fe_2MgO_4 (01-077-2367) and magnesium aluminum oxide $Mg_{0.4}Al_{2.4}O_4$ (01-084-0378); and some presence of metallic iron (00-003-1050), which are the common phases [46–48].

Among the slag foams S1–S7, the commonly identified phase is gehlenite (Figure 5). It can be concluded that monticellite and magnesium iron oxide are predominant in slag foam samples, with the highest MgO content. The presence of monticellite is detected only for slag foam samples where the lowest basicity was applied. Additionally, merwinite occurs in samples with the lowest basicity it aligns with [49].

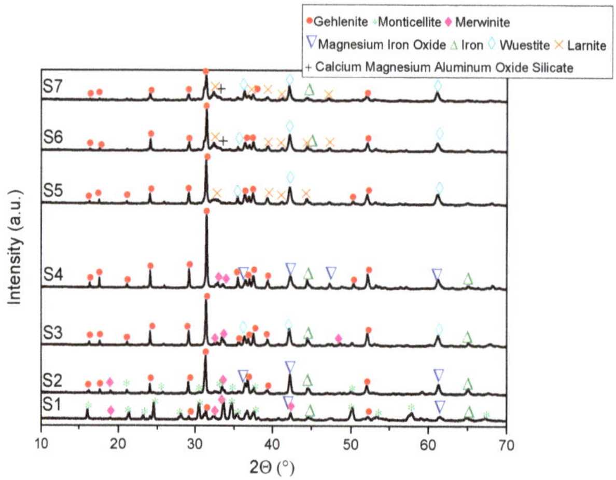

Figure 5. XRD patterns of slag foams obtained by applying slags with different basicity, stable content FeO 25 wt.%, and coke as a carbon source.

Slag foams S6 and S7 showed similar phases, including gehlenite, wuestite, metallic iron, and a phase identified as calcium magnesium aluminum oxide silicate CaO-MgO-Al_2O_3-SiO_2 (00-017-0736). This phase was detected only in slag foam samples where the highest basicity was set. Furthermore, the presence of larnite was found in samples S5–S7, where basicity from 2.7 to 3.4 was the highest among other slag foams. For slag foams S6 and S7, it could be one of the reasons for a decrease in foam height.

The slag foam samples were examined using SEM/EDS analysis to understand their chemical composition. The EDS study detected the presence of various elements, including aluminum (Al), silicon (Si), calcium (Ca), magnesium (Mg), iron (Fe), and oxygen (O) (as shown in Table 7). This analysis confirmed the findings of the XRD analysis and established correlations between these elements, confirming the identified phases. The general chemical composition range of the slag foams is as follows: 35.76–39.62 wt.% O; 2.27–7.48 wt.% Mg; 6.57–14.04 wt.% Al; 4.92–10.32 wt.% Si; 15.32–30.62 wt.% Ca; and 13.11–20.30 wt.% Fe. These elements align with oxygen distribution and are consistent with the mineral phases identified through XRD, which mainly consist of oxide compounds. An apparent decrease in the Mg content characterizes this group of slag foam. The Al content is increased, which is reasonable by using Al_2O_3 crucibles. This is observed for all samples unrelated to the slag foams or the carbon source used.

Table 7. SEM/EDS analysis results of general chemical composition for slag foam obtained by applying slags with different basicities, stable content FeO 25 wt.%, and coke as a carbon source, wt.%.

Sample	O	Mg	Al	Si	Ca	Fe
S1	39.62	7.38	6.57	10.32	15.82	20.30
S2	37.98	7.48	13.11	8.25	15.32	17.86
S3	38.26	4.50	8.56	9.92	24.35	14.41
S4	35.76	3.17	10.60	6.46	30.62	13.38
S5	36.31	2.71	10.68	5.41	27.37	17.52
S6	36.41	2.27	14.04	4.93	29.23	13.11
S7	38.16	2.27	13.72	4.92	26.08	14.85

Table 8 shows the result of slag foaming where stable basicity and Al_2O_3 content were applied. The highest value of the foam height and the foam volume of slag was achieved using 26.5 wt.% FeO, and the content of MgO 13.52 wt.% and were 8.5 cm and 264.83 cm^3, respectively. For the other two samples, the foam height and volume differ insignificantly. However, the lowest value of foaming characters is characterized by a sample of 24.0 wt.% FeO, and 16.03 wt.% MgO was used. In all three cases, stable foaming was observed during the experiments, although the foaming time varied and was about 8, 5.5, and 4 min for samples S8, S9, and S10, respectively. The foaming was faster with increasing content of FeO.

Table 8. Slag foaming characters (coke as a carbon source; stable basicity and Al_2O_3 content).

Sample	Foam Height h_{foam}, cm	Foam Volume V_{foam}, cm^3	Relative Foaming Volume, $\Delta V/V_0$	Volumetric Gas Fraction, X_{gas}
S8	7.1	221.21	6.89	0.87
S9	8.5	264.83	8.44	0.89
S10	7.9	246.14	8.29	0.89

Figure 6 illustrates detected phase patterns in slag foam samples with stable basicity and Al_2O_3 content. The relatively high content of MgO likely influenced the formation of two spinels (magnesium iron oxide and magnesium aluminum oxide) in all three slag foam samples.

Additionally, it was observed that merwinite is present in samples with low basicity, as mentioned earlier. Sample S10 is also distinguished by the presence of wuestite, which can be attributed to the fact that the highest FeO content was applied among this group.

Table 9 presents values of general chemical composition for slag foam obtained by applying slags with stable basicity and Al_2O_3 content. An increase in the content of Fe and a decrease in the content of Mg were distinctive. The rest of the elements had similar values with slight fluctuations.

Table 9. SEM/EDS analysis results of general chemical composition for slag foam obtained by applying slags with stable basicity and Al_2O_3 content and coke as a carbon source, wt.%.

Sample	O	Mg	Al	Si	Ca	Fe
S8	38.82	7.06	5.20	9.87	21.40	17.65
S9	35.80	7.80	4.79	10.69	20.71	20.21
S10	36.07	5.43	4.48	10.52	21.18	22.32

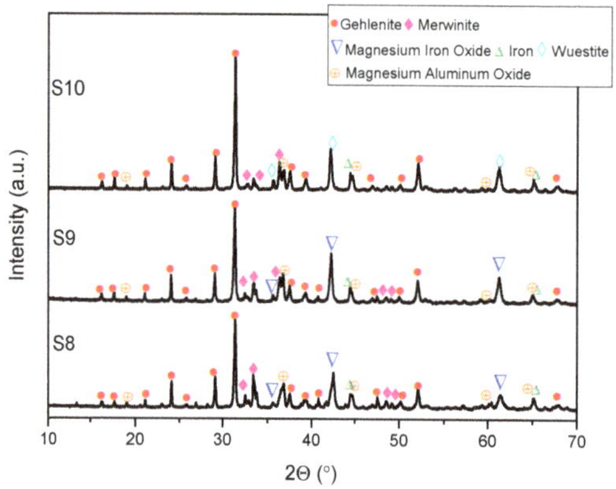

Figure 6. XRD patterns of slag foams obtained by applying slags with stable basicity and Al_2O_3 content and coke as a carbon source.

Table 10 presents the results of foaming slags with consistent basicity 3.8 but varying FeO content ranging from 10 to 50 wt.%. Increasing the FeO content from 10 wt.% to 30 wt.% increased both the slag foam height, from 2.3 cm to 10.1 cm, and the slag foam volume, ranging from 71.66 cm^3 to 314.68 cm^3. The viscosity decreased at higher FeO content, and foaming became significant, causing a decrease in the foaming index (Table 1). Additionally, increasing the FeO content within the slag results in a notable increase in slag density. These changes induce swifter drainage of the foam bubbles [50]. However, further increases in the FeO content (50 wt.%) led to a slight loss of foaming. Moreover, an increase in the FeO content leads to a decrease in the foaming index. The last two types of slag show a notable decrease in the foaming index. However, for the foaming, this was not reflected by a sharp change in the foaming characters. Based on experimental observations, it was noted that the last two types of slag exhibited a fluid-like behavior. This observation aligns with the findings mentioned in [51], which indicates that increased FeO content in the slag leads to higher fluidity. After adding coke, several slag foam collapses occurred, and a stable foam was formed.

Table 10. Slag foaming characters (coke as a carbon source; stable basicity).

Sample	Foam Height h_{foam}, cm	Foam Volume V_{foam}, cm^3	Relative Foaming Volume, $\Delta V/V_0$	Volumetric Gas Fraction, X_{gas}
S11	2.3 *	71.66	1.37	0.58
S12	8.0	249.25	7.25	0.88
S13	9.0	280.41	8.68	0.90
S14	9.3	289.76	9.33	0.90
S15	9.8	305.34	9.43	0.90
S16	10.1	314.68	9.74	0.91
S17	9.6	299.10	10.03	0.91
S18	9.4	292.87	9.80	0.91

* height of oscillation.

Conversely, at a low FeO content (10 wt.%) after adding coke to the molten slag, instead of bubbles distributed, chaotic behavior of foaming was observed where the slag surface collapsed. Indeed, the height of this slag foam represents the highest point reached by the oscillation. This could be attributed to the high viscosity of the slag, preventing significant foaming [51,52]. The foaming index decreases with increased FeO content, which aligns with this study [12].

Based on the results shown in Table 10, the critical FeO content was obtained with 30 wt.% of this oxide within slag composition, below which foaming increased with increasing FeO content and above which the slag was too fluid, the foam became less stable, and multiple collapses occurred. Nevertheless, even with an increase in FeO content from 30 wt.% to 50 wt.%, the foaming characters did not significantly deteriorate. For the slags containing 40 wt.% FeO, the foam height and volume were measured at 9.6 cm and 299.10 cm^3, respectively. Similarly, for the slags containing 50 wt.% FeO, the foam height and volume were 9.4 cm and 292.87 cm^3, respectively.

Figure 7 depicts the correlation between the ratio of the slag heights and the weight ratio of coke/slag for a series of slags with varying FeO content, with coke utilized as the carbon source. The data reveal a general tendency for the weight ratio of coke/slag to increase as the ratio of slag heights increases; for slag containing 50 wt.% FeO, a slight decline in this tendency is observed.

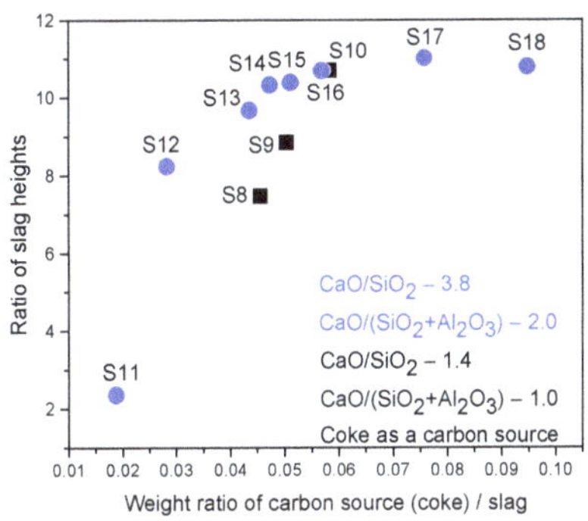

Figure 7. Correlation between the ratio of slag heights and weight ratio of coke/slag.

The main crystallized phases in slag foams S11–S13 and S16–S18 are gehlenite, wuestite, metallic iron, and a calcium magnesium aluminum oxide silicate phase (Figure 8). In the case of S11 and S13–S18, there is a correlation between the amounts of calcium and silica, suggesting the presence of dicalcium silicate or larnite phases. Moreover, when analyzing slag foam S11 and S12, S14, and S15, and S18, they were observed to contain spinel magnesium iron oxide. Slag foams S14 and S15 represent the following phases: gehlenite; wuestite; calcium magnesium aluminum oxide silicate; larnite; and magnesium iron oxide. The difference only in slag foam S15 also consists of brownmillerite.

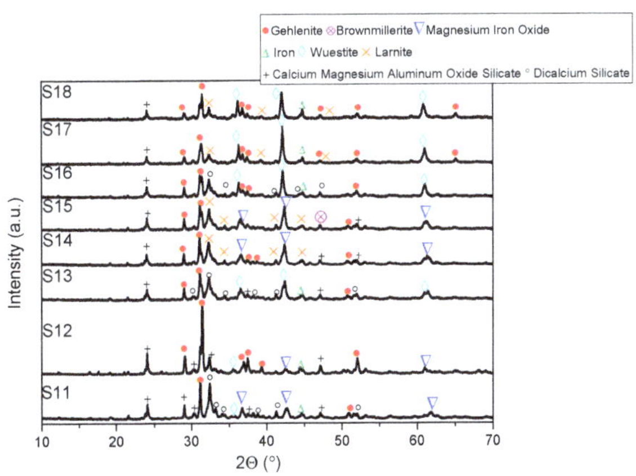

Figure 8. XRD patterns of slag foams obtained by applying slags with different FeO content, stable basicity, and coke as a carbon source.

Table 11 presents the general chemical composition range of the slag foams as follows: 31.72–39.74 wt.% O; 1.36–6.18 wt.% Mg; 5.75–13.72 wt.% Al; 2.53–8.25 wt.% Si; 15.30–37.56 wt.% Ca; and 6.59–31.81 wt.% Fe. A characteristic of this group is an increase in Fe content; however, it is somewhat irregular.

Table 11. SEM/EDS analysis results of general chemical composition for slag foam obtained by applying slags with different content FeO, stable basicity, and coke as a carbon source, wt.%.

Sample	O	Mg	Al	Si	Ca	Fe
S11	39.74	3.49	13.01	4.57	31.15	8.03
S12	38.34	1.36	12.54	3.61	37.56	6.59
S13	34.50	1.80	12.55	5.18	35.13	10.83
S14	33.94	3.61	9.43	3.29	30.44	19.28
S15	32.26	2.49	13.72	4.95	30.09	16.49
S16	37.48	6.18	5.75	8.25	22.39	19.95
S17	31.72	2.03	9.86	2.53	27.23	26.63
S18	31.98	4.81	13.24	2.87	15.30	31.81

To quantify slag foaming as a function of basicity or FeO content and foaming characters (foam height and foaming index), graphs were plotted with different basicities and FeO content, as shown in Figure 9a,b. In the case of basicity, the foaming characters initially increased as basicity increased. The group of slags with a stable FeO content demonstrated this tendency, with the S5 slag achieving maximum height and volume values. However, beyond this point, there was a subsequent decrease in the height and volume of the slag foam.

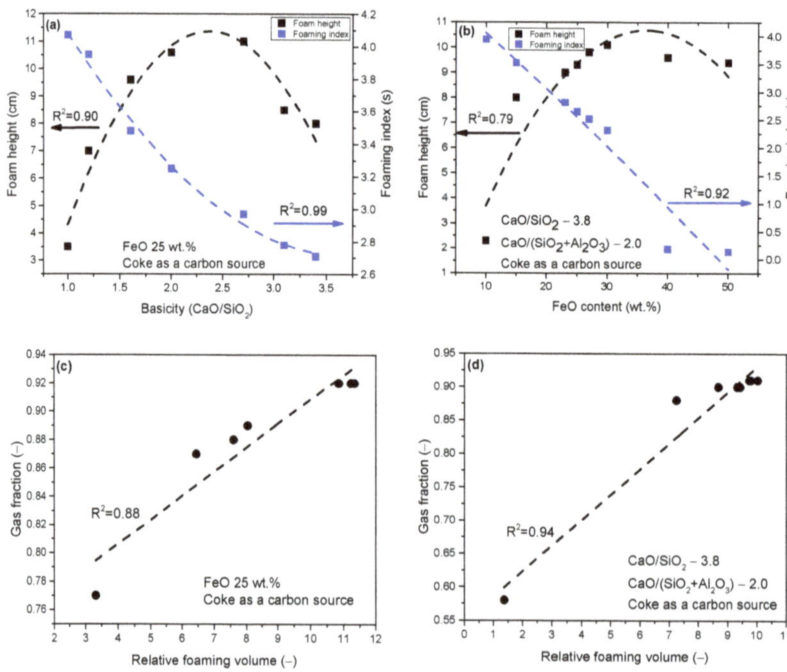

Figure 9. Foam height (coke as a carbon source) and foaming index as a function of (**a**) basicity and (**b**) FeO content of slags; relationship between relative foaming volume and gas fraction of slags with stable FeO content (**c**) and basicity (**d**).

In the case of variation of FeO content, the foaming characters initially increased until reaching maximum values for the group of slags with stable basicity. The slag with 30 wt.% FeO reached the highest height and volume values in this group. However, further increases in FeO content resulted in a decrease in the height and volume of the slag foam.

Figure 9c,d summarizes the calculation results and represents the relationship between the gas fraction indicator and the relative volume of foam.

These dependencies are presented for both groups of slags, namely, for both stable FeO content and stable basicity. It has been observed that the lower the value of the volumetric gas fraction, the lower the relative foaming volume. Additionally, when using coke with an increase in the content of FeO from 15 wt.% to 50 wt.%, a significant change in the gas fraction was not observed (Figure 9d).

When using coke as a carbon source, the most optimal foaming characters were obtained for a group of slags with different basicities from 1.2 to 2.7 and for group slags with different FeO content from 15 to 30 wt.%.

3.3. Evaluation of Slag Foaming Behavior: Case of 100 wt.% Biochar as a Carbon Source

Table 12 shows the foaming characters for 100 wt.% biochar and slag samples with different basicities. The foaming was unstable when using slag with a basicity of 1.0, and the value of 3.1 cm characterizes the highest oscillation point during the experiment. Distinctive from the coke case is that the increase in basicity did not affect the foaming behavior, and in cases with slag foams S2–S6, the foaming values did not change and were within the experimental error corresponding to the average value of the foam height of 8.7 cm (±0.12) and the average foam volume of 270.61 cm^3 (±24.5). The absence of change in foam height with increasing basicity could be due to the low interaction between highly reactive biochar and slag, as was reported in [27]. Despite some disadvantages of using biochar, in this case, stable foaming was achieved when a group of slag with different basicities from 1.2 to 3.4 was applied. The highest foaming values were achieved when using slag with a basicity of 3.4 and are 10.0 cm (foam height) and 311.56 cm^3 (foam volume).

Table 12. Slag foaming characters (100 wt.% biochar as a carbon source; stable content of FeO 25 wt.%).

Sample	Foam Height h_{foam}, cm	Foam Volume V_{foam}, cm^3	Relative Foaming Volume $\Delta V/V_0$	Volumetric Gas Fraction X_{gas}
S1	3.1 *	96.59	2.30	0.70
S2	8.7	271.06	8.70	0.89
S3	8.6	267.94	7.60	0.88
S4	8.7	271.06	8.00	0.89
S5	8.9	277.29	7.90	0.89
S6	8.6	267.94	7.60	0.88
S7	10.0	311.56	9.10	0.90

* height of oscillation.

To summarize, the typical phases among the given slag foam samples are gehlenite, calcium magnesium aluminum oxide silicate, merwinite, and magnesium iron oxide (Figure 10). The presence of larnite is typical for samples with higher basicity, which was also observed for samples of slag foam (coke case). Only the S1 slag foam is characterized by the vast majority of peaks assigned to monticellite since this phase is notable for slags with low basicity, as noted in the case of coke. Additionally, metallic iron is present only in sample S5.

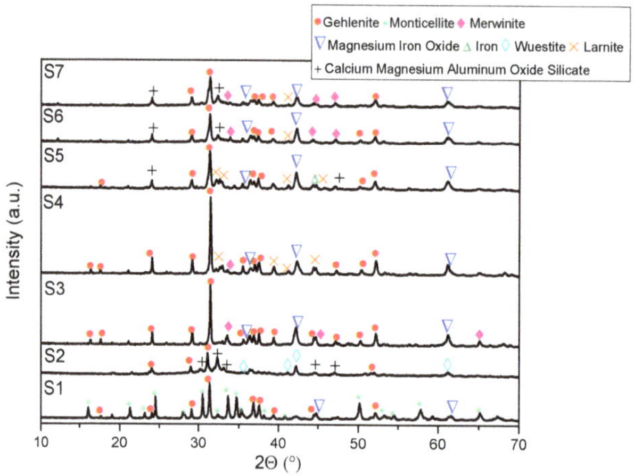

Figure 10. XRD patterns of slag foams obtained by applying slags with different basicities, stable FeO content, and biochar as a carbon source.

The general chemical composition range of the slag foams obtained using biochar (Table 13) is as follows: 35.23–38.50 wt.% O; 2.09–9.08 wt.% Mg; 5.65–16.15 wt.% Al; 3.72–12.23 wt.% Si; 20.82–29.89 wt.% Ca; and 13.33–20.59 wt.% Fe. These samples of slag foam exhibit a decrease in the levels of Mg and Si, while there is an increase in the content of Ca.

Table 13. SEM/EDS analysis results of general chemical composition for slag foam obtained by applying slags with different basicities, stable content FeO 25 wt.%, and biochar as a carbon source, wt.%.

Sample	O	Mg	Al	Si	Ca	Fe
S1	37.53	9.08	5.65	12.23	20.82	14.69
S2	35.23	2.09	10.98	3.72	27.40	20.59
S3	38.50	2.71	11.43	7.03	23.07	17.25
S4	36.15	4.69	10.68	7.62	26.72	14.15
S5	37.73	2.35	16.15	5.49	24.95	13.33
S6	38.25	2.53	13.33	4.35	23.92	17.61
S7	37.25	2.29	12.22	3.97	29.89	14.38

Table 14 shows the result of slag foaming using slags with stable basicity and content of Al_2O_3. In the case of biochar, the foaming characters are lower than with coke, and it is also noteworthy that there was no significant foaming change, and it was within experimental error. Furthermore, the foaming was fast in this case and ranged from 2 to 3 min.

Table 14. Slag foaming characters (biochar as a carbon source; stable basicity and Al_2O_3 content).

Sample	Foam Height h_{foam}, cm	Foam Volume V_{foam}, cm^3	Relative Foaming Volume, $\Delta V/V_0$	Volumetric Gas Fraction, X_{gas}
S8	6.7	208.75	6.20	0.86
S9	6.9	214.98	6.34	0.86
S10	6.8	186.94	5.67	0.85

The XRD analysis results for slag foams indicated no difference in the phase composition when using biochar as a carbon source compared to using coke. All the detected phases observed in the coke case are also present in the samples with biochar, as shown in Figure 11.

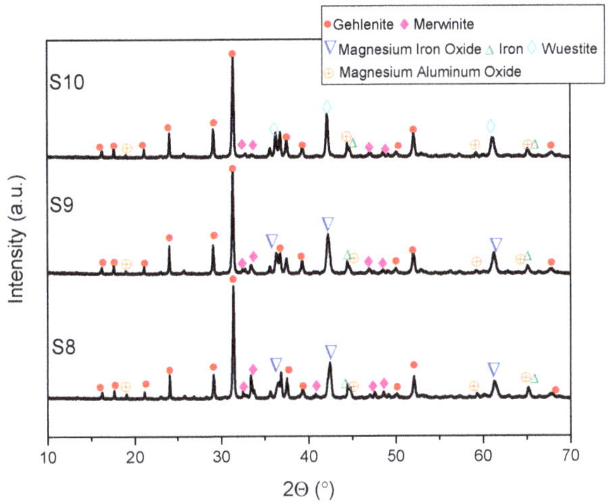

Figure 11. XRD patterns of slag foams obtained by applying slags with stable basicity and Al_2O_3 content and biochar as a carbon source.

The general chemical composition of slag foam samples (Table 15) using biochar is not much different from the point where coke was used, and the elemental contents are close to this case.

Table 15. SEM/EDS analysis results of general chemical composition for slag foam obtained by applying slags with stable basicity and Al_2O_3 content and biochar as a carbon source, wt.%.

Sample	O	Mg	Al	Si	Ca	Fe
S8	36.24	9.08	4.39	10.14	21.28	18.87
S9	37.65	6.34	4.76	10.37	20.01	20.87
S10	37.12	5.22	5.23	10.09	19.29	23.05

Table 16 presents the results of using 100 wt.% biochar for slag compositions with different FeO content. The lowest value of the foam height is distinctive for a sample with a FeO content of 10 wt.% and is 2.5 cm, as with coke. This is the value for the highest point of the oscillation height. Stable foaming was obtained for samples containing FeO from 15 to 27 wt.%. In cases with a FeO content of 30–50 wt.%, very fast, uncontrolled foaming was observed, which exceeded the sizes of the crucible.

Figure 12 shows the correlation between the ratio of slag heights and the weight ratio of biochar/slag for a series of slags with varying FeO content, with biochar utilized as the carbon source. The data reveal a general tendency for the weight ratio of biochar/slag to increase as the ratio of slag heights increases; however, for slags containing 40 and 50 wt.% FeO, a slight decrease in this trend is observed. Furthermore, there is a visible decrease for samples S8–S10. The overall behavior is similar in the case of coke.

Table 16. Slag foaming characters (100 wt.% biochar as a carbon source; stable basicity).

Sample	Foam Height h_{foam}, cm	Foam Volume V_{foam}, cm^3	Relative Foaming Volume $\Delta V/V_0$	Volumetric Gas Fraction, X_{gas}
S11	2.5 *	77.89	2.60	0.72
S12	6.7	208.75	5.80	0.85
S13	10.1	314.68	9.20	0.90
S14	8.0	249.25	7.90	0.89
S15	9.5	295.99	9.00	0.90
S16	11.0	343.66	11.70	0.92
S17	10.4	324.03	10.60	0.91
S18	10.8	336.49	10.60	0.91

* height of oscillation.

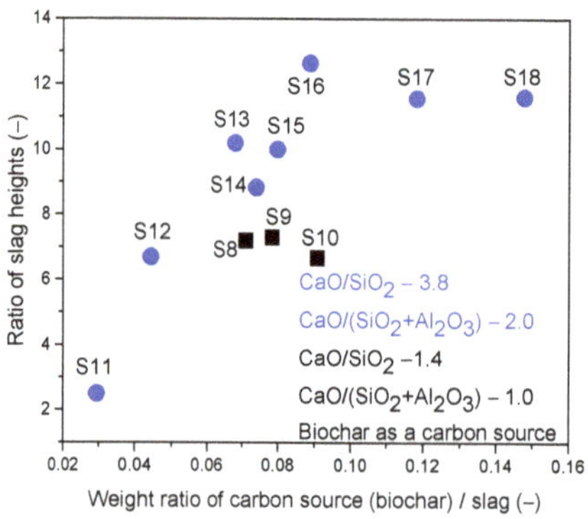

Figure 12. Correlation between the ratio of slag heights and weight ratio of biochar/slag.

The phase composition analysis (Figure 13) showed that most samples contained gehlenite, except for the slag foam sample with 10 wt.% FeO. Larnite was detected in samples S11 (10 wt.% FeO), S13 (23 wt.% FeO), S15 (27 wt.% FeO), S16 (30 wt.% FeO), and S17 (40 wt.% FeO). Remarkably, metallic iron was only found in samples S14 (25 wt.% FeO), S15, S17, and S18 (50 wt.% FeO). Out of all the samples, only sample S14 exhibited the presence of magnesium aluminum oxide along with magnesium iron oxide. Spinel magnesium iron oxide was detected in the slag foam samples with FeO content ranging from 10 to 30 wt.%.

Additionally, the phase of brownmillerite was detected in slag foams S11, S13, S15, S16, and S18. Some samples, specifically S12 and S14, were characterized by the presence of monticellite and merwinite. Peaks corresponding to wuestite were observed in samples S13, S15, S17, and S18. Furthermore, calcium magnesium aluminum oxide silicate was present in samples S11, S13, and S16–S18.

The general chemical composition range of the slag foams obtained using biochar by applying slags with different content FeO, stable basicity (Table 17) is as follows: 31.72–39.74 wt.% O; 1.36–6.18 wt.% Mg; 5.75–13.72 wt.% Al; 2.53–8.25 wt.% Si; 15.30–37.56 wt.% Ca; and 6.59–31.81 wt.% Fe. Nonetheless, with the carbon source applied, these slag foams are characterized by an increased content of Fe, as previously observed for this kind of slag group.

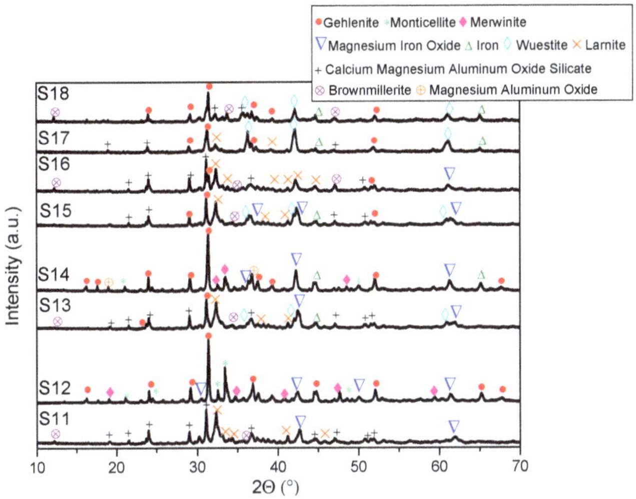

Figure 13. XRD patterns of slag foams obtained by applying slags with different FeO content, stable basicity, and biochar as a carbon source.

Table 17. SEM/EDS analysis results of general chemical composition for slag foam obtained by applying slags with different content FeO, stable basicity, and biochar as a carbon source, wt.%.

Sample	O	Mg	Al	Si	Ca	Fe
S11	39.74	3.49	13.01	4.57	31.15	8.03
S12	38.34	1.36	12.54	3.61	37.56	6.59
S13	34.50	1.80	12.55	5.18	35.13	10.83
S14	33.94	3.61	9.43	3.29	30.44	19.28
S15	32.26	2.49	13.72	4.95	30.09	16.49
S16	37.48	6.18	5.75	8.25	22.39	19.95
S17	31.72	2.03	9.86	2.53	27.23	26.63
S18	31.98	4.71	13.24	2.87	15.30	31.81

Figure 14a–d shows the relationship between foam height, foaming index, and basicity or FeO content of slag samples. Notably, gas formation increases despite the absence of a change in the foam height in samples S2–S6 (Figure 14c,d). It can be concluded that when applying biochar as a carbon source, the most optimal foaming characters were obtained for a group of slags with different basicities from 1.2 to 3.4.

In the case of using biochar in the group of slags in which the content of FeO changed (Figure 14b), it was noticed that with an increase in the content of this oxide and with the addition of biochar, foaming occurred faster. Most likely, this is explained by the fact that biochar is a highly reactive carbon material (judging by FTIR and 1H NMR results) with a high content of volatile matter compared to coke. These experiments were characterized by a rather high biochar reactivity with air and CO_2 and, as a result, a high reactivity with FeO, which is confirmed by the results, where FeO from 30–50 wt.% were applied, and uncontrolled foaming occurred. These conclusions have also been made in [32].

A significant disadvantage when using biochar was that when this material was added, some parts burned and did not contribute to the foaming process, which indicates that this material should be granulated [32] or should be subjected to a higher pyrolysis temperature.

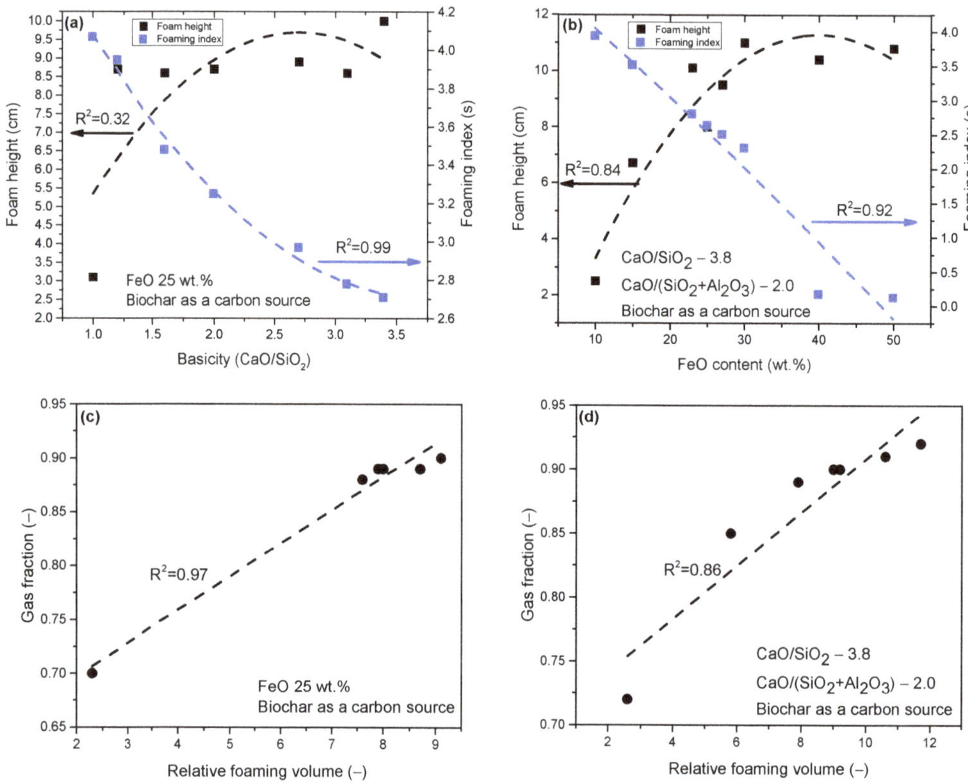

Figure 14. Foam height (biochar as a carbon source) and foaming index as a function of (**a**) basicity and (**b**) FeO content of slags; relationship between relative foaming volume and gas fraction of slags with stable: FeO content (**c**) and basicity (**d**).

3.4. Evaluation of Slag Foaming Behavior: The Case of Coke and Biochar Mixture with a Ratio of 1:1 as a Carbon Source

Table 18 shows the results of slag foaming with a stable content of FeO 25 wt.% and using a mixture of coke and biochar with a ratio of 1:1 as the carbon source. Unexpectedly, the foaming behavior differed from coke or biochar when the mixture was applied as a carbon source. With increasing basicity, the height and volume of the slag foam increased from 6.4 cm and 199.40 cm^3, respectively, and reached a maximum value for slag with basicity of 2.7 at 10.5 cm for the slag foam height and 327.14 cm^3 for the slag height volume. A subsequent increase in basicity led to a slight decrease in these indicators for the last two slag samples. However, from experimental observations, the last two samples, S6 and S7, were characterized by fast foaming compared to other slag samples. As with the use of coke or biochar, for slag with basicity 1.0, the lowest foaming characters for this group of slags were obtained. However, foaming characters' values are higher than coke or biochar cases. This foam was stable during the test, which, in turn, leads to the conclusion that, somehow, the use of a mixture of coke and biochar mitigates the disadvantages of using coke or biochar as individual carbon sources.

Table 18. Slag foaming characters (mixture of coke and biochar with a ratio of 1:1 as a carbon source; stable content of FeO 25 wt.%).

Sample	Foam Height h_{foam}, cm	Foam Volume V_{foam}, cm^3	Relative Foaming Volume, $\Delta V/V_0$	Volumetric Gas Fraction, X_{gas}
S1	6.4	199.40	6.11	0.86
S2	7.3	227.44	6.40	0.87
S3	9.7	302.22	9.66	0.91
S4	10.6	330.26	10.91	0.92
S5	10.5	327.14	8.55	0.90
S6	9.7	302.22	8.70	0.90
S7	9.3	289.76	8.30	0.89

For samples of slag foam in the case of using slags with different basicities and stable FeO content, the following main phases were identified and presented in Figure 15. The typical phases among the given samples are gehlenite, metallic iron, merwinite, and larnite. Magnesium iron oxide is also present in samples S1, S2, and S6. Monticellite is present in samples S1 and S2, as previously detected for slag foam samples with the lowest basicity and high content of MgO. Wuestite is present in samples S3–S5 and S7. Calcium magnesium aluminum oxide silicate is present in samples S6 and S7. Notably, this phase was seen mainly for samples where higher basicity prevails. However, the exception is XRD patterns of slag foams by applying slags with different basicities, stable FeO content, and, in the case of using biochar, as a carbon source. The phase of larnite is detected for slag foams with high basicity. Lastly, brownmillerite is present only in sample S5.

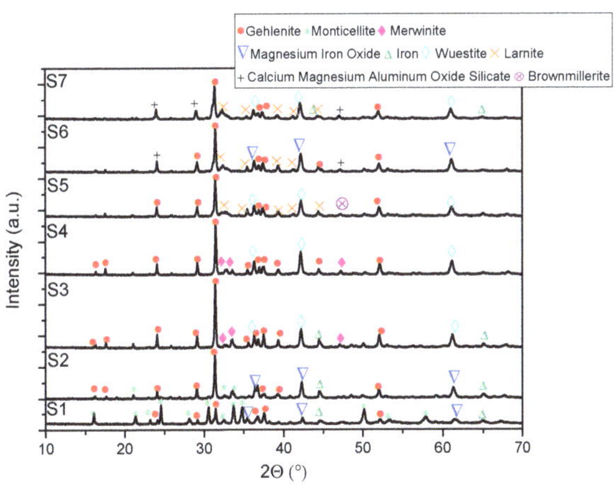

Figure 15. XRD patterns of slag foams obtained by applying slags with different basicities, stable FeO 25 wt.%, and a mixture of coke and biochar with a ratio of 1:1 as a carbon source.

The general chemical composition range of the slag foams obtained using a mixture of coke and biochar by applying slags with different basicities and stable FeO (Table 19) is as follows: 31.90–40.70 wt.% O; 2.16–6.16 wt.% Mg; 7.05–15.42 wt.% Al; 4.55–9.25 wt.% Si; 21.39–30.71 wt.% Ca; and 14.61–18.05 wt.% Fe. It should be noted that the Mg content is relatively stable, and only for the slag foam sample, its content is 6.16 wt.%.

Table 19. SEM/EDS analysis results of general chemical composition for slag foam obtained by applying slags with different basicities, stable content FeO 25 wt.%, and mixture of coke and biochar with a ratio of 1:1 as a carbon source, wt.%.

Sample	O	Mg	Al	Si	Ca	Fe
S1	37.80	3.75	15.42	4.62	22.99	15.42
S2	38.24	6.16	10.35	9.25	21.39	14.61
S3	34.72	3.48	7.05	8.81	27.89	18.05
S4	40.70	2.16	12.69	5.32	25.08	14.06
S5	34.06	2.55	11.21	7.35	26.98	17.84
S6	31.90	2.68	13.24	5.52	30.71	15.95
S7	34.66	2.32	12.13	4.55	29.66	16.68

Table 20 presents the results of slag foaming with stable basicity and Al_2O_3 content. When biochar was used along with coke in the mixture, a foaming tendency was observed to be analogous compared to using only coke. It is worth noting that the foaming characters of the slag were practically similar to those of when coke was used as the individual carbon source. The foaming time was the longest, lasting approximately 9 min when using S8 slag. This longer duration could be attributed to the higher viscosity of the S8 slag compared to the other samples. In contrast, the foaming time for the remaining two samples was around 5 min.

Table 20. Slag foaming characters (mixture of coke and biochar with a ratio of 1:1 as a carbon source; stable basicity and Al_2O_3 content).

Sample	Foam Height h_{foam}, cm	Foam Volume V_{foam}, cm^3	Relative Foaming Volume, $\Delta V/V_0$	Volumetric Gas Fraction, X_{gas}
S8	7.5	233.67	7.06	0.88
S9	8.6	267.95	8.15	0.89
S10	8.0	249.25	8.52	0.90

In all three samples, the following common phases were identified: gehlenite; merwinite; magnesium iron oxide; magnesium aluminum oxide; and metallic iron (Figure 16). The slag sample, which had the highest amount of FeO at 30.7 wt.%, also contained wuestite in the slag foam. This tendency was observed in all samples with stable basicity and Al_2O_3.

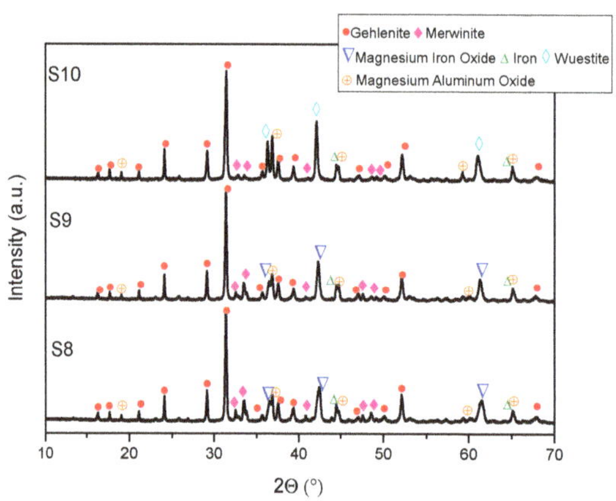

Figure 16. XRD patterns of slag foams obtained by applying slags with stable basicity and Al_2O_3 content and a mixture of coke and biochar with a ratio of 1:1 as a carbon source.

Table 21 shows the content of elements for slag foams obtained using a mixture of coke and biochar as a carbon source. As in the previous cases, this group is characterized by a decrease in the content of Mg and an increase in Fe.

Table 21. SEM/EDS analysis results of general chemical composition for slag foam obtained by applying slags with stable basicity and Al_2O_3 content and a mixture of coke and biochar with a ratio of 1:1 as a carbon source, wt.%.

Sample	O	Mg	Al	Si	Ca	Fe
S8	38.86	9.88	5.08	9.16	19.15	17.87
S9	39.79	7.83	4.60	10.84	18.78	18.16
S10	40.76	5.05	4.82	9.09	20.22	20.06

Table 22 shows the foaming characters of slag with different FeO content. It should be noted that when using a mixture of coke and biochar with a ratio of 1:1, foaming characters were obtained that were either close to or slightly higher than those obtained using coke. A higher FeO content led to a proportional increase in the height and volume of the slag foam. Precisely when the slag contained 40 or 50 wt.%, FeO showed a maximum foam height of 10.0 cm and 10.1 cm and foam volume of 311.57 cm^3 and 314.68 cm^3, respectively. However, it is important to note that using 50 wt.% FeO reached a critical point in the experiment. Only this specific test lasted approximately 3 min because, after adding a mixture of coke and biochar, the foaming exceeded the capacity of the crucible in which the slag was initially placed. This behavior can be explained by the fact that, in contrast to the use of 100 wt.% coke, in this case, part of the coke was replaced by biochar, which, in terms of its physicochemical properties, is more reactive, has a greater porosity and specific surface area in comparison with high-temperature coke, which is influenced by excessive and non-controlled foaming. Along with this, according to [52], an increase in FeO content enhances gas generation, thereby increasing foam height. Combining the increase in FeO content with the use of biochar could have a notable impact on the foaming process.

Table 22. Slag foaming characters (mixture of coke and biochar with a ratio of 1:1 as a carbon source; stable basicity).

Sample	Foam Height h_{foam}, cm	Foam Volume V_{foam}, cm^3	Relative Foaming Volume, $\Delta V/V_0$	Volumetric Gas Fraction, X_{gas}
S11	3.4 *	105.93	2.58	0.72
S12	8.1	252.37	7.35	0.88
S13	8.2	255.48	7.37	0.88
S14	8.4	261.72	8.33	0.89
S15	9.6	299.10	8.60	0.90
S16	9.6	299.10	9.67	0.91
S17	10.0	311.57	9.75	0.91
S18	10.1	314.68	11.16	0.92

* height of oscillation.

The utilization of 50 wt.% FeO may be unfavorable due to two main reasons. Firstly, it results in the significant generation of a gas fraction, which may not be desirable for the specific application. Secondly, the low viscosity of the liquid resulting from this composition allows for the easy removal of the formed bubbles [53]. Slag S11 containing the smallest amount of FeO 10 wt.% had a limited foam height, as it was too viscose.

Figure 17 shows the correlation between the ratio of slag heights, the weight ratio of coke and biochar/slag, and a set of slags with different FeO content. Generally, as the ratio of slag heights increases, there is a tendency for the weight ratio of coke and biochar/slag to slag heights to also increase. However, it is noteworthy that for slags containing 50 wt.% FeO, unlike the behavior observed with coke, the weight ratio sharply increases in this case.

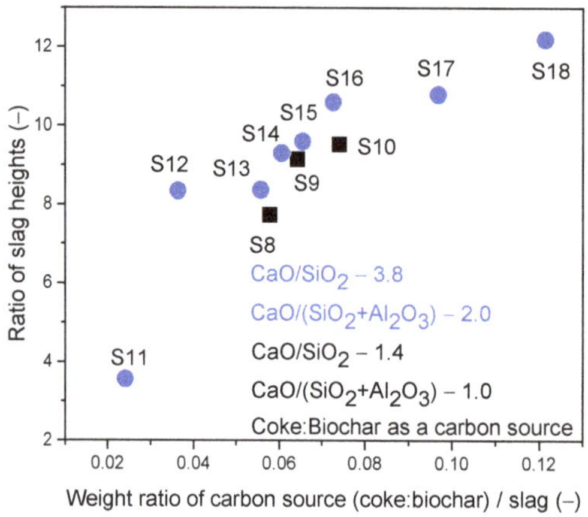

Figure 17. Correlation between the ratio of slag heights and weight ratio of coke and biochar/slag.

For samples of slag foam in the case of using slags with different content FeO and stable basicity, it was identified that slag foams S11, S13, S14, S16, and S17 consist of gehlenite, calcium magnesium aluminum oxide silicate, larnite, magnesium iron oxide, and brownmillerite (Figure 18). Additionally, the presence of the wuestite phase was detected for samples S12, S13, and S18. The peak corresponding to metallic iron was detected for slag samples S14, S15, and S18.

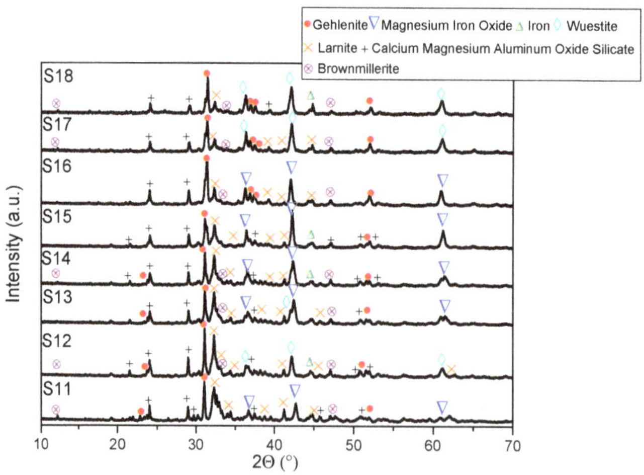

Figure 18. XRD patterns of slag foams obtained by applying slags with different FeO content, stable basicity, and a mixture of coke and biochar with a ratio of 1:1 as a carbon source.

Noteworthy, the phase magnesium iron oxide was not found for samples of slag foam, where the FeO content was set to 40 and 50 wt.% both in the case of using biochar and a mixture of coke and biochar. These samples are characterized mainly by the presence of brownmillerite, wuestite, and metallic iron.

The general chemical composition range of the slag foams obtained by applying slags with different content FeO, stable basicity (Table 23) is as follows: 26.16–47.17 wt.% O;

1.16–4.35 wt.% Mg; 7.86–13.64 wt.% Al; 1.77–4.27 wt.% Si; 14.83–41.59 wt.% Ca; and 7.31–32.12 wt.% Fe. Unlike the previous results observed for this group, where the iron content showed a consistent change for samples S13–S17, the pattern of change was irregular in this case. Notably, slag sample S11 shows the lowest value among all the samples.

Table 23. SEM/EDS analysis results of general chemical composition for slag foam obtained by applying slags with different content FeO, stable basicity, and mixture of coke and biochar with a ratio of 1:1 as a carbon source, wt.%.

Sample	O	Mg	Al	Si	Ca	Fe
S11	41.58	3.35	11.47	3.88	32.39	7.31
S12	31.85	1.42	8.86	2.78	41.59	13.49
S13	37.24	4.35	10.16	4.27	26.14	17.83
S14	47.17	3.03	13.64	3.39	14.83	17.93
S15	28.49	1.43	9.58	2.43	35.43	22.63
S16	26.16	1.16	13.04	2.81	30.05	26.77
S17	33.61	2.67	10.94	2.65	24.81	25.33
S18	27.11	1.69	7.86	1.77	29.44	32.12

Figure 19a–d shows the summarization of the results of slag foaming using a mixture of coke and biochar. For a group of slags with a stable FeO content, the maximum values of foaming character (foam height) were achieved for slags with basicity 2.0 and 2.7, the values of which are nearly similar. Then, there was a decrease in foam height.

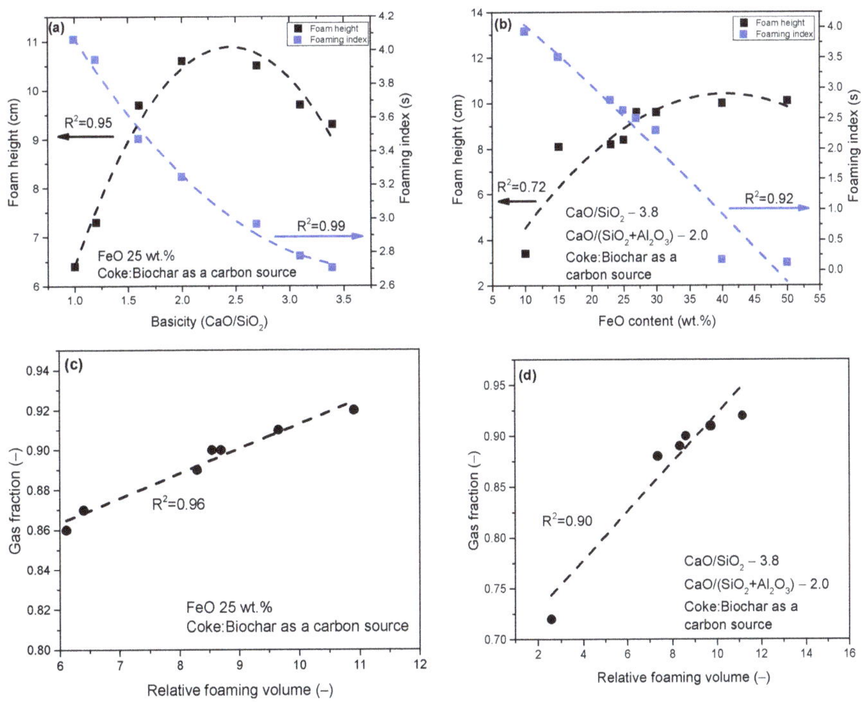

Figure 19. Foam height (mixture of coke and biochar with a ratio of 1:1 as a carbon source) and foaming index as a function of (**a**) basicity and (**b**) FeO content of slags; relationship between relative foaming volume and gas fraction of slags with stable FeO content (**c**) and basicity (**d**).

In contrast to the coke case, a different foaming behavior was observed for a set of slags with consistent basicity but varying FeO content (as shown in Figure 19b). As the FeO content increased, the foaming characters showed a corresponding increase, with the maximum values reached at 50 wt.% FeO. Notably, unlike the tests conducted with coke, gas formation (as shown in Figure 19d) continued to increase.

Similarly, the third group, comprising slags with different FeO content of 15 to 30 wt.%, showed optimal foaming properties. Particularly, the partial substitution of coke with biochar in the mixture showed that in the case of basicity variations, stable foaming was achieved with a minimum basicity value of 1.0 and a FeO content of 25 wt.%. For different FeO content, stable foaming was obtained from a range of 15 wt.% to 40 wt.% of this oxide, with a basicity of 3.8.

The results showed that adding biochar to the coke positively affects the foaming process. When a mixture of coke and biochar with a ratio of 1:1 was used, it produced a different foaming behavior than coke or biochar as carbon sources. Firstly, the reaction FeO + C = Fe + CO is of great importance in the slag foaming [12,53]; therefore, after adding the mixture to the molten slag, the biochar component, possessing greater porosity, surface area [34], and higher reactivity relative to coke, as evidenced by FTIR results, could react fast enough at the initial stages of foaming. It should be taken into account that the volatile matters of biochar (41.04 wt.% daf) could also contribute to the gasification $C(s) + CO_2(g) = 2CO(g)$ via the Boudouard reaction [54], as well as in the reaction of the water vapor [30] in accordance with $H_2O(g) + C(s) = CO(g) + H_2$. However, experimental results showed that the foaming time with 100 wt.% biochar was 3–4 min, indicating its fast reactivity. On the other hand, when using the mixture, the foaming time ranged from 6 to 9 min depending on the slag composition (the only exception is one test, which lasted about 3 min in the case of 50 wt.% FeO and was disregarded). This leads to the conclusion that even with a partial replacement of coke, it could have a major effect throughout the entire foaming process. Additionally, using a mixture of biochar and coke probably generated more CO/CO_2 than using 100 wt.% coke or 100 wt.% biochar, resulting in a longer reaction time compared to biochar use and enhancing the foaming process.

4. Conclusions

The foaming behavior of slags with chemical composition variations and three carbon sources, namely, 100 wt.% high-temperature coke as a reference, 100 wt.% biochar and a mixture of coke and biochar with a ratio of 1:1 has been evaluated both quantitively and qualitatively. The main findings can be emphasized:

(1) FTIR analysis indicates clear chemical distinctions between biochar and coke, with different functional groups in each carbon source. The reduction in certain functional groups of coke can be attributed to the higher carbonization temperature, leading to the loss of volatile and oxygen-containing functional groups. These results were confirmed judging by the 1H NMR spectra;

(2) Employing 100 wt.% coke as the carbon source resulted in optimal foaming characters for slags with different basicity values ranging from 1.2 to 2.7 and FeO content from 15 to 30 wt.%. For stable basicity (B2) at 1.4 and consistent Al_2O_3 content, the highest foam height and volume were achieved with 26.5 wt.% FeO and 13.52 wt.% MgO;

(3) When using 100 wt.% biochar as a carbon source, the most optimal foaming characters were obtained for a group of slags with different basicities from 1.2 to 3.4. Notably, the influence on foam height was not observed with the increase in basicity B2 from 1.2 to 3.1. For group slags with different FeO content, stable foaming was obtained for samples containing this oxide from 15 to 27 wt.%. In the slag group, where stable basicity of 1.4 and Al_2O_3 content were used, foaming characters were lower than coke, there were no significant changes, and the characters were within experimental error;

(4) A mixture of coke and biochar led to stable foaming with a minimum basicity value of 1.0 and 25 wt.% FeO content. Unlike the other carbon sources, this mixture provided stable foaming across all slag compositions with varying basicity values. Stable

foaming was observed in the presence of 15 wt.% to 40 wt.% FeO, presenting a broader range compared to individual applications of coke or biochar. For slags with stable basicity and Al_2O_3 content, similar foaming tendencies were observed compared to using coke.

(5) XRD analysis revealed consistent trends in the composition of slag foam samples unrelated to the carbon source used. Higher MgO content was associated with predominant phases such as monticellite and magnesium iron oxide. Slag foam samples with the lowest basicity showed the presence of monticellite and merwinite, while those with the highest basicity contained calcium magnesium aluminum oxide silicate, larnite, or dicalcium silicate;

(6) When a mixture of coke and biochar was used, it produced different foaming behavior than coke or biochar. From experimental observation, biochar reacts fast enough at the initial stages of foaming, and, probably, coke has a major effect throughout the entire foaming process. It could be assumed that this mixture might generate more CO/CO_2 than using 100 wt.% coke or 100 wt.% biochar individually. Consequently, this increased gas generation could lead to a prolonged reaction time compared to using 100 wt.% biochar, thereby enhancing the overall foaming process.

In summary, the utilization of different carbon sources influenced the foaming behavior, with the coke and biochar mixture demonstrating unique characteristics and potential for improved foaming outcomes.

Author Contributions: Conceptualization, L.K. and J.S.; methodology, L.K.; investigation, L.K.; writing—original draft preparation, L.K.; writing—review and editing, J.S.; supervision, J.S. All authors have read and agreed to the published version of the manuscript.

Funding: This research received no external funding.

Data Availability Statement: Not applicable.

Acknowledgments: Support by the scholarship program "Scholarship of the Scholarship Foundation of the Republic of Austria, Postdocs," [MPC-2022-02241], financed by the Federal Ministry of Education, Science and Research of Austria, is gratefully acknowledged. Additionally, Voestalpine Stahl GmbH, Linz, Austria, is gratefully acknowledged for providing samples for this research. The support from Heng Zheng (Montanuniversitaet Leoben) during this research is gratefully appreciated.

Conflicts of Interest: The authors declare no conflict of interest.

References

1. Zhu, T.X.; Coley, K.S.; Irons, G.A. Progress in Slag Foaming in Metallurgical Processes. *Metall. Mater. Trans. B* **2012**, *43*, 751–757. [CrossRef]
2. Matsuura, H.; Fruehan, R.J. Slag Foaming in an Electric Arc Furnace. *ISIJ Int.* **2009**, *49*, 1530–1535. [CrossRef]
3. Chychko, A.; Seetharaman, S. Foaming in Electric Arc Furnace Part I: Laboratory Studies of Enthalpy Changes of Carbonate Additions to Slag Melts. *Metall. Mater. Trans. B* **2011**, *42*, 20–29. [CrossRef]
4. Vieira, D.; Almeida, R.A.M.D.; Bielefeldt, W.V.; Vilela, A.C.F. Slag Evaluation to Reduce Energy Consumption and EAF Electrical Instability. *Mater. Res.* **2016**, *19*, 1127–1131. [CrossRef]
5. Ji, F.-Z.; Barati, M.; Coley, K.; Irons, G.A. Kinetics of Coal Injection into Iron Oxide Containing Slags. *Can. Metall. Q.* **2005**, *44*, 85–94. [CrossRef]
6. Agnihotri, A.; Singh, P.K.; Singh, D.; Gupta, M. Foamy Slag Practice to Enhance the Energy Efficiency of Electric Arc Furnace: An Industrial Scale Validation. *Mater. Today Proc.* **2021**, *46*, 1537–1542. [CrossRef]
7. de Almeida, R.A.M.; Vieira, D.; Bielefeldt, W.V.; Vilela, A.C.F. Slag Foaming Fundamentals—A Critical Assessment. *Mater. Res.* **2017**, *20*, 474–480. [CrossRef]
8. Kieush, L.; Koveria, A.; Schenk, J.; Rysbekov, K.; Lozynskyi, V.; Zheng, H.; Matayev, A. Investigation into the Effect of Multi-Component Coal Blends on Properties of Metallurgical Coke via Petrographic Analysis under Industrial Conditions. *Sustainability* **2022**, *14*, 9947. [CrossRef]
9. Ogawa, Y.; Katayama, H.; Hirata, H.; Tokumitsu, N.; Yamauchi, M. Slag Foaming in Smelting Reduction and Its Control with Carbonaceous Materials. *ISIJ Int.* **1992**, *32*, 87–94. [CrossRef]
10. Wu, K.; Qian, W.; Chu, S.; Niu, Q.; Luo, H. Behavior of Slag Foaming Caused by Blowing Gas in Molten Slags. *ISIJ Int.* **2000**, *40*, 954–957. [CrossRef]
11. Zhang, Y.; Fruehan, R.J. Effect of Carbonaceous Particles on Slag Foaming. *Metall. Mater. Trans. B* **1995**, *26*, 813–819. [CrossRef]

12. Bhoi, B.; Jouhari, A.K.; Ray, H.S.; Misra, V.N. Smelting Reduction Reactions by Solid Carbon Using Induction Furnace: Foaming Behaviour and Kinetics of FeO Reduction in CaO–SiO_2–FeO Slag. *Ironmak. Steelmak.* **2006**, *33*, 245–252. [CrossRef]
13. Zhang, Y.; Fruehan, R.J. Effect of Gas Type and Pressure on Slag Foaming. *Metall. Mater. Trans. B* **1995**, *26*, 1088–1091. [CrossRef]
14. Kieush, L.; Schenk, J.; Koveria, A.; Hrubiak, A.; Hopfinger, H.; Zheng, H. Evaluation of Slag Foaming Behavior Using Renewable Carbon Sources in Electric Arc Furnace-Based Steel Production. *Energies* **2023**, *16*, 4673. [CrossRef]
15. European Commission. Directorate General for Research and Innovation. In *Sustainable EAF Steel Production (GREENEAF)*; Publications Office: Luxembourg, 2013.
16. Cirilli, F.; Baracchini, G.; Bianco, L. EAF Long Term Industrial Trials of Utilization of Char from Biomass as Fossil Coal Substitute. *La Metall. Ital.* **2017**, *109*, 13–17.
17. Nwachukwu, C.M.; Wang, C.; Wetterlund, E. Exploring the Role of Forest Biomass in Abating Fossil CO_2 Emissions in the Iron and Steel Industry—The Case of Sweden. *Appl. Energy* **2021**, *288*, 116558. [CrossRef]
18. Funke, A.; Demus, T.; Willms, T.; Schenke, L.; Echterhof, T.; Niebel, A.; Pfeifer, H.; Dahmen, N. Application of Fast Pyrolysis Char in an Electric Arc Furnace. *Fuel Process. Technol.* **2018**, *174*, 61–68. [CrossRef]
19. Kalde, A.; Demus, T.; Echterhof, T.; Pfeifer, H. Determining the Reactivity of Biochar-Agglomerates to Replace Fossil Coal in Electric Arc Furnace Steelmaking. In Proceedings of the EUBCE 2015 Online Conference Proceedings, Online, 1–4 June 2015; pp. 497–507.
20. Chen, W.-H.; Peng, J.; Bi, X.T. A State-of-the-Art Review of Biomass Torrefaction, Densification and Applications. *Renew. Sustain. Energy Rev.* **2015**, *44*, 847–866. [CrossRef]
21. Ru, B.; Wang, S.; Dai, G.; Zhang, L. Effect of Torrefaction on Biomass Physicochemical Characteristics and the Resulting Pyrolysis Behavior. *Energy Fuels* **2015**, *29*, 5865–5874. [CrossRef]
22. Yang, H.; Yan, R.; Chen, H.; Zheng, C.; Lee, D.H.; Liang, D.T. In-Depth Investigation of Biomass Pyrolysis Based on Three Major Components: Hemicellulose, Cellulose and Lignin. *Energy Fuels* **2006**, *20*, 388–393. [CrossRef]
23. Kieush, L.; Koveria, A.; Boyko, M.; Yaholnyk, M.; Hrubiak, A.; Molchanov, L.; Moklyak, V. Influence of Biocoke on Iron Ore Sintering Performance and Strength Properties of Sinter. *Min. Miner. Depos.* **2022**, *16*, 55–63. [CrossRef]
24. Shukla, I. Potential of Renewable Agricultural Wastes in the Smart and Sustainable Steelmaking Process. *J. Clean. Prod.* **2022**, *370*, 133422. [CrossRef]
25. Fidalgo, B.; Berrueco, C.; Millan, M. Chars from Agricultural Wastes as Greener Fuels for Electric Arc Furnaces. *J. Anal. Appl. Pyrolysis* **2015**, *113*, 274–280. [CrossRef]
26. Echterhof, T. Review on the Use of Alternative Carbon Sources in EAF Steelmaking. *Metals* **2021**, *11*, 222. [CrossRef]
27. Huang, X.; Ng, K.W.; Giroux, L.; Duchesne, M.; Li, D.; Todoschuk, T. Interaction Behavior of Biogenic Material with Electric Arc Furnace Slag. *Fuels* **2021**, *2*, 420–436. [CrossRef]
28. Robinson, R.; Brabie, L.; Pettersson, M.; Amovic, M.; Ljunggren, R. An Empirical Comparative Study of Renewable Biochar and Fossil Carbon as Carburizer in Steelmaking. *ISIJ Int.* **2022**, *62*, 2522–2528. [CrossRef]
29. Demus, T.; Reichel, T.; Schulten, M.; Echterhof, T.; Pfeifer, H. Increasing the Sustainability of Steel Production in the Electric Arc Furnace by Substituting Fossil Coal with Biochar Agglomerates. *Ironmak. Steelmak.* **2016**, *43*, 564–570. [CrossRef]
30. Yunos, N.F.M.; Zaharia, M.; Idris, M.A.; Nath, D.; Khanna, R.; Sahajwalla, V. Recycling Agricultural Waste from Palm Shells during Electric Arc Furnace Steelmaking. *Energy Fuels* **2012**, *26*, 278–286. [CrossRef]
31. Hoikkaniemi, E.; Sulasalmi, P.; Visuri, V.-V.; Fabritius, T. Biochar as a Slag Foaming Agent in EAF—A Novel Experimental Setup. In Proceedings of the 5th European Academic Symposium on EAF Steelmaking, Oulu, Finland, 5 June 2023; pp. 1–5.
32. Di Giovanni, C.; Li, D.; Ng, K.W.; Huang, X. Ranking of Injection Biochar for Slag Foaming Applications in Steelmaking. *Metals* **2023**, *13*, 1003. [CrossRef]
33. Kieush, L.; Rieger, J.; Schenk, J.; Brondi, C.; Rovelli, D.; Echterhof, T.; Cirilli, F.; Thaler, C.; Jaeger, N.; Snaet, D.; et al. A Comprehensive Review of Secondary Carbon Bio-Carriers for Application in Metallurgical Processes: Utilization of Torrefied Biomass in Steel Production. *Metals* **2022**, *12*, 2005. [CrossRef]
34. Kieush, L.; Schenk, J.; Koveria, A.; Rantitsch, G.; Hrubiak, A.; Hopfinger, H. Utilization of Renewable Carbon in Electric Arc Furnace-Based Steel Production: Comparative Evaluation of Properties of Conventional and Non-Conventional Carbon-Bearing Sources. *Metals* **2023**, *13*, 722. [CrossRef]
35. Menad, N.-E.; Kana, N.; Seron, A.; Kanari, N. New EAF Slag Characterization Methodology for Strategic Metal Recovery. *Materials* **2021**, *14*, 1513. [CrossRef]
36. Heo, J.H.; Park, J.H. Assessment of Physicochemical Properties of Electrical Arc Furnace Slag and Their Effects on Foamability. *Metall. Mater. Trans. B* **2019**, *50*, 2959–2968. [CrossRef]
37. Jiang, R.; Fruehan, R.J. Slag Foaming in Bath Smelting. *Metall. Mater. Trans. B* **1991**, *22*, 481–489. [CrossRef]
38. Lahiri, A.K.; Seetharaman, S. Foaming Behavior of Slags. *Metall. Mater. Trans. B* **2002**, *33*, 499–502. [CrossRef]
39. Hong, L.; Hirasawa, M.; Sano, M. Behavior of Slag Foaming with Reduction of Iron Oxide in Molten Slags by Graphite. *ISIJ Int.* **1998**, *38*, 1339–1345. [CrossRef]
40. Xiang, J.; Wang, X.; Yang, M.; Wang, J.; Shan, C.; Fan, G.; Qiu, G.; Lv, X. Slag-Foaming Phenomenon Originating from Reaction of Titanium-Bearing Blast Furnace Slag: Continuous Monitoring of Foaming Height and Calibration. *J. Mater. Res. Technol.* **2021**, *11*, 1184–1192. [CrossRef]

41. Ren, X.; Cai, H.; Chang, J.-M.; Fan, Y.-M. TG-FTIR Study on the Pyrolysis Properties of Lignin from Different Kinds of Woody Biomass. *Mater. Sci.* **2018**, *3*, 1–7. [CrossRef]
42. Ben, H.; Hao, N.; Liu, Q.; Ragauskas, A.J. Solid-State NMR Investigation of Bio-Chars Produced from Biomass Components and Whole Biomasses. *Bioenerg. Res.* **2017**, *10*, 1036–1044. [CrossRef]
43. Jennings, K.R. Spectrometric Identification of Organic Compounds (Fifth Edition). Wiley, New York. *Org. Mass Spectrom.* **1991**, *26*, 813. [CrossRef]
44. Vidacak, B.; Arvanitidis, I.; Jönsson, P.G.; Sjöberg, P. Observation on Foaming of EAF Slags in the Production of Stainless Steel: Observation on Foaming of EAF Slags. *Scand. J. Metall.* **2002**, *31*, 321–327. [CrossRef]
45. Zhang, Y.; Fruehan, R.J. Effect of the Bubble Size and Chemical Reactions on Slag Foaming. *MMTB* **1995**, *26*, 803–812. [CrossRef]
46. Herbelin, M.; Bascou, J.; Lavastre, V.; Guillaume, D.; Benbakkar, M.; Peuble, S.; Baron, J.-P. Steel Slag Characterisation—Benefit of Coupling Chemical, Mineralogical and Magnetic Techniques. *Minerals* **2020**, *10*, 705. [CrossRef]
47. Mombelli, D.; Mapelli, C.; Barella, S.; Gruttadauria, A.; Le Saout, G.; Garcia-Diaz, E. The Efficiency of Quartz Addition on Electric Arc Furnace (EAF) Carbon Steel Slag Stability. *J. Hazard. Mater.* **2014**, *279*, 586–596. [CrossRef] [PubMed]
48. Riboldi, A.; Borgese, L.; Vassalini, I.; Cornacchia, G.; Gelfi, M.; Boniardi, M.V.; Casaroli, A.; Depero, L.E. Micro-Raman Spectroscopy Investigation of Crystalline Phases in EAF Slag. *Appl. Sci.* **2020**, *10*, 4115. [CrossRef]
49. Gaye, H.; Lehmann, J. Steelmaking Slags. In *Slag Atlas*; Verlag Stahleisen GmbH: Dusseldorf, Germany, 1995.
50. Corbari, R.; Matsuura, H.; Halder, S.; Walker, M.; Fruehan, R.J. Foaming and the Rate of the Carbon-Iron Oxide Reaction in Slag. *Metall. Mater. Trans. B* **2009**, *40*, 940–948. [CrossRef]
51. Luz, A.P.; Avila, T.A.; Bonadia, P.; Pandolfelli, V.C. Slag Foaming: Fundamentals, Experimental Evaluation and Application in the Steelmaking Industry. *Refract. Worldforum* **2011**, *3*, 91–98.
52. Ito, K.; Fruehan, R.J. Study on the Foaming of CaO-SiO_2-FeO Slags: Part I. Foaming Parameters and Experimental Results. *Metall. Trans. B* **1989**, *20*, 509–514. [CrossRef]
53. Luz, A.P.; Tomba Martinez, A.G.; López, F.; Bonadia, P.; Pandolfelli, V.C. Slag Foaming Practice in the Steelmaking Process. *Ceram. Int.* **2018**, *44*, 8727–8741. [CrossRef]
54. Teasdale, S.L.; Hayes, P.C. Observations of the Reduction of FeO from Slag by Graphite, Coke and Coal Char. *ISIJ Int.* **2005**, *45*, 634–641. [CrossRef]

energies

MDPI

Article

Technological Prospects of Biochar Derived from Viticulture Waste: Characterization and Application Perspectives

Veronica D'Eusanio [1,2,*], Antonio Lezza [1], Biagio Anderlini [1], Daniele Malferrari [1], Marcello Romagnoli [3,4] and Fabrizio Roncaglia [1,2,3,*]

1 Department of Chemical and Geological Sciences, University of Modena and Reggio Emilia, 41125 Modena, Italy; antonio.lezza@unimore.it (A.L.); biagio.anderlini@unimore.it (B.A.); daniele.malferrari@unimore.it (D.M.)
2 National Interuniversity Consortium of Materials Science and Technology (INSTM), 50121 Firenze, Italy
3 Interdepartmental Centre H2-MORE, University of Modena and Reggio Emilia, Via Università 4, 41121 Modena, Italy; marcello.romagnoli@unimore.it
4 Department of Engineering "Enzo Ferrari", University of Modena and Reggio Emilia, Via Vivarelli 10, 41125 Modena, Italy
* Correspondence: veronica.deusanio@unimore.it (V.D.); fabrizio.roncaglia@unmore.it (F.R.)

Abstract: The increasing demand for sustainable agricultural practices aimed at reducing carbon dioxide emissions has driven the exploration of converting viticulture residues into biochar. This study investigates the potential technological applications of biochar as a filler for the production of electrically conductive composite materials, suitable to Bipolar Plate (BP) manufacturing. Grape seeds (GSs), defatted grape seeds (DGSs), wood stems (WSs), and whole grape seeds (WGSs) were converted into biochar samples through low-temperature (300 °C) pyrolysis for 3 or 24 h. The composition and thermal stability of biochar were evaluated through thermogravimetric analysis (TG), which provided valuable insights into interpreting the in-plane conductivity (IPC) values of the BP samples. Pyrolyzed GS and DGS biochar samples demonstrated enhanced thermal stability and conferred higher IPC values compared to WS counterparts. This indicates a clear correlation between the formation of carbon-rich structures during pyrolysis and overall electrical conductivity. In contrast, pyrolyzed WGSs produced BP samples with lower IPC values due to the presence of lipids, which were not effectively degraded by the low-temperature pyrolysis.

Keywords: biochar; biomass pyrolysis; viticulture waste; TGA; electro-conductive materials

1. Introduction

Wine production, deeply rooted in the cultural fabric of Europe, is a fundamental pillar of social, commercial, and economic life in major producing countries such as Italy, France, and Spain. The wine industry in Italy, for instance, boasts a significant output, producing approximately 5.1 million tonnes of wine and 8.1 million tonnes of grapes in 2021 alone [1]. However, this substantial production volume also generates considerable waste. Vinification processes inherently produce byproducts such as grape pomace, lees, and residues from cleaning agents and chemicals used in the production process. Grape pomace, mainly composed of skins, seeds, and some stems, generally accounts for ~20–30% (*w*/*w*) of the original grape weight [2]. Managing these byproducts poses environmental challenges, as improper disposal or treatment can lead to pollution and ecosystem degradation. For instance, incineration or landfilling can result in adverse environmental consequences, including surface and groundwater pollution, CO_2 emissions, attraction of agricultural pests and flies, and the production of unpleasant odors [2–4].

The considerable amounts of waste byproducts suggest the importance of devising strategies aimed at minimizing the environmental impact within the wine industry. This necessitates a focus on circular-economy principles, wherein high-value byproducts are effectively utilized to create additional value and reduced waste. Examples of such practices include the extraction of grape seed oil or polyphenols and conversion into biofuels. These approaches not only contribute to waste reduction but also offer economic and environmental benefits to wineries, aligning with sustainability principles in viticulture and oenology. In pursuit of zero waste in the viticulture sector, converting solid residues from grape pomace into biochar through pyrolysis represents a promising option [2,5]. This approach offers several advantages, including the potential to incorporate secondary waste generated from previous extractions, such as defatted grape seeds, as well as waste materials devoid of significant extractable components, such as woody stems or canes.

The pyrolysis process is easily implemented but intrinsically complex, involving diverse and interrelated physical and chemical transformations that result in the production of numerous volatile and non-volatile species [6]. Several process parameters, such as residence time, heating rate, and feedstock composition [7], influence the yield and morphology of the obtained biochar. However, pyrolysis temperature remains the most important parameter [8]. The characteristics of the obtained biochar also depend significantly on the specific biomass substrate, solid-volume-to-surface-area ratio, and moisture content [9]. The explicit quantities of cellulose, hemicellulose, and lignin in the biomass significantly influence the nature and yield of pyrolysis products. For instance, cellulose tends to produce more volatile organic compounds (VOCs) and tars, whereas lignin contributes to the formation of char and aromatic compounds. Additionally, the presence of proteins and lipids within the biomass leads to the formation of nitrogen- and oxygen-containing compounds. The moisture content of the biomass also plays a crucial role, affecting the thermal degradation pathways and the overall energy efficiency of the process. Other factors, such as the presence of minerals and inorganic elements, further modulate the pyrolysis reactions and the quality of the resulting bio-oil, gas, and biochar.

Several studies have highlighted the potential of biochar production from biomass via slow pyrolysis, detecting an annual net energy production ranging from 26,008 to 49,903 MJ/ha in the form of syngas [5]. The energy yield from syngas ranges between 2 and 7 MJ/MJ, surpassing the energy output of ethanol production, which typically ranges from 0.7 to 2.2 MJ/MJ [10,11]. Furthermore, biochar production yields substantial heat, offering opportunities to reduce reliance on fossil fuels. Beyond energy benefits, the conversion of biomass into biochar holds promise in mitigating net greenhouse gas emissions [12], improving soil properties [13], and acting as a remediation agent for wastewater contaminants [14–16], thus underscoring its potential as an environmentally sustainable solution.

Grape skins, seeds, and stems are substrates rich in dietary fibers, comprising major constituents such as cellulose, hemicellulose, and lignin [17–19]. This composition prompts an evaluation of grape waste biochar as electrically conducting fillers, given that biochar derived from lignin and cellulose can achieve conductivities comparable to carbon black [20,21], a commonly used filler in antistatic composite materials. This enhanced conductivity arises from the high-carbon polyaromatic structure developed during pyrolysis, a property that, in some instances, was exploited to obtain graphene-like structures through post-modification [22].

Bipolar Plates (BPs) play a crucial role in ensuring the proper distribution of reactants and products during the operation of fuel cells and electrolyzers [23]. They are instrumental in both the production of clean hydrogen and its effective utilization. Compared to metallic BPs, graphite-based BPs offer notable advantages, including lower corrosion and reduced weight. While bulk graphite boasts exceptional electrical and thermal conductivity, as well as high chemical inertness, it does exhibit brittleness. This limitation can be addressed with composite BPs [24], composed of powdered graphite and an organic polymeric binder. High-filled composite BPs can retain most of the graphite's beneficial properties, achieving excellent in-plane electrical conductivities (IPCs) as required by the technical standards (>100 S/cm) [23].

In this study, the valorization of different winery byproducts, specifically woody stems (WSs), grape skins (GSs), whole grape seeds (WGSs), and defatted grape seeds (DGSs) were evaluated through biochar production. Two pyrolysis conditions were investigated, all conducted at a consistent temperature of 300 °C but varying in pyrolysis duration (3 h and 24 h). The selection of these specific pyrolysis times was intended to thoroughly investigate the impact of thermal processing duration on the yield and properties of the resulting biochar. Slow pyrolysis, characterized by lower temperatures and extended processing times, generally leads to higher biochar yields [25]. By examining both shorter (3 h) and longer (24 h) durations, we aimed to assess how the composition of biochar evolves over pyrolysis times. The 3 h pyrolysis allows us to study the initial stages of biomass decomposition and biochar formation, corresponding to the minimum processing time (energy input), useful for industrial applications. Conversely, the 24 h pyrolysis process allows us to observe the effects of prolonged heat exposure on the final composition and quality of biochar. Longer pyrolysis times are expected to facilitate more complete carbonization, potentially enhancing electrical conductivity. The biochar samples were characterized by thermogravimetric (TG) analysis. Moreover, the obtained biochars were evaluated as a substitute for fossil graphite within composite BPs, potentially enhancing both its cost-effectiveness and sustainability by improving the renewable carbon footprint. The findings could contribute to the development of more environmentally friendly hydrogen economy while leveraging agricultural waste for valuable applications.

2. Materials and Methods

2.1. Chemicals

Diglycidylether of bisphenol A (DGEBA) resin DER 331® (187 g/eq) was purchased from Dow Chemicals (Midland, MI, USA); isophoronediamine (IPDA) hardener (42.57 g/eq) and dichloromethane (DCM) were purchased from Merck (Darmstadt, Germany); GraphCOND® 45/98 X, D50 75 mm, 98% carbon (G45) and GraphTERM® 23/99.9, D50 23 mm, 99.9% carbon (G23) were purchased from Georg H. LUH GmbH (Walluf, Germany); carbon black (CB) (PBX51) D50 17 mm was purchased from Cabot Corporation (Alpharetta, GA, USA).

2.2. Sample Preparation

Woody stems (WSs) of *Vitis vinifera* cultivar Graspa Rossa were collected from a farm in the Modena (Italy) area, alongside the grapes utilized for obtaining grape skin (GS) and whole grape seed (WGS) samples. Defatted grape seeds (DGSs) were sourced from Randi Group, a local company based in Faenza, Italy, specializing in grape seed oil production (https://www.randi-group.com/it/randi-group/, accessed on 14 February 2024). DGSs are derived from a blend of grape seeds collected from wineries across northern Italy, irrespective of cultivar or geographic origin. According to the company, DGSs constitute ~90% of the waste generated from grape seed oil production, consistent with findings from previous studies [26–28].

All samples were pre-dried in a laboratory oven (Argo LAB, Carpi, MO, Italy) (oven-dried samples, OD) to decrease moisture levels to below 8 wt%. This low moisture content contributes to the long-term stability of the matrices and inhibits the growth of mold and undesirable microorganisms.

Pyrolysis was conducted in a laboratory muffle under N_2 flow, carbonizing the samples at 300 °C for 3 and 24 h. Figure 1 presents the DGS samples pyrolyzed for 3 and 24 h, by way of example.

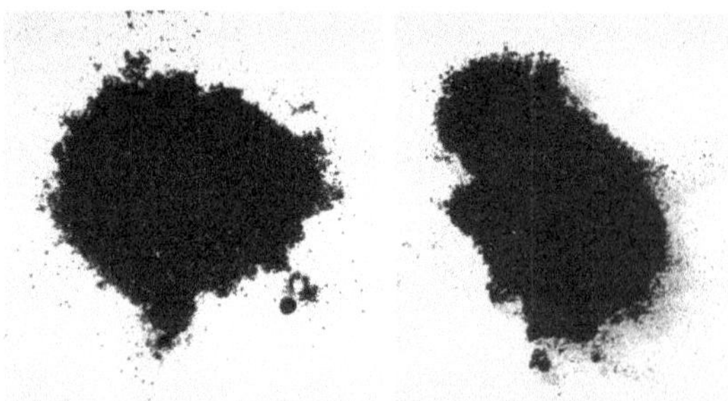

Figure 1. Defatted grape seed (DGS) samples pyrolyzed at 300 °C for 3 h (**left**) and 24 h (**right**).

2.3. Proximate Composition

The moisture and ash content were assessed following protocols recommended by the Association of Official Analytical Chemists [29]. The moisture content was determined by drying the samples at 105 °C until a constant weight was achieved. The ash content was determined using a laboratory furnace with a gradual temperature ramp up to 550 °C. Each measurement was performed in triplicate, and the results were averaged. CHN elemental analysis was conducted with a CHNS Analyzer Flash2000 (Thermo-Scientific, Waltham, MA, USA).

2.4. TGA

Thermogravimetric analysis was conducted using a Seiko SSC 5200 thermal analyzer (Seiko Instruments Inc., Chiba, Japan) under inert atmosphere conditions. The heating protocol employed a gradient of 10 °C /min over the temperature range of 25–1000 °C, with ultrapure helium purging at a flow rate of 100 μL/min.

2.5. Powder Preparation

All samples were prepared maintaining a fixed 85/15 filler-to-binder ratio [24]. In a 100 mL round-bottom glass flask, equipped with a magnetic stirring bar, CB, G23, G45, biochar, and DCM (35 mL) were introduced, and the suspension was stirred at 360 rpm for 5 min. Subsequently, DGEBA epoxy resin (0.56 g, dissolved in 2 mL of DCM) was added under stirring, and 2 min later, IPDA (0.12 g, dissolved in 2 mL of DCM) was slowly dropped under stirring. The mixture was further stirred for 5 min. Then, the solvent was removed via a rotary evaporator, resulting in 4.5 g of powder, which was divided into four replicates of approximately 1.0 g per sample. For each type of the obtained biochar samples (WSs, GSs, WGSs, and DGSs), six different loadings (wt%) were included within the BP formulations, as shown in Table 1. Additionally, two different pyrolysis conditions (3 h at 300 °C or 24 h at 300 °C) were evaluated.

Table 1. Compositions of biochar conductive samples.

Biochar Loading (wt%)	Biochar (g)	G45 (g)	G23 (g)	CB (g)	DGEBA (g)	IPDA (g)
10	0.46	2.65	0.57	0.19	0.56	0.12
15	0.68	2.47	0.54	0.18	0.56	0.12
20	0.91	2.29	0.50	0.17	0.56	0.12
25	1.14	2.12	0.46	0.16	0.56	0.12
30	1.37	1.94	0.42	0.14	0.56	0.12
35	1.59	1.77	0.38	0.13	0.56	0.12

2.6. Molding

Each dried powder sample (approximately 1.0 g) was pressed at 160 °C and 100 MPa for 12 min in a 2 cm i.d. mold. The obtained discs (see Figure 2), having a thickness close to 2 mm, were subsequently post-cured in oven at 60 °C for 48 h.

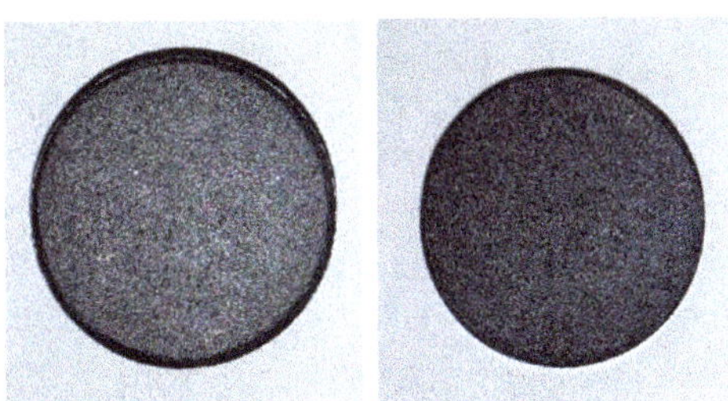

Figure 2. WS_3h prepared BP samples with biochar loading of 10 wt% (**left**) and 35 wt% (**right**).

2.7. In-Plane Conductivity (IPC) Measurements

IPC was measured by means of a digital sourcemeter (Keithley 2400 SMU, Keithley Instruments, Inc., Cleveland, OH, USA) equipped with a four-point probe, as described earlier [24]. Each sample was measured in 10 different points (five per side), and the mean IPC (S/cm) value was obtained as the inverse of the resistivity ρ (Ω·cm), as depicted in Equation (1). The mean resistivity value was corrected accordingly to the sample geometry, by means of the F(t/s) factor of Equation (2), tabulated by F.M. Smits [30].

$$\mathrm{IPC} = \frac{1}{\rho} \tag{1}$$

$$\rho = \left(\frac{\pi}{\ln 2} \cdot \frac{V}{I}\right) \cdot t \cdot F\left(\frac{t}{s}\right) \tag{2}$$

where V is the measured voltage (mV), I is the measured current (mA), t is the sample thickness (cm), and s is the spacing between two probes (cm).

2.8. Density Measurements

The density was calculated with standard weight/volume formula. Samples masses were measured with an analytical balance. The sample thickness was obtained as the mean value of four measurements carried out with a 0.01 mm resolution micrometer, while the samples' area was almost constant (=3.141 cm^2) due to the mold's internal dimensions.

2.9. Statistical Analysis

The results are expressed as means ± standard deviations of the mean and statistically analyzed using Matlab R2023a software. For comparison of the results, a one-way analysis of variance (ANOVA) test followed by a Tukey–Kramer post hoc ANOVA test was used, and p values of less than 0.05 were considered statistically significant. PCA analysis was carried out by using the PLS-Toolbox 2.1 for Matlab© (distributed by Eigenvector Research, Wenatchee, WA, USA).

3. Results

3.1. Proximate Composition

The geographical origin, degree of maturation, cultivar, and other factors exert a substantial and meaningful influence on the chemical composition of plant-based matrices [31,32]. In this study, proximate analysis offers a comprehensive insight into the variations in chemical composition of the matrices following the pyrolysis procedure. The results of the proximate analysis are reported in Table 2.

The data reported in Table 2 show an increase in carbon content (C %) and a decrease in the hydrogen (H %) and oxygen (O %) contents with pyrolysis time. Therefore, O/C and H/C ratios decreased accordingly. This trend is primarily attributed to the substantial loss of oxygenated and other functionalized compounds during heating, along with the release of VOCs and removal of inherent water. The observed increase in nitrogen content (N %) during pyrolysis can be attributed to several interrelated factors associated with the differential thermal stability and decomposition rates of various biomass components. Specifically, proteins exhibit a longer half-life compared to other components, reflecting their higher stability. Complete protein degradation during pyrolysis only occurs at temperatures above 500 °C; at lower temperatures, proteins are transformed into amines (through decarboxylation) or *N*-heterocycles [33,34] through Maillard reactions with sugars.

Thermochemical conversion of organic material involves numerous physical and chemical changes that yield solids (char), liquids, and gases, which may exhibit improved properties as materials or as fuels. Pyrolysis or carbonization, conducted in the absence of oxidants (e.g., oxygen), primarily result in the formation of solids as primary products. Slow pyrolysis is particularly effective for biochar production, typically yielding around 35% biochar from the dry biomass weight [7]. During heating, free moisture is first lost, and from approximately 120 °C onward, the chemical bonds of primary biomass constituents—hemicellulose, cellulose, and lignin—are progressively rearranged, forming new functional groups (e.g., carboxyl, lactone, quinine, chromene, anhydride, phenol, ether, pyrone, pyridine, pyridone, and pyrrole) [7], and releasing VOCs [35]. These volatile matters include CO_2, CO, CH_4 [36], H_2O, and various condensable organic species under ambient conditions. The remaining carbon-rich, non-volatile solid residue of biomass is referred to as biochar. Biochar produced at low temperatures (300–400 °C) exhibits various oxygenated functional groups on its surface, derived from partial degradation of initial biopolymers. Conversely, biochar produced at higher temperatures (600–700 °C) exhibits a highly hydrophobic nature with well-organized carbon-rich layers reminiscent of graphite [37]. A lower O/C ratio enhances the stability of biochar, making it similar to carbon black [25]. The H/C ratio serves as an indicator of biochar structure; values below 0.1 suggest a high abundance of fused aromatic ring structures [38]. Biochar also retains a significant portion of the mineral matter originally present in the biomass, resulting in an increase in ash percentage content compared to the non-pyrolyzed samples.

Table 2. Proximate analysis of the oven-dried biomass samples, before and after pyrolysis.

	Pyrolysis Conditions	WSs	GSs	WGSs	DGSs
Moisture	-	6.97 ± 0.04 [ab]	6.89 ± 0.07 [a]	6.12 ± 0.03 [c]	7.06 ± 0.09 [b]
	300 °C, 3 h	2.47 ± 0.02 [a]	2.56 ± 0.05 [a]	1.01 ± 0.06 [b]	0.97 ± 0.04 [b]
	300 °C, 24 h	2.44 ± 0.02 [a]	2.60 ± 0.02 [b]	1.03 ± 0.04 [c]	1.04 ± 0.03 [c]
Ash	-	3.07 ± 0.02 [a]	5.71 ± 0.04 [b]	3.13 ± 0.03 [a]	2.07 ± 0.02 [c]
	300 °C, 3 h	3.68 ± 0.02 [a]	6.65 ± 0.04 [b]	3.55 ± 0.02 [c]	2.94 ± 0.03 [d]
	300 °C, 24 h	4.23 ± 0.04 [a]	7.61 ± 0.03 [b]	3.71 ± 0.01 [c]	3.44 ± 0.04 [d]
C	-	45.0 ± 0.3 [a]	47.2 ± 0.6 [b]	51.5 ± 0.4 [c]	51.2 ± 0.5 [c]
	300 °C, 3 h	51.0 ± 0.1 [a]	57.8 ± 0.4 [b]	63.9 ± 0.4 [c]	58.9 ± 0.1 [d]
	300 °C, 24 h	61.7 ± 0.1 [a]	63.1 ± 0.2 [b]	68.8 ± 0.2 [c]	64.9 ± 0.1 [d]
H	-	6.99 ± 0.21 [a]	6.72 ± 0.53 [ab]	6.01 ± 0.31 [b]	6.08 ± 0.11 [b]
	300 °C, 3 h	5.57 ± 0.14 [a]	5.08 ± 0.01 [b]	5.98 ± 0.15 [c]	5.28 ± 0.02 [b]
	300 °C, 24 h	3.10 ± 0.04 [a]	3.62 ± 0.03 [b]	5.62 ± 0.02 [c]	3.72 ± 0.07 [b]
N	-	0.50 ± 0.05 [a]	1.46 ± 0.08 [b]	1.22 ± 0.02 [c]	1.43 ± 0.07 [b]
	300 °C, 3 h	0.58 [*a]	1.67 ± 0.05 [b]	1.90 ± 0.36 [b]	2.26 ± 0.39 [b]
	300 °C, 24 h	1.19 ± 0.03 [a]	2.31 ± 0.01 [b]	2.38 ± 0.03 [b]	2.89 ± 0.15 [c]
S	-	<0.1	<0.1	<0.1	<0.1
	300 °C, 3 h	<0.1	<0.1	<0.1	<0.1
	300 °C, 24 h	<0.1	<0.1	<0.1	<0.1
O (by difference)	-	44.4 ± 0.3 [a]	38.9 ± 0.6 [b]	38.1 ± 0.3 [b]	39.2 ± 0.5 [b]
	300 °C, 3 h	39.2 ± 0.2 [a]	28.8 ± 0.4 [b]	24.7 ± 0.9 [c]	30.6 ± 0.5 [d]
	300 °C, 24 h	29.8 ± 0.1 [a]	23.4 ± 0.2 [b]	19.5 ± 0.2 [c]	25.0 ± 0.4 [d]
O/C	-	0.99	0.82	0.74	0.77
	300 °C, 3 h	0.77	0.50	0.39	0.52
	300 °C, 24 h	0.48	0.37	0.28	0.39
H/C	-	0.16	0.14	0.12	0.12
	300 °C, 3 h	0.11	0.09	0.09	0.09
	300 °C, 24 h	0.05	0.06	0.08	0.06

Biomass was oven-dried before pyrolysis. Data are expressed as mean ± SD of three replicates; * SD < 0.01; [a,b,c,d] differences between means indicated by the same letter within the same row are not statistically significant ($p < 0.05$) according to the Tukey–Kramer HSD post hoc test.

3.2. TGA

Each biomass constituent decomposes at different rates and within distinct temperature ranges. The TGA profiles of WS, GS, WGS and DGS samples at a 10 °C/min heating rate are shown in Figure 3a–d, respectively. The TGA data are represented by the dashed lines, while the differential thermal gravimetric, DTG, and percentage mass losses are shown by solid lines and the percentage values on the left-hand scale.

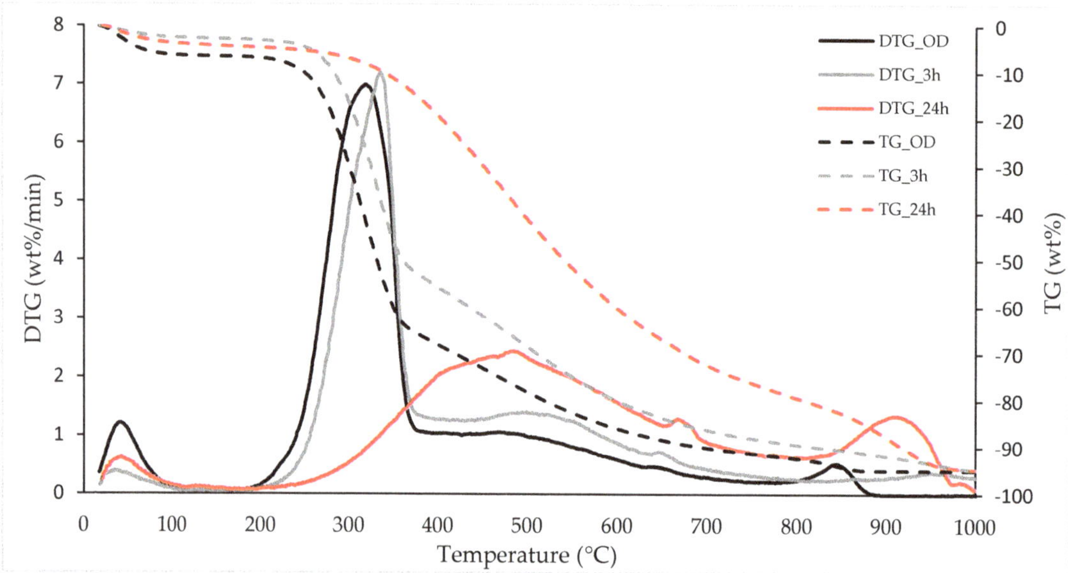

(a)

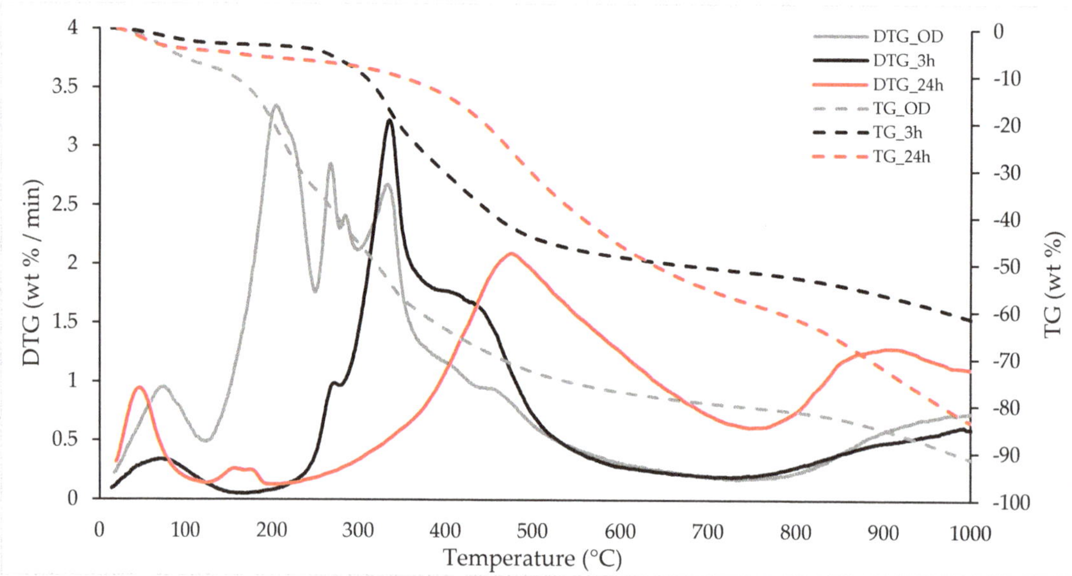

(b)

Figure 3. *Cont.*

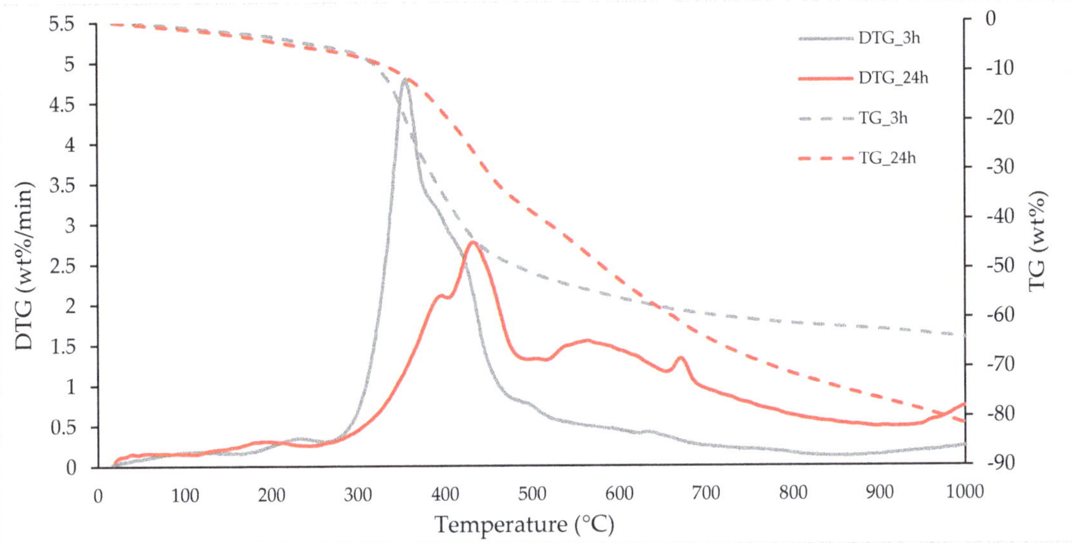

(c)

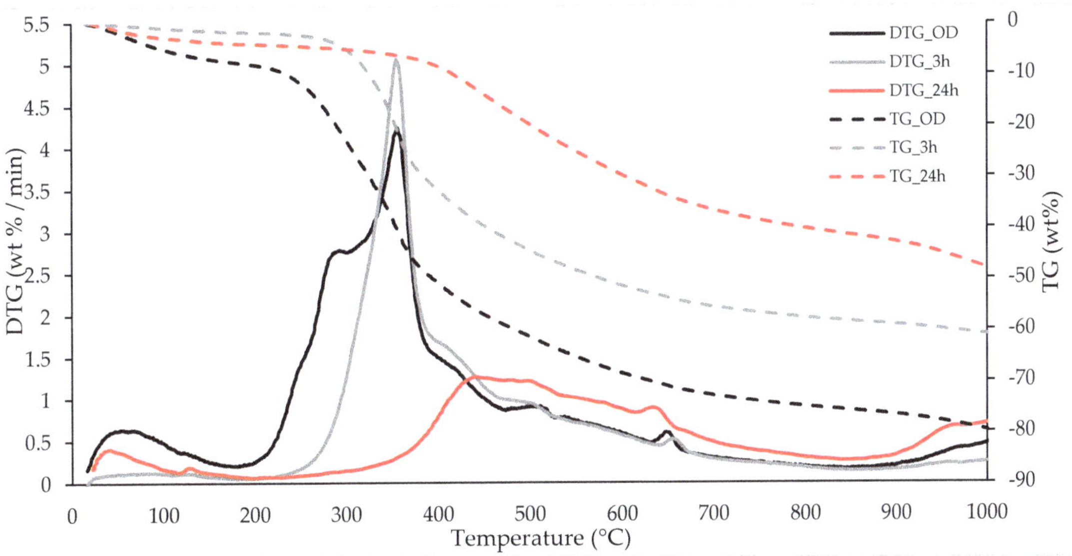

(d)

Figure 3. TG (dashed lines) and DTG (solid lines) curves of the OD and pyrolyzed (for 3 or 24 h) WS (**a**), GS (**b**), WGS (**c**), and DGS (**d**) samples at a heating rate of 10 °C/min in helium atmosphere.

Table 3 summarizes the thermal degradation processes observed in the TGA/DTG profiles. This analysis provides information on the various degradative processes involving the main biomass constituents, such as hemicellulose, cellulose, lignin, proteins (2–15%) [28,31,39,40], and, in specific samples like WGSs, lipids (8–20%) [39,41,42]. These degradation processes can also occur simultaneously within specific temperature ranges, which are clearly identifiable in the thermograms of Figure 3. Comparing the TG and DTG profiles of non-pyrolyzed and pyrolyzed biomasses provides insights into the extent of degradation of their constituents.

Table 3. Thermally activated processes.

Region	Thermally Activated Process
20–120 °C	Removal of moisture and VOCs.
120–250 °C	Removal of bound water, low-boiling VOCs, and caramelization of sugars. Degradation of pectin and proteins.
250–400 °C	Degradation of hemicellulose, cellulose, and lipids.
400–800 °C	Slow volatilization of lignin and volatilization of intermediate compounds formed after pyrolysis.
800–1000 °C	Volatilization of carbon residues (C20–C40 fragments).

The main volatilization fraction of the biomass took place between 200 and 400 °C for the oven-dried and 3 h pyrolyzed samples, while the samples pyrolyzed for 24 h showed a constant mass loss between 200 °C and 1000 °C. The significant release of VOCs can be attributed to the thermal decomposition of the three primary biomass constituents: hemicellulose, cellulose, and lignin.

The early mass losses observed in the DTG curves (Figure 3, temperature range of 30–120 °C) are attributed to the evaporation of moisture. As the treated samples were all pre-dried, the relative mass loss is limited (1–7%), with higher values for non-pyrolyzed samples. Some VOC loss also occurred within this temperature range.

In the temperature range of 120 °C to 250 °C, several thermal processes occur. Bound water, primarily associated with the inorganic fraction like mineral salts' crystallization water, and structural water, formed from condensation reactions of -OH groups in simple non-cellulosic carbohydrates, are removed. Moreover, semi-volatile compounds with medium-to-low vapor pressure (SVOCs), initially present or formed during heating, are completely eliminated. Thermal degradation of free amino acids begins around 180 °C, while proteins degrade more slowly, persisting up to 200–220 °C. Notably, the non-pyrolyzed GS sample (Figure 3b) exhibits significant mass loss (−34.0% at 250 °C), distinguishing it from other biomasses in this temperature region. This is mainly due to the degradation of pectin, which is absent in the WS, DGS, and WGS samples. Pectin typically presents two DTG signals, approximately at 200 and 230 °C, and is associated with decarboxylation and dehydration events in lateral chains [43–45].

The events occurring at higher temperatures (250–400 °C) are mainly attributed to the degradation of hemicelluloses and cellulose. Typically, hemicellulose degradation in an inert atmosphere occurs in the range of 220–315 °C [46–49], followed by the pyrolysis of cellulose (315–400 °C) [46–49]. These signals were observed in the DTG curves of OD and 3 h pyrolyzed samples. In the WS_OD sample (Figure 3a), a single broad peak extending from approximately 200 °C to 400 °C is observed, with a maximum at 318 °C. This peak likely includes the degradation of simple non-cellulosic carbohydrates, hemicelluloses, and cellulose. Accordingly, the mass loss relative to hemicellulose was absent in the WS_3h sample, as this biomass fraction was degraded during the thermal treatment (3 h at 300 °C), while the degradation of cellulose was still detectable (starting from ~230 °C and peaking at 335 °C). The DGS_3h sample (Figure 3d) is similar to WS_3h in this region, with a single peak in the DTG curve at 358 °C attributable to cellulose. However, the DGS_OD sample showed a different behavior, presenting a shoulder with a maximum rate of mass loss at 289 °C related to hemicelluloses, followed by a peak at 358 °C related to cellulose. The hemicellulose decomposition step often appears as a shoulder instead of a well-defined peak, a result that is in line with previous studies [48,50]. The GS samples (Figure 3b) showed two peaks in the DTG curves with maximum mass loss rates at 268 °C (GS_OD) and 272 °C (GS_3h), and 333 °C (GS_OD) and 335 °C (GS_3h), attributable to the degradation of hemicelluloses and cellulose, respectively. Additionally, the GS_OD sample showed a third small peak with a maximum loss rate at 285 °C, whose origin is unclear and requires further investigation. The absence of a clear distinction between the two peaks for hemicellulose and cellulose in the WS_OD sample indicates a higher hemicellulose content compared to the DGS and GS samples. The DTG curve of the WGS_3h sample (Figure 3c) showed a peak with a maximum at 355 °C, attributable to cellulose degradation. However, a shoulder emerged starting from approximately 380 °C and extending up to approximately 480 °C. A peak at 381 °C was linked to the degradation of unsaturated fatty acids by other authors [51].

Given the high content of unsaturated fatty acids in grape seeds [52], we can attribute this peak to their degradation. This signal is shifted at higher temperature (397 °C) in the WGS_24h sample. During extended pyrolysis, the breakdown of unsaturated fatty acids can result in the formation of intermediate compounds that may decompose at higher temperatures compared to the shorter pyrolysis duration.

Lignin, the first biomass constituent to undergo decomposition at lower temperatures, degrades slowly with a very low reaction rate extending throughout the pyrolysis process up to 800 °C. Its pyrolysis occurs over a broad temperature range due to the heterogeneous composition of lignin [53], leaving a higher residue at 1000 °C [49]. Significant lignin degradation occurs around 400 °C [46], marked by peaks or shoulders observed in the DTG curves of all analyzed samples. The main signal observed in all the samples pyrolyzed for 24 h up to 800 °C can be mainly attributed to the volatilization of the lignin fraction and other intermediate compounds formed after pyrolysis. In particular, depolymerization of hemicelluloses and celluloses leads to the formation of char residues that are stable up to 600–700 °C [49,54], while protein denaturation and degradation leads to the formation of *N*-heterocycles and nitriles, which are stable up to 800 °C [34].

Above 800 °C and up to the final temperature of 1000 °C, the residual stages of biomass degradation were observed. This temperature range represents the typical window for carbon pyrolysis, characterized by the thermal decomposition of low-volatile matter, including carbon fragments C20–C40, in the presence of mineral ashes. The total mass losses at 1000 °C vary among the samples, ranging from −48.9% for DGS_24h to −95.2% for WS_3h. For the GS and DGS samples pyrolyzed for 3 or 24 h, the mass loss at 1000 °C is significantly lower compared to their respective non-pyrolyzed samples. This can be attributed to the removal of VOCs and the formation of a more thermally stable char during the pyrolysis process. Additionally, the ash content of these pyrolyzed biomasses contributes to a lower mass loss at 1000 °C. In contrast, the WS samples showed a mass loss at 1000 °C of around 95% for all three samples, regardless of the pyrolysis process. This phenomenon can be attributed to several factors. The lower ash content in WS means that a greater proportion of the sample is composed of organic material that can decompose and volatilize at high temperatures. Additionally, the DTG curve of the WS_OD sample (Figure 3a) and some literature data indicated a higher content of hemicelluloses and cellulose in wood stems compared to grape skin and seeds [39,55–57], which typically leave less residue at 1000 °C [49].

3.3. Bipolar Plates

Several considerations are required to evaluate the use of biochar as a substitute for fossil graphite in the manufacture of composite BPs. Since the extraction and processing of fossil graphite are energy-intensive, it is essential to prefer biochars obtained through a low-energy pyrolysis process. This approach aligns with reducing carbon loss and corresponding CO_2 emissions, though it might impact the intrinsic electrical conductivity typically achieved during pyrolysis.

Several composite formulations including 10, 15, 20, 25, or 30 wt% of each of the available biochars (WSs, GSs, WGSs, and DGSs) were prepared, as shown in Table 1. Additionally, two alternative low-temperature pyrolysis conditions (300 °C for 3 h or 24 h) were considered. The obtained BP samples were then measured in terms of density (Table 4) and IPC (Table 5).

Table 4. Density values (g/cm^3) of the BP samples.

Sample \ wt%	10	15	20	25	30	35
GS3	1.78 ± 0.04 [a]	1.72 ± 0.02 [ac]	1.69 ± 0.04 [ac]	1.62 *[ac]		
WGS3	1.59 ± 0.04 [b]	1.56 ± 0.02 [b]	1.53 ± 0.05 [b]	1.49 ± 0.02 [b]		
DGS3	1.72 ± 0.07 [a]	1.67 ± 0.02 [a]	1.64 ± 0.03 [bc]	1.57 ± 0.01 [bc]	1.52 ± 0.07 [a]	1.50 ± 0.01 [a]
WD3	1.74 ± 0.04 [a]	1.73 ± 0.03 [cd]	1.70 ± 0.02 [cd]	1.68 ± 0.02 [ad]		
GS24	1.77 ± 0.04 [a]	1.74 ± 0.01 [cd]	1.70 ± 0.03 [cd]	1.64 ± 0.01 [ac]		
WGS24	1.74 ± 0.03 [a]	1.71 ± 0.03 [ad]	1.70 ± 0.01 [cd]	1.66 ± 0.01 [ac]	1.60 ± 0.01 [ab]	
DGS24	1.84 ± 0.04 [a]	1.77 ± 0.0[2] [c]	1.75 ± 0.03 [ad]	1.68 ± 0.01 [ad]	1.64 ± 0.02 [b]	1.57 ± 0.04 [b]
WD24	1.82 ± 0.06 [a]	1.76 ± 0.02 [cd]	1.72 ± 0.06 [cd]	1.64 ± 0.08 [ac]		

Data are expressed as mean ± standard deviation of three replicates; * SD < 0.01; [a,b,c,d] differences between means indicated by the same letters within the same column are not statistically significant ($p < 0.05$) according to the Tukey–Kramer HSD post hoc test.

Table 5. IPC values (S/cm) of the BP samples.

Sample \ wt%	10	15	20	25	30	35
GS3	127.6 ± 9.6 [ab]	113.1 ± 10.2 [a]	96.7 ± 7.1 [abc]	87.5 ± 8.8 [ab]		
WGS3	102.6 ± 11.4 [a]	96.4 ± 7.2 [a]	89.7 ± 8.7 [ac]	73.5 ± 9.1 [a]		
DGS3	132.5 ± 10.1 [b]	120.7 ± 9.9 [a]	114.2 ± 8.1 [bc]	95.4 ± 11.8 [ab]	81.7 ± 10.6 [ab]	74.4 ± 10.6 [a]
WD3	120.5 ± 8.6 [ab]	110.9 ± 11.7 [a]	99.8 ± 6.2 [abc]	89.4 ± 5.9 [ab]		
GS24	127.8 ± 9.6 [ab]	114.2 ± 10.5 [a]	94.1 ± 7.7 [abc]	82.3 ± 11.5 [ab]		
WGS24	103.4 ± 9.9 [a]	95.0 ± 10.3 [a]	86.1 ± 7.2 [a]	75.0 ± 8.9 [ab]	66.6 ± 5.9 [a]	
DGS24	128.7 ± 10.2 [ab]	117.1 ± 10.6 [a]	109.4 ± 8.7 [c]	100.0 ± 10.0 [b]	88.8 ± 8.5 [b]	74.6 ± 9.9 [a]
WD24	113.5 ± 9.6 [ab]	104.2 ± 10.2 [a]	96.2 ± 8.8 [abc]	79.6 ± 6.9 [ab]		

Data are expressed as mean ± standard deviation of three replicates; [a,b,c] differences between means indicated by the same letters within the same column are not statistically significant ($p < 0.05$) according to the Tukey–Kramer HSD post hoc test.

Almost all composite samples featuring 10 wt% of biochar inclusion (first row of Tables 4 and 5) resulted in small deviations compared to the biochar-free reference (IPC = 153 S/cm; d = 1.83 g/cm^3) [24]. This suggests that the biochar developed a respectable intrinsic conductivity. However, the inclusion of greater amounts of biochar resulted in progressively degraded performance in terms of both density and IPC. Densities values (Table 4) showed a much more pronounced influence from the pyrolysis time, presumably due to the relatively slow mass loss at 300 °C (see Figure 3). Overall, some interesting trends were evident from this study:

- Prolonged pyrolysis: Extending the pyrolysis duration from 3 h to 24 h provided only a small improvement in the intrinsic conductivity of the biochar. This indicates that the additional energy and time investment for longer pyrolysis may not yield substantial benefits in terms of electrical performance.
- Fat removal: The removal of fats from the grape seeds had a significant positive impact on the performance of the biochar. Defatted biochar (DGS3 and DGS24), compared to GS biochar, produced excellent composite materials, maintaining good performance even at higher biochar loadings. Specifically, composites with 25 wt% of DGS biochar (30 wt% of substituted G45 graphite) demonstrated superior results.
- Carbon content (C %): Biochars featuring higher carbon content (lower O/C ratio) such as DGS and GS (see Table 2) led to samples with higher IPC values, suggesting a higher intrinsic conductivity occurring in biochars with higher carbon content.

- Thermal stability: GS and DGS biochars exhibited higher thermal stability compared to WS samples, as shown by the relative TGA/DTG curves (Figure 3). Therefore, a higher degree of carbonization corresponds to a more stable structure.

4. Conclusions

The results of this study demonstrate that converting wine production waste into biochar through low-energy pyrolysis is a promising strategy to enhance the carbon neutrality of this activity. This option not only provides a simple way to retain most of the carbon present in the waste as biochar but also highlights its favorable application in the preparation of conductive composite materials suitable for BP manufacture. This, in turn, synergistically promotes the sustainable production (water electrolyzers) and use (fuel cells) of clean hydrogen.

These findings also provide a strong basis for further development. The TG analysis yielded valuable insights for interpreting the IPC values of the BP samples. Specifically, the pyrolyzed GS and DGS samples demonstrate enhanced thermal stability compared to their WS counterparts, along with higher IPC values. This greater thermal stability, indicative of a more graphite-like behavior, correlates positively with increased conductivity. Vegetable oil extraction is a low-energy input process and is highly desired for this application. After the separation of this valuable fraction, the resulting waste biomass can be easily processed through a short pyrolysis cycle (e.g., 3 h at 300 °C) to produce high-value biochar. The promising preliminary results for DGS biochar as a substitute for fossil graphite in conductive composite materials highlight the need for further research. By evaluating additional technical parameters such as hydrogen permeability and flexural strength, future studies can provide a more comprehensive understanding of the biochar-based composites' suitability for practical applications. This ongoing research will be crucial in optimizing the materials for performance, durability, and sustainability, thereby advancing the development of cleaner energy solutions.

Author Contributions: Conceptualization, F.R. and V.D.; methodology, V.D. and D.M.; software, V.D. and D.M.; validation, F.R., B.A. and A.L.; formal analysis, F.R., M.R. and V.D.; investigation, V.D., B.A., and A.L.; resources, F.R. and M.R.; data curation, V.D. and A.L.; writing—original draft preparation, V.D., M.R. and B.A.; writing—review and editing, F.R., D.M. and A.L.; visualization, V.D.; supervision, F.R. and M.R.; project administration, F.R. and M.R.; funding acquisition, F.R. All authors have read and agreed to the published version of the manuscript.

Funding: This research received no external funding.

Data Availability Statement: All data are contained within the article.

Acknowledgments: It is a pleasure to gratefully thank Giovanni Randi S.p.A. (Randi Group, Faenza, Italy) for providing us with the defatted grape seed (DGS) sample used in this article.

Conflicts of Interest: The authors declare no conflicts of interest.

References

1. FAOSTAT. Available online: https://www.fao.org/faostat/en/#data (accessed on 17 October 2022).
2. Jin, Q.; Wang, Z.; Feng, Y.; Kim, Y.-T.; Stewart, A.C.; O'Keefe, S.F.; Neilson, A.P.; He, Z.; Huang, H. Grape Pomace and Its Secondary Waste Management: Biochar Production for a Broad Range of Lead (Pb) Removal from Water. *Environ. Res.* **2020**, *186*, 109442. [CrossRef]
3. Beres, C.; Costa, G.N.S.; Cabezudo, I.; Da Silva-James, N.K.; Teles, A.S.C.; Cruz, A.P.G.; Mellinger-Silva, C.; Tonon, R.V.; Cabral, L.M.C.; Freitas, S.P. Towards Integral Utilization of Grape Pomace from Winemaking Process: A Review. *Waste Manag.* **2017**, *68*, 581–594. [CrossRef]
4. Devesa-Rey, R.; Vecino, X.; Varela-Alende, J.L.; Barral, M.T.; Cruz, J.M.; Moldes, A.B. Valorization of Winery Waste vs. the Costs of Not Recycling. *Waste Manag.* **2011**, *31*, 2327–2335. [CrossRef] [PubMed]
5. Gaunt, J.L.; Lehmann, J. Energy Balance and Emissions Associated with Biochar Sequestration and Pyrolysis Bioenergy Production. *Environ. Sci. Technol.* **2008**, *42*, 4152–4158. [CrossRef]
6. Li, Y.; Xing, B.; Ding, Y.; Han, X.; Wang, S. A Critical Review of the Production and Advanced Utilization of Biochar via Selective Pyrolysis of Lignocellulosic Biomass. *Bioresour. Technol.* **2020**, *312*, 123614. [CrossRef] [PubMed]

7. Tomczyk, A.; Sokołowska, Z.; Boguta, P. Biochar Physicochemical Properties: Pyrolysis Temperature and Feedstock Kind Effects. *Rev. Environ. Sci. Biotechnol.* **2020**, *19*, 191–215. [CrossRef]
8. Vaiškūnaitė, R.; Mažeikienė, A.; Mohammadi, K. Effects of Pyrolysis Temperature on Biochar Physicochemical and Microbial Properties for H2S Removal from Biogas. *Sustainability* **2024**, *16*, 5424. [CrossRef]
9. Li, Y.; Gupta, R.; Zhang, Q.; You, S. Review of Biochar Production via Crop Residue Pyrolysis: Development and Perspectives. *Bioresour. Technol.* **2023**, *369*, 128423. [CrossRef] [PubMed]
10. Patzek, T.W.; Pimentel, D. Thermodynamics of Energy Production from Biomass. *Crit. Rev. Plant Sci.* **2005**, *24*, 327–364. [CrossRef]
11. Metzger, J.O. Production of Liquid Hydrocarbons from Biomass. *Angew. Chem. Int. Ed.* **2006**, *45*, 696–698. [CrossRef]
12. Kung, C.-C.; McCarl, B.A.; Cao, X. Economics of Pyrolysis-Based Energy Production and Biochar Utilization: A Case Study in Taiwan. *Energy Policy* **2013**, *60*, 317–323. [CrossRef]
13. Chan, K.Y.; Van Zwieten, L.; Meszaros, I.; Downie, A.; Joseph, S. Agronomic Values of Greenwaste Biochar as a Soil Amendment. *Soil Res.* **2007**, *45*, 629. [CrossRef]
14. Ding, W.; Dong, X.; Ime, I.M.; Gao, B.; Ma, L.Q. Pyrolytic Temperatures Impact Lead Sorption Mechanisms by Bagasse Biochars. *Chemosphere* **2014**, *105*, 68–74. [CrossRef] [PubMed]
15. Shen, Z.; Hou, D.; Jin, F.; Shi, J.; Fan, X.; Tsang, D.C.W.; Alessi, D.S. Effect of Production Temperature on Lead Removal Mechanisms by Rice Straw Biochars. *Sci. Total Environ.* **2019**, *655*, 751–758. [CrossRef] [PubMed]
16. Jia, M.; Wang, F.; Bian, Y.; Jin, X.; Song, Y.; Kengara, F.O.; Xu, R.; Jiang, X. Effects of pH and Metal Ions on Oxytetracycline Sorption to Maize-Straw-Derived Biochar. *Bioresour. Technol.* **2013**, *136*, 87–93. [CrossRef] [PubMed]
17. Venkitasamy, C.; Zhao, L.; Zhang, R.; Pan, Z. Grapes. In *Integrated Processing Technologies for Food and Agricultural By-Products*; Elsevier: Amsterdam, The Netherlands, 2019; pp. 133–163. ISBN 978-0-12-814138-0.
18. Deng, Q.; Penner, M.H.; Zhao, Y. Chemical Composition of Dietary Fiber and Polyphenols of Five Different Varieties of Wine Grape Pomace Skins. *Food Res. Int.* **2011**, *44*, 2712–2720. [CrossRef]
19. Barros, A.; Gironés-Vilaplana, A.; Texeira, A.; Baenas, N.; Domínguez-Perles, R. Grape Stems as a Source of Bioactive Compounds: Application towards Added-Value Commodities and Significance for Human Health. *Phytochem. Rev.* **2015**, *14*, 921–931. [CrossRef]
20. Kane, S.; Ulrich, R.; Harrington, A.; Stadie, N.P.; Ryan, C. Physical and Chemical Mechanisms That Influence the Electrical Conductivity of Lignin-Derived Biochar. *Carbon Trends* **2021**, *5*, 100088. [CrossRef]
21. Weber, K.; Quicker, P. Properties of Biochar. *Fuel* **2018**, *217*, 240–261. [CrossRef]
22. Plenča, K.; Cvetnić, S.; Prskalo, H.; Kovačić, M.; Cvetnić, M.; Kušić, H.; Matusinović, Z.; Kraljić Roković, M.; Genorio, B.; Lavrenčič Štangar, U.; et al. Biomass Pyrolysis-Derived Biochar: A Versatile Precursor for Graphene Synthesis. *Materials* **2023**, *16*, 7658. [CrossRef]
23. Jeong, K.I.; Oh, J.; Song, S.A.; Lee, D.; Lee, D.G.; Kim, S.S. A Review of Composite Bipolar Plates in Proton Exchange Membrane Fuel Cells: Electrical Properties and Gas Permeability. *Compos. Struct.* **2021**, *262*, 113617. [CrossRef]
24. Roncaglia, F.; Romagnoli, M.; Incudini, S.; Santini, E.; Imperato, M.; Spinelli, L.; Di Bona, A.; Biagi, R.; Mucci, A. Graphite-Epoxy Composites for Fuel-Cell Bipolar Plates: Wet vs Dry Mixing and Role of the Design of Experiment in the Optimization of Molding Parameters. *Int. J. Hydrogen Energy* **2021**, *46*, 4407–4416. [CrossRef]
25. Spokas, K.A. Review of the Stability of Biochar in Soils: Predictability of O:C Molar Ratios. *Carbon Manag.* **2010**, *1*, 289–303. [CrossRef]
26. Bordiga, M.; Travaglia, F.; Locatelli, M. Valorisation of Grape Pomace: An Approach That Is Increasingly Reaching Its Maturity—A Review. *Int. J. Food Sci. Technol.* **2019**, *54*, 933–942. [CrossRef]
27. Shinagawa, F.B.; Santana, F.C.D.; Torres, L.R.O.; Mancini-Filho, J. Grape Seed Oil: A Potential Functional Food? *Food Sci. Technol.* **2015**, *35*, 399–406. [CrossRef]
28. D'Eusanio, V.; Malferrari, D.; Marchetti, A.; Roncaglia, F.; Tassi, L. Waste By-Product of Grape Seed Oil Production: Chemical Characterization for Use as a Food and Feed Supplement. *Life* **2023**, *13*, 326. [CrossRef] [PubMed]
29. Associataion of Official Analytical Chemist. *AOAC Official Methods of Analysis of the Association of Official's Analytical Chemists*, 14th ed.; Associataion of Official Analytical Chemist: Washington, DC, USA, 1990; pp. 223–225+992–995.
30. Smits, F.M. Measurement of Sheet Resistivities with the Four-Point Probe. *Bell Syst. Tech. J.* **1958**, *37*, 711–718. [CrossRef]
31. D'Eusanio, V.; Genua, F.; Marchetti, A.; Morelli, L.; Tassi, L. Characterization of Some Stilbenoids Extracted from Two Cultivars of Lambrusco—Vitis Vinifera Species: An Opportunity to Valorize Pruning Canes for a More Sustainable Viticulture. *Molecules* **2023**, *28*, 4074. [CrossRef] [PubMed]
32. D'Eusanio, V.; Genua, F.; Marchetti, A.; Morelli, L.; Tassi, L. Exploring the Mineral Composition of Grapevine Canes for Wood Chip Applications in Alcoholic Beverage Production to Enhance Viticulture Sustainability. *Beverages* **2023**, *9*, 60. [CrossRef]
33. Chen, D.; Mei, J.; Li, H.; Li, Y.; Lu, M.; Ma, T.; Ma, Z. Combined Pretreatment with Torrefaction and Washing Using Torrefaction Liquid Products to Yield Upgraded Biomass and Pyrolysis Products. *Bioresour. Technol.* **2017**, *228*, 62–68. [CrossRef]
34. Leng, L.; Yang, L.; Chen, J.; Leng, S.; Li, H.; Li, H.; Yuan, X.; Zhou, W.; Huang, H. A Review on Pyrolysis of Protein-Rich Biomass: Nitrogen Transformation. *Bioresour. Technol.* **2020**, *315*, 123801. [CrossRef] [PubMed]
35. Neves, D.; Thunman, H.; Matos, A.; Tarelho, L.; Gómez-Barea, A. Characterization and Prediction of Biomass Pyrolysis Products. *Prog. Energy Combust. Sci.* **2011**, *37*, 611–630. [CrossRef]

36. Klasson, K.T. Biochar Characterization and a Method for Estimating Biochar Quality from Proximate Analysis Results. *Biomass Bioenergy* **2017**, *96*, 50–58. [CrossRef]
37. Ahmad, M.; Rajapaksha, A.U.; Lim, J.E.; Zhang, M.; Bolan, N.; Mohan, D.; Vithanage, M.; Lee, S.S.; Ok, Y.S. Biochar as a Sorbent for Contaminant Management in Soil and Water: A Review. *Chemosphere* **2014**, *99*, 19–33. [CrossRef] [PubMed]
38. Leng, L.; Huang, H.; Li, H.; Li, J.; Zhou, W. Biochar Stability Assessment Methods: A Review. *Sci. Total Environ.* **2019**, *647*, 210–222. [CrossRef] [PubMed]
39. Spinei, M.; Oroian, M. The Potential of Grape Pomace Varieties as a Dietary Source of Pectic Substances. *Foods* **2021**, *10*, 867. [CrossRef] [PubMed]
40. Sagdic, O.; Ozturk, I.; Yetim, H.; Kayacier, A.; Dogan, M. Mineral Contents and Nutritive Values of the Pomaces of Commercial Turkish Grape (*Vitis vinifera* L.) Varieties. *Qual. Assur. Saf. Crops Foods* **2014**, *6*, 89–93. [CrossRef]
41. Rombaut, N.; Savoire, R.; Thomasset, B.; Castello, J.; Van Hecke, E.; Lanoisellé, J.-L. Optimization of Oil Yield and Oil Total Phenolic Content during Grape Seed Cold Screw Pressing. *Ind. Crops Prod.* **2015**, *63*, 26–33. [CrossRef]
42. Beveridge, T.H.J.; Girard, B.; Kopp, T.; Drover, J.C.G. Yield and Composition of Grape Seed Oils Extracted by Supercritical Carbon Dioxide and Petroleum Ether: Varietal Effects. *J. Agric. Food Chem.* **2005**, *53*, 1799–1804. [CrossRef]
43. Kozioł, A.; Środa-Pomianek, K.; Górniak, A.; Wikiera, A.; Cyprych, K.; Malik, M. Structural Determination of Pectins by Spectroscopy Methods. *Coatings* **2022**, *12*, 546. [CrossRef]
44. Einhorn-Stoll, U.; Kastner, H.; Fatouros, A.; Krähmer, A.; Kroh, L.W.; Drusch, S. Thermal Degradation of Citrus Pectin in Low-Moisture Environment—Investigation of Backbone Depolymerisation. *Food Hydrocoll.* **2020**, *107*, 105937. [CrossRef]
45. Zhou, S.; Xu, Y.; Wang, C.; Tian, Z. Pyrolysis Behavior of Pectin under the Conditions That Simulate Cigarette Smoking. *J. Anal. Appl. Pyrolysis* **2011**, *91*, 232–240. [CrossRef]
46. Díez, D.; Urueña, A.; Piñero, R.; Barrio, A.; Tamminen, T. Determination of Hemicellulose, Cellulose, and Lignin Content in Different Types of Biomasses by Thermogravimetric Analysis and Pseudocomponent Kinetic Model (TGA-PKM Method). *Processes* **2020**, *8*, 1048. [CrossRef]
47. Pardo, R.; Taboada-Ruiz, L.; Fuente, E.; Ruiz, B.; Díaz-Somoano, M.; Calvo, L.F.; Paniagua, S. Exploring the Potential of Conventional and Flash Pyrolysis Methods for the Valorisation of Grape Seed and Chestnut Shell Biomass from Agri-Food Industry Waste. *Biomass Bioenergy* **2023**, *177*, 106942. [CrossRef]
48. Khiari, B.; Jeguirim, M. Pyrolysis of Grape Marc from Tunisian Wine Industry: Feedstock Characterization, Thermal Degradation and Kinetic Analysis. *Energies* **2018**, *11*, 730. [CrossRef]
49. Yang, H.; Yan, R.; Chen, H.; Lee, D.H.; Zheng, C. Characteristics of Hemicellulose, Cellulose and Lignin Pyrolysis. *Fuel* **2007**, *86*, 1781–1788. [CrossRef]
50. Jeguirim, M.; Bikai, J.; Elmay, Y.; Limousy, L.; Njeugna, E. Thermal Characterization and Pyrolysis Kinetics of Tropical Biomass Feedstocks for Energy Recovery. *Energy Sustain. Dev.* **2014**, *23*, 188–193. [CrossRef]
51. Khalil, I.K.; Mustafa, A.S.; Mustafa, A.M. Study of Fatty Acid Composition, Physiochemical Properties and Thermal Stability of Broad Beans (Vicia Faba) Seed Oil. *Agric. Biol. J. N. Am.* **2017**, *8*, 141–146. [CrossRef]
52. Tangolar, S.G.; Özoğul, Y.; Tangolar, S.; Torun, A. Evaluation of Fatty Acid Profiles and Mineral Content of Grape Seed Oil of Some Grape Genotypes. *Int. J. Food Sci. Nutr.* **2009**, *60*, 32–39. [CrossRef]
53. Valente, M.; Brillard, A.; Schönnenbeck, C.; Brilhac, J.-F. Investigation of Grape Marc Combustion Using Thermogravimetric Analysis. Kinetic Modeling Using an Extended Independent Parallel Reaction (EIPR). *Fuel Process. Technol.* **2015**, *131*, 297–303. [CrossRef]
54. Shen, D.K.; Gu, S.; Bridgwater, A.V. The Thermal Performance of the Polysaccharides Extracted from Hardwood: Cellulose and Hemicellulose. *Carbohydr. Polym.* **2010**, *82*, 39–45. [CrossRef]
55. Prozil, S.O.; Evtuguin, D.V.; Lopes, L.P.C. Chemical Composition of Grape Stalks of *Vitis vinifera* L. from Red Grape Pomaces. *Ind. Crops Prod.* **2012**, *35*, 178–184. [CrossRef]
56. Mendes, J.A.S.; Xavier, A.M.R.B.; Evtuguin, D.V.; Lopes, L.P.C. Integrated Utilization of Grape Skins from White Grape Pomaces. *Ind. Crops Prod.* **2013**, *49*, 286–291. [CrossRef]
57. D'Eusanio, V.; Morelli, L.; Marchetti, A.; Tassi, L. Aroma Profile of Grapevine Chips after Roasting: A Comparative Study of Sorbara and Spergola Cultivars for More Sustainable Oenological Production. *Separations* **2023**, *10*, 532. [CrossRef]

energies

MDPI

Article

High-Energy-Density Hydrochar and Bio-Oil from Hydrothermal Processing of Spent Coffee Grounds—Experimental Investigation

Mariusz Wądrzyk *, Jakub Katerla, Rafał Janus, Marek Lewandowski, Marek Plata and Łukasz Korzeniowski

Faculty of Energy and Fuels, AGH University of Krakow, 30-059 Kraków, Poland; katerla@agh.edu.pl (J.K.); rjanus@agh.edu.pl (R.J.); lewandowski@agh.edu.pl (M.L.); mplata@agh.edu.pl (M.P.); lukkor@agh.edu.pl (Ł.K.)
* Correspondence: wadrzyk@agh.edu.pl; Tel.: +48-12-617-38-90

Abstract: Spent coffee grounds (SCGs), a by-product of coffee brewing, have high application potential. However, their high moisture content complicates conventional conversion without energy-intensive drying. This study explores a new route to convert SCGs to high-carbon bioproducts, such as hydrochar and bio-oil, through hydrothermal processing. The effect of the processing variables, i.e., temperature, residence time, and the application of the binary solvent as a reaction medium, on the distribution of the resultant bioproducts was investigated. The quality of the fabricated bioproducts was analyzed by means of instrumental techniques such as EA, ATR-FTIR, GC-MS, and GC-TCD-FID. Two dominant fractions were liquid bio-oil and solid hydrochar. The highest char yield (39 wt.%) was observed under milder conditions (low T and short residence times), while more severe conditions led to an increase in bio-oil formation, which reached a maximum of 46 wt.%. The resulting bio-oils were of similar quality, presenting high carbon content (71–74 wt.%) and energetic values (approximately 35 MJ/kg). Also, hydrochars showed a noticeable energy densification compared to raw materials, where the C content and HHV reached up to 73.8 wt.% and 30 MJ/kg, respectively. The addition of co-solvent to water improves the bio-oil yield as a result of the enhanced stabilization of reactive intermediates.

Keywords: spent coffee grounds; hydrothermal treatment; solvothermolysis; bio-oil; hydrochar

1. Introduction

The food industry generates abundant amounts of various types of residues at each stage of the product life cycle, i.e., production, processing, and consumption. The number of residues is gradually growing year on year. It is the result of the constant growth of the global population, urbanization, a high demand for goods, and intensive social development. Consequently, the disposal of the increasing amounts of waste is one of the biggest global challenges we face, and its solution is one of the fundamentals of the development of the circular economy. One of the goods for which production is permanently noticeably increasing is coffee, which is the second most popular hot beverage globally, following tea. Generally, in 2023, the coffee market represented ca. 39% of the revenue connected with the hot-beverages market [1]. According to the International Coffee Organization [2], global coffee production for the period 2021/22 exceeded 10.1 million tons and has generally had a growing trend over the last two decades. For example, the consumption of coffee in Poland increased from 98 to 150 thousand tons between 1995 and 2020. Consequently, SCGs are an abundantly available waste stream, which is predicted to increase in the forthcoming years and consequently could be valorized to obtain value-added green chemicals and energy carriers, or even drop-in biocomponents.

According to Kim et al. [3], the residue remaining after conventional coffee brewing accounts for approximately 90% of the total input mass. This is consistent with the data of Blinová et al. [4], who indicated that the production of spent coffee grounds (SCGs) is twice that of ground coffee when it comes to instant coffee production. Roughly estimated, the global generation of spent coffee grounds could be more than 9 million tons of waste each year. Nowadays, the resulting spent coffee grounds do not have a reasonable application route, and the majority are disposed of through unsustainable methods such as landfilling, incineration, or composting. The mentioned disposal methods involve environmental issues due to the release of greenhouse gases (GHGs) such as carbon dioxide (CO_2), methane (CH_4), and nitrous oxide (N_2O). It was reported by Zhang et al. [5] that the food loss sector could even be responsible for 8% of global GHG emissions, releasing 4.2 tCO_2/t of food leftovers to the environment. On the one hand, the incineration of biomass fuels is considered CO_2 neutral, but processing items with high moisture content like SCGs is energy-intensive. On the other hand, landfilling food waste (like SCGs) is deemed the main source of greenhouse gas emissions (particularly methane), releasing numerous active biochemicals into the environment, e.g., tannins and polyphenols [6]. Moreover, wet-type feedstocks exhibit low biochemical stability and, consequently, there is a tendency toward the rapid growth of various hazardous pathogens. The natural consequence is its negative impact on the environment and human health. Thus, considering the scale of coffee production, there is a need for new technologies for sustainable development for SCG utilization, taking into account environmental, societal, and energy aspects. It should be mentioned that, nowadays, the research and development sector is actively investigating various routes of effective conversion of SCGs towards value-added products. This is the result of the huge potential of SCGs as a feedstock that contains numerous valuable chemicals such as polyphenols, lignin, cellulose, hemicellulose, and lipids, which can be further extracted and transformed accordingly [7]. Furthermore, a lack of competition dilemmas in the food market and a large supply of coffee residuals make this raw material attractive in the context of the production of useful bioproducts. Recently, Atabani et al. [8] reviewed different routes of valorization of SCGs for fuels and value-added chemicals like bioactive compounds, adsorbents, polymers, and nanocomposites, and they emphasized that this kind of waste offers a big application potential. Also, the utilization of SCGs to produce polymers and catalysts has been reviewed by other researchers [9]. In particular, the conversion of SCGs into fuels and energy carriers was an interesting alternative. For instance, in the literature, there exist reports concerning the production of hydrochar from SCGs by hydrothermal carbonization, which was tested as an alternative to coal for energy applications [10], in the production of a coke substitute [11], as briquettes and pellets [12,13], via valorization through pyrolysis process towards liquid and solid bioproducts [14–16], for manufacturing esters for fueling compression ignition (CI) engines [17,18], and for co-processing with other wastes [19]. Moreover, some research also focused on the implementation of SCGs in other applications such as organic amendments of agricultural soil [20], activated carbons devised for the electrochemical energy storage materials [21] or adsorption/purification purposes [22]. Furthermore, the feasibility of the conversion of residual coffee waste to methane production through anaerobic digestion was also reported [23]. Arauzo et al. [24] tested the possibility of recovering phenolic compounds from SCGs by combining hydrothermal delignification with ultrasound-assisted extraction. Such technologies that focus on the transformation of residual matter generated by the food industry sector (or any other type of biomass) are in line with the sustainability approach that is developed worldwide by countries and leading R&D institutions. These types of smart technologies perfectly address the challenges of the sustainable development philosophy and bring forward the implementation of the assumptions of the circular economy.

Among the mentioned transformation methods, one that is particularly prospective is the liquefaction of SCGs towards liquid bio-oil as a source of fuels and chemicals. In the case of raw residuals with a high moisture content such as spent coffee grounds, the conversion that occurs under near-critical conditions of water is a particularly interesting opportunity. Recently, this route was thoroughly investigated for different kinds of wet raw materials like algae [25], vegetable leftover [26,27], fruit pomace [28], lignocellulosic biomass [29], municipal wastes [30] or industrial wastes like brewer spent grains [31]. The main product of biomass transformation is a viscous bio-oil that exhibits a relatively high heating value. The bio-oil yield and composition are highly dependent on the biochemical composition of the raw materials, i.e., the content of proteins, carbohydrates, lipids, lignin, etc. In principle, the composition of raw materials and processing conditions of hydrothermal liquefaction result in a wide range of bio-oil yields (varying in the range of ca. 20 wt.% to 60 wt.%) for various raw materials [32,33]. Hydrothermal liquefaction is commonly performed in the presence of water at elevated temperatures and under high pressure. Water, due to unique changes in its physicochemical properties, allows for the solubilization of various groups of organic compounds, offers a high reaction rate, and could act as a catalyst for numerous reactions. With this, it poses an excellent solvent under near-critical conditions. In the case of coffee spent grounds, Yang et al. [34] investigated the possibility of hydrothermal liquefaction in an aqueous environment, and yield of obtained bio-oils ranged between 15 and 47 wt.%. The average bio-oil yield varied between 25 and 30 wt.%, and the highest yield was noted for the very low concentration of biomass in the slurry, which is not economically justified in view of implementation on a commercial scale. Among the tested processing variables, the authors noted the most pronounced effect of processing temperature. The bio-oil yield was found to initially increase continuously from 200 to 275 °C, but it started to decrease after a further temperature increase to 300 °C. However, Liakos [35] investigating the HTL of SCGs between 280 and 350 °C reported the lowest bio-oil yield for the lowest temperature (280 °C) and the highest bio-oil yield (34.0 wt.%) at 300 °C as the optimal temperature. The further increase in temperature up to 350 °C did not cause a noticeable change in the bio-oil yield. The residence time effect depends very often on the severity of processing conditions. In the mentioned report of Liakos et al. [35], the optimal residence time for SCGs was noted as 30 min and 15 min for 300 and 320 °C, respectively. Prolongation of the processing period resulted in a gradual decrease in the bio-oil yield. It is worth noting that the different optimal processing conditions can be caused by differences in the structure of each type of SCGs. In another paper [36], the authors reported on the effect of different extraction methods on the recovery of bio-oil after the HTL of SCGs. They noted the bio-oil yield ranging from 15.2 wt.% to a maximum 39.3 wt.% for different extraction methods. Another reported approach was the co-processing of spent coffee grounds with lignocellulosic feedstocks like a paper filter, corn stalk and white pine bark in subcritical water [37] as a medium, but the bio-oil yield did not exceed 30 wt.%.

The aspects of economic feasibility and environmental impact are critically important for assessing hydrothermal liquefaction technology in practice. Today, HTL is considered one of the most used methods for processing wet-type feedstock with moisture reaching even 90 wt.%, which could supersede non-sustainable methods. Compared to pyrolysis, the second most widely tested method of waste valorization, HTL has a great advantage because it uses water present in the residual matter as media and does not need prior drying (in contrast to pyrolysis), which is an energy sink [38]. Furthermore, HTL is highly flexible regarding the type of feedstock that can process both wet and dry organic matter, which can overcome issues with the seasonal availability of some raw materials. Furthermore, the quality of the HTL bio-oil is higher than that of the pyrolysis bio-oil [38]. HTL also has some drawbacks that affect the feasibility and scalability of technology: (i) the necessity to use feedstock quite fast because of possible spontaneous biochemical processes or storage under refrigeration conditions; (ii) limited applications of HTL liquid bio-oil because of high heteroatoms content and undesirable physicochemical properties (e.g., viscosity, acidity and low oxidative stability); (iii) advanced applications of HTL bio-oil requiring deep downstream processing, which could be problematic (e.g., due to catalyst deactivation), and (iv) high corrosivity of the subcritical water as a reaction environment [39]. Nevertheless, the development of HTL technology has occurred rapidly in the last decade, resulting in few pilot/demonstration plants developed by several companies and numerous ongoing projects around the world. According to Dhalsamant et al. [39], who focused on the techno-economic analysis of HTL, an economically feasible HTL process is highly dependent on the bio-oil yield. The authors reported that the minimum selling price (MSP) of the bio-oil was found to be \$3.5 and \$1.8 per GGE (gallon of gasoline-equivalent) when its yield was 47 wt.% and 80 wt.%, respectively. In another paper [38], it was reported that the minimum fuel selling price (MFSP) of finished fuel achieved using HTL technology was in the range of 0.6–1.3 USD/L. It is worth noting that there is a possibility of using existing crude oil infrastructure (refineries) for processing the resultant HTL bio-oil, which could reduce the capital cost. Also, location, tax rates and government support can affect technology, making it more competitive. HTL also allows us to provide noticeable environmental benefits. Lozano et al. [38] reported that for HTL technology, GHG emissions can be reduced by 85% compared to its fossil counterparts (19.4 kg CO_2eq/GJ_{fuel} versus 126 kg CO_2eq/GJ_{fuel}). Furthermore, the integration of HTL technology with carbon capture has the potential to reduce CO_2 by 102% compared to the fossil baseline. Nowadays, food waste disposed through non-environmentally sustainable approaches (e.g., landfilling, incineration) can even be responsible for the emission of ~8% of total anthropogenic GHG emissions that constitute 4.4 $GtCO_2eq$ [39].

Despite the richness of the literature regarding the investigation of the potential conversion of wet-type feedstock using near-critical conditions of water toward solid hydrochar and liquid bio-oil, there is a lack of reports on the conversion of SCGs in a medium other than a purely hydrothermal one, which can affect the maximization yield of bio-oil and, consequently, HTL technology feasibility. In recent reports, it can be found that the introduction of an additional solvent to water could be beneficial for increasing the yield and quality of the resultant bio-oil [40]. For instance, the conversion of fruit pomace in a binary solvent system was recently reported to be more favorable towards the formation of liquid products than processing in a purely aqueous medium [41,42]. The addition of tested organic solvents to water under near-critical conditions causes a change in the chemical nature of the reaction medium, and therefore, this supports the hydrolysis processes and promotes the stabilization of the formed intermediate products. The presence of solvents with hydrogen donor ability such as isopropanol, ethanol, or tetralin could stabilize the products of primary conversion and reduce the incidence of secondary repolymerization. Moreover, the water–organic solvent mixture increases the solvent penetration capability of the pristine structure of biomass while providing a better solubility of intermediates, consequently leading to the deeper decomposition of building blocks. The binary solvent system with alcohols also favors ionic reactions over the radical ones, which together enhance bio-oil formation [40].

Herein, we investigated the efficiency of the conversion of spent coffee grounds to high-energy density products through hydrothermal liquefaction under various processing conditions, including the effect of conversion in the mixture of water and isopropanol (1:1 m/m). Monohydroxy alcohol serves as a source of hydrogen in situ. To the best of our knowledge, no scientific work has been reported to date on the thermochemical liquefaction of spent coffee grounds in binary solvent systems composed of a mixture of monohydroxy alcohol and water. We have found that the addition of isopropanol as an oxygen-containing solvent creates favorable conditions for the efficient conversion of biomacromolecules present in the processed waste biomass. It is pertinent to mention that we studied the feasibility of direct utilization of the as-received wet raw material. In the majority of research, the feedstock was preliminarily dried and then mixed with deionized water to prepare the starting slurry. Here, we made an effort to compare the results obtained for the initially dried feedstock with the loading of as-received (wet) biomass, which brings new insights and could be more relevant for real conditions due to the simplification of the entire processing procedure as well as the reduction of the overall technology energy input.

2. Materials and Methods

2.1. Feedstock

The raw material was obtained by brewing ground coffee (*Coffea arabica*) using an automatic home-use coffee machine working at a pressure of 15 bar. Herein, we test the as-received wet spent coffee grounds. The raw material was kept sealed in plastic bags and stored frozen at −18 °C. The moisture content was determined by drying the samples at 105 °C until a constant mass was achieved. The mean moisture content was equal to 62 wt.%.

2.2. Hydrothermal Treatment Procedure

Hydrothermal tests of SCGs were conducted in a 500 cm^3 stainless steel Parr 4575A batch reactor (Parr Instrument Company, Moline, IL, USA), allowing us to record the processing parameters (temperature, pressure, stirring time). In the present contribution, as-received waste was used for studies without any pretreatment operations. Therefore, the slurry was prepared using a raw sample (wet), and considering the moisture content of the feedstock, we calculated the appropriate amount of distilled water to be added to achieve a constant dry biomass/solvent ratio (1:9 by mass). We selected 10 wt.% dry biomass in the slurry based on our previous research and other reports in the literature [32,33], taking into account the high bio-oil yield and economic feasibility aspects of the technology. Next, ca. 100 g of the slurry of spent coffee grounds was put into the reactor. First, a series of hydrothermal tests were performed in a purely aqueous medium with the aim of determining the effect of process variables (temperature, residence time) on the yield distribution of the obtained bioproduct. Hydrothermal experiments were carried out at processing temperatures between 260 and 320 °C and for two residence times of 30 and 60 min. The experiments were carried out at an autogenous pressure, which corresponded to the real operational pressure between ca. 5 and 12 MPa. The heating step to the set experimental temperature took 31 to 44 min depending on the process temperature, which corresponded to the heating rate between 6.8 and 7.7 °C min^{-1}. The cooling was initiated after the desired reaction time at the selected temperature and lasted for ca. 60 min. The temperature and pressure profiles as a function of time for various runs are provided in the Supplementary Information section (Figure S1). Next, we studied the effect of reaction media on the performance of SCG conversion. For this purpose, additional comparative tests were carried out in a mixture of water and isopropyl alcohol, for which the proportion was equal to 50:50 by mass. The comparative analysis between the water and water–isopropanol systems was carried out under conditions found to be optimal in the first experimental series concerning bio-oil yield, i.e., 320 °C and 30 min.

The recovery of the bioproducts from the reaction mixture was carried out by a standard procedure employing solvent extraction with dichloromethane (DCM), followed by filtration under reduced pressure and solvent evaporation. During filtration, the solid product (hydrochar) was additionally washed with dichloromethane to extract all soluble chemicals and finally dried at 105 °C. The yield of gas products was calculated using the ideal gas law, based on the pressure gain measured after the process and the gas composition. Each phase of the liquefaction product was recovered quantitatively to estimate the yield. Details of procedure protocols were described in a previous report [43]. Isopropanol and dichloromethane (both analytical-grade) were purchased from Avantor Performance Materials Poland.

2.3. Analytical Procedure for Bioproducts Characterization

2.3.1. Elemental Analysis

The elemental composition (C, H, N) of the raw spent coffee grounds, as well as the produced bio-oils and hydrochars, was determined using an EuroVector EA3100 CHNS analyzer (EuroVector, Pavia, Italy). Calibration was performed using 2,5-bis(5-tert-butyl-2-benzo-oxazol-2-yl) thiophene (BBOT) as a reference material and the standard protocol recommended by the producer. The oxygen content was calculated by the difference. Each run was repeated at least twice, and the actual result is the mean value. The relative standard deviation was below 2%. Based on the elemental composition of the HTL bioproducts, the energy recovery (ER) in the form of HTL bio-oil as a designed component group was calculated according to Equation (1).

$$ER_{bio-oil,i} = \frac{Y_{bio-oil,i} \cdot HHV_{bio-oil,i}}{HHV_{feedstock}}, \quad (1)$$

- $ER_{bio\text{-}oil,i}$—energy recovery in the form of resultant bio-oil as desired bioproduct; i means bio-oils obtained under various processing conditions (%).
- $Y_{bio\text{-}oil,i}$—yield of the i-bio-oil (wt.%).
- $HHV_{bio\text{-}oil,i}$—higher heating value of i-bio-oil (MJ/kg).
- $HHV_{feedstock}$—higher heating value of the dry feedstock (22.5 MJ/kg).

2.3.2. Fourier Transform Infrared Spectroscopy (FT-IR)

The qualitative composition of the obtained bio-oils was investigated by means of mid-infrared Fourier transform spectroscopy. Spectra were collected on a Nicolet iS5 (Thermo Scientific, Waltham, MA, USA) spectrometer equipped with an attenuated total reflectance device (ATR) with a diamond crystal and DLaTGS detector (Thermo Scientific, Waltham, MA, USA). The raw material was analyzed using the DRIFT technique (using an EasiDiff™ kit; Pike Technologies, Fitchburg, WI, USA). In this case, the sample was prepared as a 1 wt.% mixture in KBr. For each spectrum, sixteen scans were recorded in a wavenumber region of 4000–400 cm^{-1} at a resolution of 4 cm^{-1}.

2.3.3. Gas Chromatography Coupled to Mass Spectrometry (GC-MS)

A qualitative analysis of the volatile components of the obtained bio-oils was carried out using a GC-MS Agilent apparatus model GC 7890B equipped with a MS 5977A mass spectrometer (Agilent Technologies, Santa Clara, CA, USA) and a DB-17 capillary column (30 m × 0.32 mm × 0.5 μm; Agilent Technologies, Santa Clara, CA, USA). The samples were prepared as 5 wt.% solutions in acetone (Avantor Performance Materials Poland; HPLC grade). The identification of compounds was carried out using a deconvolution algorithm. The MS spectra were identified using the reference MS library, specifically the NIST14 chemical database (chemical base G1034C). The shares of the group of compounds were calculated as a quotient of the peak area of a particular compound in relation to the cumulative peak area of all the components detected. Detailed information on the conditions of GC-MS analyses was previously reported [44]. For better legibility, the identified compounds were divided into five groups according to their functional groups, i.e., (i) esters, (ii) carboxylic acids, (iii) phenol derivatives, (iv) other oxygen compounds (cyclic oxygenates, ketones, aldehydes, sterols, etc.) and (v) nitrogen compounds.

2.3.4. Gas Chromatography (GC-TCD-FID)

The composition of the chosen gas phases was analyzed using a gas chromatograph (GC-SRI, model 302, SRI Instruments, Torrance, CA, USA) equipped with TCD and FID detectors coupled in a cascade. The separation of the constituents was carried out using two packed columns, namely HayeSep D of a length of 2 m and molecular sieve (13X), L = 2 m.

3. Results and Discussion

3.1. Effect of Process Variables on Liquefaction Products Yield Distribution

3.1.1. Temperature

The effect of the processing temperature on the hydrothermal liquefaction of the spent coffee grains is depicted in Figure 1. As can be seen, the yield distribution of the resultant bioproducts changes noticeably with a change in processing temperature. Two dominant group fractions were liquid bio-oil and a solid residue called hydrochar. The highest hydrochar yield (38.9 wt.%) was observed at the lowest tested temperature (260 °C), while the highest bio-oil yield was denoted under the most severe conditions of the tested range (320 °C). Generally, the variation in process temperature caused noticeable changes, especially for both mentioned groups. The bio-oil yield ranged from 15.9 wt.% to 35.7 wt.% at 260 and 320 °C, respectively. These results are in line with previous reports by Yang et al. [34,37] on the transformation of SCGs in a subcritical aqueous medium. Higher temperatures led to a gradual decomposition of the macromolecules building the raw material through hydrolysis reactions, as well as subsequent decomposition and the repolymerization reactions of intermediates, thus enhancing the formation of the compounds contained in bio-oil. The yield of the aqueous-phase organics decreased slightly with temperature, while the gaseous products increased. This observation also confirms the more intensive decomposition reactions of the hydrolysis products under harsher conditions. In the case of the HTL process, the optimal conditions seem to be around 300–320 °C, allowing for the high yield of bio-oil with considerable heating value (due to high carbon content and low oxygen content). The processing in more severe conditions additionally involves high energy demand to reach the processing temperature, difficulties linked with high operational pressure, and the high corrosivity of media forcing us to use the most resistible materials.

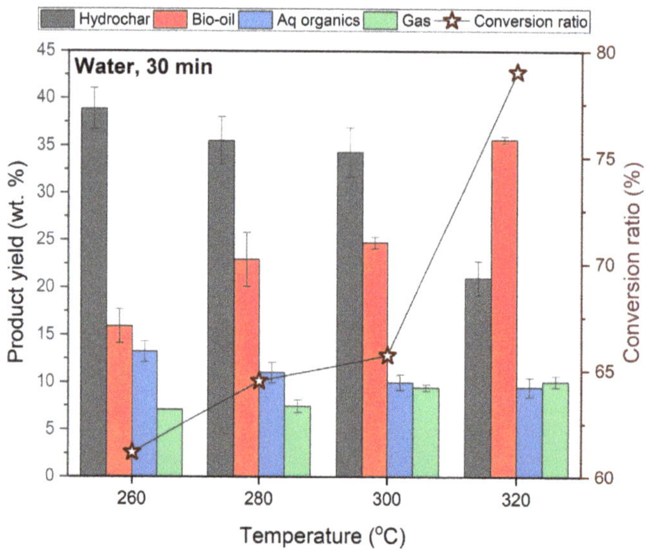

Figure 1. Effect of processing temperature ranging between 260 and 320 °C on the distribution of HTL bioproducts yields (residence time of 30 min and water as solvent).

3.1.2. Residence Time

The effect of the reaction time depends on the processing temperature, which is presented in Figure 2. At a lower processing temperature (280 °C), prolonging the duration of the process results in a slight decrease in hydrochar yield and an increase in gas production. On the other hand, a longer processing time at 300 °C resulted in a decrease in bio-oil yield and an increase in hydrochar formation. This can be assigned to the formation of secondary solids as a consequence of the intensification of the intermediates' repolymerization reactions. The kinetics of the reaction network between different groups of biopolymers (and its derivatives) could also vary. For example, according to [45], the rate constant for biocrude formation from protein and lipids was higher than that from carbohydrates. Higher processing temperatures involve higher rates of decomposition reactions that occur and reach the activation energy of more possible reactions. Consequently, primary decomposition products are formed faster, and the prolongation of the process leads to secondary reactions, i.e., the cracking of liquid products and the repolymerization/ rearrangement of intermediates forming secondary highly condensed structures. Generally, it is worth noting that, compared to processing temperature, the impact of residence time is rather minor. The diversity of yields with processing conditions also affected the conversion ratio. The latter for processing in a water medium in both series varied within the range of 61–79%.

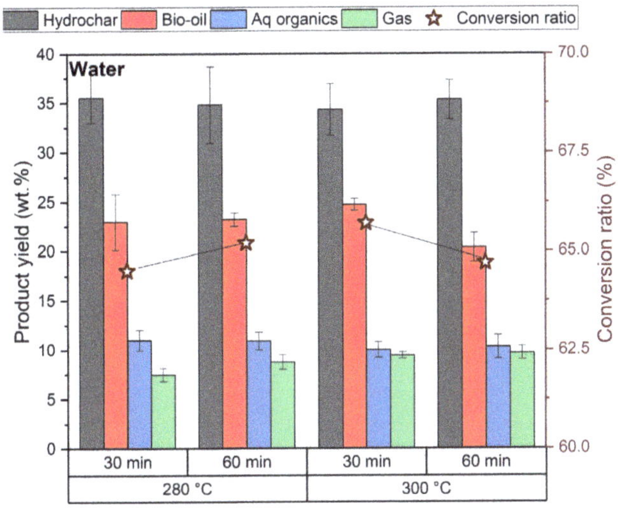

Figure 2. Effect of residence time ranging between 30 and 60 min on the distribution of HTL bioproducts yields (processing temperatures at 280 and 300 °C, and water as solvent).

3.1.3. Reaction Media Type

Recent reports on the hydrothermal processing of biomass have confirmed that applying a binary solvent system may enhance the conversion of raw materials. Therefore, this contribution focuses also on the additional comparative assessment of the effect of processing SCGs in purely hydrothermal media with binary solvent systems, under the previously established optimal conditions. Figure 3 comparatively presents the distribution of the thermochemical liquefaction bioproducts yields for binary solvent systems and pure water. It can be seen at a glance that the addition of oxygen-containing solvents, such as monohydroxy alcohols, can be beneficial for the increase in the yield of the desired product, i.e., bio-oil. The transformation of spent coffee grounds using a water–isopropanol mixture as a thermochemical liquefaction medium increased bio-oil yield, i.e., reaching 46.1 wt.%. A noticeably higher bio-oil yield compared to transformation using a purely aqueous medium (~35 wt.%) can be related to various reaction mechanisms that occur under different conditions. It can be mentioned here that the observed phenomenon can be ascribed to the better stabilization of intermediates formed during depolymerization in alcohols. Isopropanol, as a secondary alcohol, can act as a hydrogen donor and is prone to mentioned reactions. The enhancement of bio-oil yield in the binary solvent system can be assigned to both physical and chemical effects [40]. The presence of monohydroxy alcohol can improve the solubility of the higher range of organic molecules, i.e., some of the formed intermediates. Consequently, the repolymerization reactions between them could be hindered to some extent. Moreover, it is known that isopropyl alcohol features a hydrogen-donation ability, which is beneficial for the yield and composition of the bio-oil. For instance, it can be active in hydrogenolysis reactions and radical scavenging, leading to depolymerization and the reduction of secondary char formation, respectively. Thus, co-solvents added to the system can act not only as solvents but also as reagents in occurring reactions during thermochemical liquefaction. It can also be seen that for reaction mixtures containing oxygen compounds, the yield of solid residue was noticeably lower, which additionally proved that it is a more favorable environment for the profound degradation of the SCG building blocks. This can be due to the change of reaction environment after the addition of an oxygen-containing co-solvent and therefore enhanced hydrolysis. The presented investigation demonstrated that water alone is a less effective solvent compared to its mixture with isopropanol. For pure water, some of the raw material did not undergo liquefaction under the tested processing conditions.

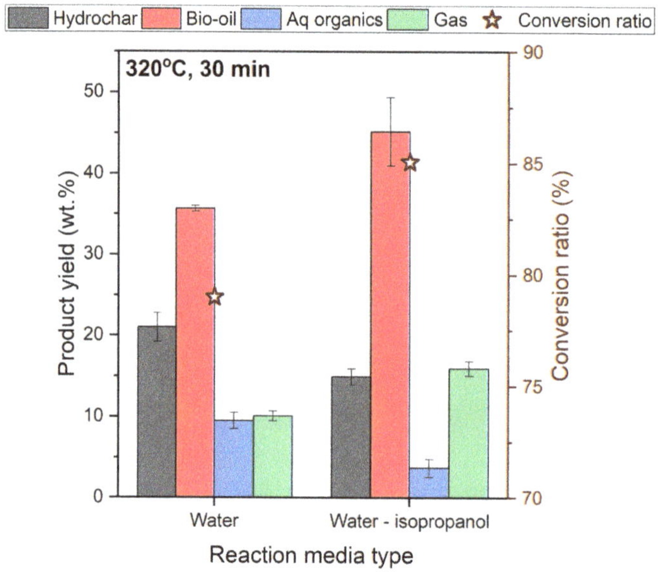

Figure 3. Yields of liquefaction products for water and water–isopropanol as reaction media (processing temperature: 320 °C; residence time: 30 min).

3.2. Effect of Processing Conditions on Bioproducts Composition

3.2.1. Elemental Composition

All bio-oils obtained through hydrothermal liquefaction in water and binary solvent systems were physically similar, featuring a high viscosity, a dark color and an intensive odor that resembled the smell of strongly roasted coffee beans. In Table 1, the results of the elemental compositions for raw material, bio-oils, and hydrochars are compared. A notably higher carbon content in the bio-oils is observed compared to the raw SCGs. In the case of the water series, the carbon content accounted for ca. 70.8–74.1 wt.%, while for bio-oil formed after applying a water–isopropanol mixture, it was around 71.6 wt.%. For the water medium, the highest values were observed for moderate processing temperatures (280–300 °C). Under more severe conditions, a noticeably higher yield of bio-oil was observed, so slightly lower relative carbon content could be the result of the presence of some oxygenates originating from the degradation of the most resistant compounds, e.g., lignin. What is worth noting here, the bio-oil elemental composition for binary solvent systems is very close to the reference bio-oil obtained in a water medium. Thus, the noticeable increase in the yield of bio-oil does not cause a decrease in the quality of the resultant bioproduct, which is highly beneficial. We also investigated the quality of the resultant hydrochar (cf. Table 1). Also, in this case, the quality of the solid residue for various processing conditions is similar. The increase in processing temperature causes gradual deoxygenation and the growth of the carbon content.

Table 1. Elemental composition of raw SCGs, together with resultant bio-oils and hydrochars obtained by hydrothermal liquefaction under various processing conditions (processing temperature: 260–320 °C, reaction media: water vs. water–isopropanol).

HTL Processing Conditions			C	H	N	O
Raw SCGs			50.8	7.5	2.4	37.5
Bio-oil	water	260 °C	72.9	9.6	1.9	15.6
		280 °C	74.8	9.8	2.3	13.1
		300 °C	74.1	9.9	2.0	14.1
		320 °C	70.8	9.5	2.1	17.7
	water–isopropanol	320 °C	71.6	9.6	2.5	16.2
Hydrochar	water	260 °C	67.6	5.6	3.9	22.9
		280 °C	70.0	5.0	4.2	20.9
		300 °C	72.0	5.1	4.5	18.4
		320 °C	73.8	5.1	4.6	16.5
	water–isopropanol	320 °C	68.4	5.0	6.1	20.6

For the sake of comparing the effect of process conditions on the change in the O/C and H/C molar ratios of the bio-oils, a van Krevelen diagram was plotted, which is depicted in Figure 4. It is worth noting that the O/C ratio for both groups of products (bio-oils, hydrochars) is noticeably lower compared to the feedstock. The H/C molar ratio for bio-oils remained at a comparable level, while that of hydrochars decreased by around 43–58% compared to SCGs. For comparison, we present these data together with the HHV ranges estimated based on the elemental (H, C, and O) composition. As can be seen, all the obtained products exhibit high energy density, and the HHVs of the resultant liquids and hydrochars were above 35 MJ/kg and around 30 MJ/kg, respectively.

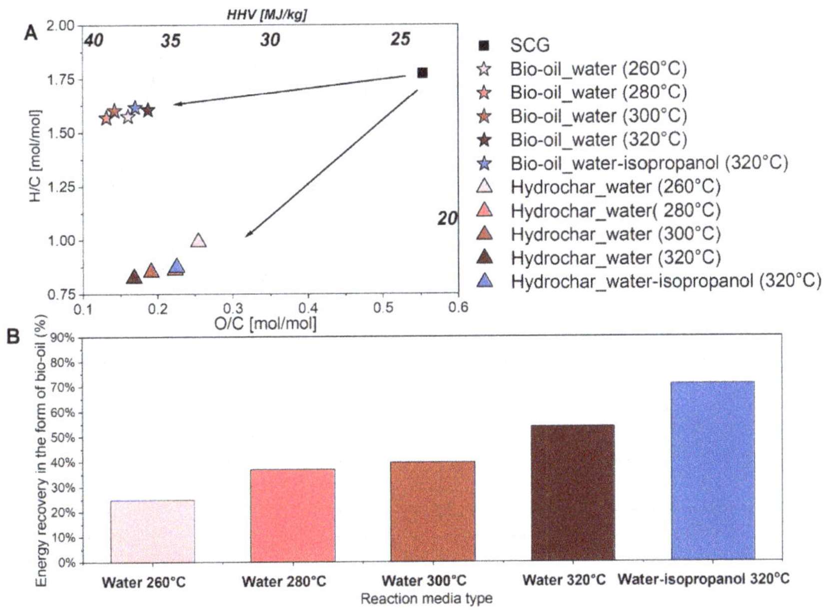

Figure 4. Comprehensive van Krevelen diagram for the studied bio-oils and hydrochars obtained at different processing conditions, i.e., processing temperature between 260 and 320 °C and single (water) and binary solvent system (water–isopropanol) (**A**) together with energy recovery in the form of bio-oil as a target group of products (**B**).

To evaluate the efficiency of biomass conversion under various processing conditions, we also calculated the energy recovery ratio in the form of bio-oils. As can be seen in Figure 4B, the highest energy recovery revealed the liquids produced in binary solvent systems. In this case, the energy recovery increased by ca. 20% compared to the reference sample obtained in a purely aqueous medium.

3.2.2. FT-IR

The FT-IR spectrum collected for as-received SCGs discloses the complexity of its composition (see Figure 5). The broad band at 3700–3000 cm^{-1} may be ascribed to the inter-and intramolecular H-bonding in alcohols, carboxylic acids, carbohydrates as well as phenolic species. The sharp band of poor intensity at 3007 cm^{-1} is assigned to the unsaturated compounds, while the quartet of overlapping modes at 2958, 2923, 2870, and 2851 cm^{-1} is attributed to the stretching C–H vibrations in methyl and methylene species. Namely, the first two are asymmetric $-CH_3$ and $-CH_2$ stretches, respectively, whereas the latter two are the corresponding symmetric ones. Considering the relative intensities thereof (i.e., the dominance of the intensity of the $-CH_2$ modes), one may conjecture that the sample is rich in long-chain aliphatic species. Interestingly, the broad band of low intensity at 2600–2500 cm^{-1} may be ascribed to the thiol compounds, which may be inherited from the proteins present in crude coffee grains. The broad absorption at ca. 2280–1900 cm^{-1} is assigned to the combination of deformation vibrations in aromatics (i.e., phenolic species). The two sharp and intensive bands at 1742 and 1661 cm^{-1} are characteristic of C=O stretching in carbonyl compounds (caffeine), esters and aldehydes, and carboxylic acids. The bands observed at 1456 and 1357 cm^{-1} are ascribed to the C–H bending modes.

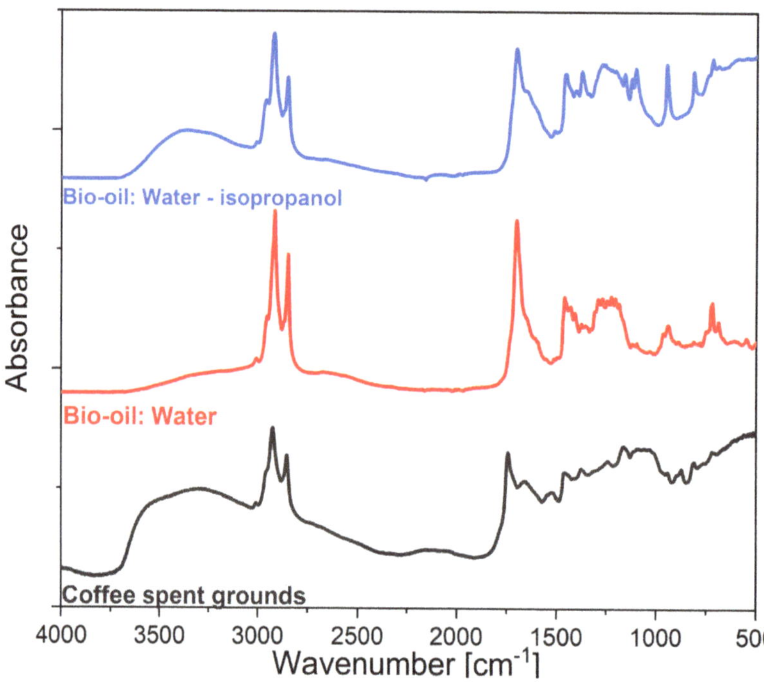

Figure 5. FT-IR spectra of coffee spent grounds and bio-oils obtained by its liquefaction using single—(water) and binary solvent system (water–isopropanol) (processing temperature: 320 °C, residence time: 30 min).

The spectra recorded for the HTL bio-oils manufactured in water are mutually similar. It is pertinent to mention that compared to coffee grounds, they feature a noticeable decrease in the intensity of the 3700–3000 cm^{-1} band and the complete elimination of the 2280–1900 cm^{-1} mode. Simultaneously, a substantial increase in the intensity of the aliphatic bands (3000–2800 cm^{-1}) and the carbonyl stretch is observed. This suggests a shift in the type of compounds, as well as a general decrease in the concentration of oxygenates like alcohols and phenols, towards long-chain aliphatic species, possibly esters, and other carbonyl compounds. Interestingly, the thiol compounds are retained in these bio-oils.

A slightly different scenario has been observed for the liquids obtained in the water–isopropyl alcohol mixtures. Namely, in this case, the intensity of the 3700–3000 cm^{-1} band is more pronounced than for a purely hydrothermal process. For both bio-oils, the shift of the C=O band from about 1750 cm^{-1} for CSG towards lower wavenumbers (approximately 1700 cm^{-1}) can be observed. This could suggest that the hydrolysis of esters contained in raw material (triglycerides) occurs intensively under subcritical conditions of water (shift from the esters band to the carboxylic acids band). It is plausible that, in the case of the water–isopropanol system, the intensity of the carbonyl band for carboxylic acid is less intensive. This can be linked to the possible transesterification between triglycerides and isopropanol, as well as the esterification of some free fatty acids (released after the hydrolysis mentioned) in a medium containing monohydroxy alcohol. Thus, this suggests the occurrence of secondary reactions between the constituents of the bio-oils and the organics present in the reaction media.

3.2.3. GC-MS

The qualitative analysis of the composition of the resultant bio-oils was conducted by means of gas chromatography coupled to mass spectrometry (GC-MS). The measurements allowed us to undertake a comparative analysis of the light fractions of the obtained liquid products. In Figure 6, the group composition of the light fractions of the resultant bio-oils obtained by using various types of reaction media is presented. Among the identified compounds, carboxylic acids and esters were the dominant groups, while phenolic compounds, along with other oxygenates and nitrogen-containing compounds, constituted a minor proportion. For all samples, the total share of the acids and esters was ca. 90% of all compounds identified. The presence of these compounds could be the result of the degradation of the lipids contained in the raw material. Interestingly, the addition of isopropanol as a co-solvent resulted in the noticeable conversion of carboxylic acid to the corresponding esters, which accounted for ca. 31.3%. In the excess presence of isopropyl alcohol, esters are formed via esterification between carboxylic acids (formed through hydrolysis of triglycerides) and the utilized monohydroxy alcohol. Once separated, carboxylic acids and esters can be utilized as drop-in biocomponents for compression ignition (CI) engines. The use of a binary solvent system affects the partial conversion of fatty acids into propyl esters, demonstrating superior oxidative stability and physicochemical properties (e.g., viscosity, cetane number) compared to commercially used fatty acid methyl esters [46]. Lipid derivatives (fatty acids and esters) can also be catalytically hydrogenated towards hydrocarbons, commercially called 'renewable diesel'. However, it should be emphasized that the most challenging operation will be the selective separation of this group of compounds from the as-received biocrude. Such a group of biocomponents can be considered as advanced biofuels, which are highly demanded in the fuel market.

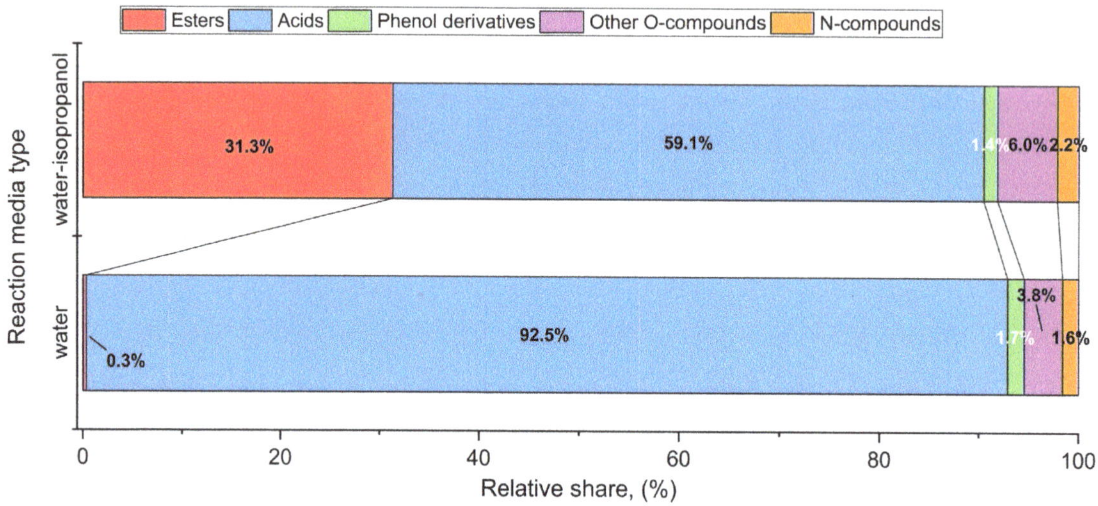

Figure 6. Group composition of the bio-oils obtained through thermochemical liquefaction using single—(water) and binary solvent system (water–isopropanol) (processing temperature: 320 °C, residence time: 30 min) (based on GC-MS analysis).

Isopropanol as a secondary alcohol is prone to hydrogen donation; thus, it is highly active in stabilizing other groups of intermediates, which eventually leads to a higher share of bio-oil yield. The presence of phenolics confirms the degradation of certain lignin structures in SCGs, while oxygen compounds indicate the degradation of polysaccharide species.

In Table 2, we comparatively gather the list of exemplary compounds detected in bio-oils with a sufficiently high threshold level of fitting to the MS library (75%) for tested reaction media types. The highest share was noticed for hexadecanoic, octadecanoic, and linoleic acids accompanied by the corresponding esters. This is in line with the results previously reported in the literature [36]. Interestingly, in the resultant liquids, we found isopropanol derivatives, which ultimately evidences our hypothesis that added monohydroxy alcohol could act not only as a solvent but also as a reagent. Thus, this proves that it is beneficial in stabilizing various reactive intermediates. The occurrence of other secondary reactions can be confirmed by the lower-molecular esters. For instance, organic acids derived from the decomposition of carbohydrates can undergo esterification, resulting in the presence of, e.g., isopropyl lactate. Additionally, the dehydration of isopropyl alcohol can occur to a small extent, which is confirmed by the presence of 2-propanol, 1-(1-methylethoxy). The presence of cyclic ketones and furan derivatives indicates that the polysaccharides (e.g., hemicellulose and cellulose) underwent depolymerization to hexoses and pentoses, followed by recombination to cyclic structures. The detected phenol derivatives are typical for the degradation of lignin structures. Nitrogen compounds are common products of protein degradation. Furthermore, the presence of some heterocyclic nitrogen compounds (i.e., pyrrolidinone or pyrrolidine) can confirm the Maillard reaction between carbohydrates and protein derivatives. Interestingly, the caffeine structure was preserved in the bio-oil throughout the HTL run. Furthermore, its content in the bio-oils roughly remains constant, notwithstanding the processing conditions.

Table 2. Exemplary compounds identified in bio-oils obtained via thermochemical liquefaction using single—(water) and binary solvent system (water–isopropanol) (processing temperature: 320 °C, residence time: 30 min).

Group	RT (min)	Compound Name	Share of Products Depending on Reaction Media Type [% of Cumulative Integral]	
			Water	Water–Isopropanol
Esters	58.2	Hexadecanoic acid, methyl ester	0.34	—
	60.4	Hexadecanoic acid, iso-propyl ester	—	18.38
	66.3	Octadecanoic acid, iso-propyl ester	—	6.39
	66.9	9,12-Octadecadienoic acid, iso-propyl ester	—	5.43
	71.3	Eicosanoic acid, isopropyl ester	—	0.60
Carboxylic acids	60.8	n-Hexadecanoic acid	49.07	31.20
	66.6	n-Octadecanoic acid	18.18	7.14
	67.3	9,12-Octadecadienoic acid	22.27	17.01
	71.7	n-Eicosanoic acid	3.22	2.14
Phenol derivatives	24.7	Phenol	1.10	0.54
	28.5	Phenol, 3-methyl-	0.05	0.06
	29.5	p-Cresol	0.14	0.13
	30.8	Mequinol	0.10	0.11
	34.1	Phenol, 2-ethyl-	0.05	—
	39.1	Phenol, 4-ethyl-2-methoxy-	0.21	0.33
Other O-compounds	12.7	Ethanol, 1-methoxy-, acetate	0.10	0.28
	13.4	(S)-Isopropyl lactate	—	0.27
	17.6	2-Cyclopenten-1-one	0.08	—
	20.8	2-Cyclopenten-1-one, 2-methyl-	0.23	0.18
	22.9	2-Propanol, 1-(1-methylethoxy)-	—	0.19
	23.2	2,5-Hexanedione	0.05	—
	25.0	2,4-Dimethylfuran	0.11	0.17
	25.5	2-Cyclopenten-1-one, 2,3-dimethyl-	0.04	0.06
	27.3	2-Cyclopenten-1-one, 3,4-dimethyl-	0.05	0.06
	27.7	1,2-Cyclopentanedione, 3-methyl-	0.18	0.23
	28.2	1-Methoxy-1,4-cyclohexadiene	0.13	0.13
	30.5	1-Methoxy-1,3-cyclohexadiene	0.09	0.41
	31.9	2-Cyclopenten-1-one, 3-ethyl-2-hydroxy-	0.08	0.17
	32.1	1,2-Cyclohexanedicarboxaldehyde	0.05	—
	40.1	4-Hydroxy-2,4,5-trimethyl-2,5-cyclohexadien-1-one	0.04	—
	40.7	Cyclohexanone, 2-(2-butynyl)-	—	0.07
	41.6	2-Cyclohexen-1-one, 4-(1-methylethyl)-	0.04	—
	42.2	4-Isopropenyl-1-methylbicyclo[4.1.0]heptan-2-one	—	0.06
	43.1	Glycerol 1,2-diacetate	0.04	0.06
	63.3	Estra-1,3,5(10)-trien-17.beta.-ol	0.07	0.05
	80.9	Phorbol	0.10	0.11
N-compounds	33.7	2-Pyrrolidinone	0.05	0.14
	37.2	Pyrrolidine, 1-acetyl-	0.10	0.19
	50.3	2-Cyclohexylpiperidine	—	0.09
	66.8	Caffeine	1.41	1.64

3.2.4. Gas-Phase Composition

We also investigated the composition of the gaseous products that evolved during the liquefaction of SCGs in an aqueous medium and its mixture with isopropanol (see Table 3). For both cases, the dominant component of the gas phase was carbon dioxide, which is indicative of the decomposition of the raw material through intensive decarboxylation. Carbon monoxide accounted for a few percentages, which proves that certain parts of macromolecules and intermediate by-products also decompose via decarbonylation. For hydrothermal liquefaction in pure water at lower concentrations, certain amounts of flammable components such as methane, ethane, propane, n-propane, isopropane, and n-propane were also detected. It is noteworthy that incorporating isopropanol as a co-solvent markedly increased the concentrations of propane and propene, with their concentration sum rising from 0.87 vol.% in water alone to 12.62 vol.% in a water and isopropanol mixture. This increase clearly confirms that isopropanol acts in this case not only as a solvent but also as a reagent. Moreover, the hydrogen content increased by more than double. This is most likely due to the reactions of isopropanol under near-critical solvent conditions. Namely, alcohol may undergo dehydrogenation towards acetone (the effect is usually observed over base catalysts, according to Equation (2)), while in the presence of acid sites, it dehydrates to propene (Equation (3)) [47]. Subsequently, propene may react with hydrogen, eventually forming propane (viz. Equation (4)). Therefore, the observed increase in hydrogen, propene, and propane in the gas phase is expected.

Table 3. Comparison of the composition of gaseous phase evolved during liquefaction of SCGs using single—(water) and binary solvent system (water–isopropanol) (processing temperature: 320 °C, residence time: 30 min).

Reaction Medium	Gas Components									
	CO_2	CO	H_2	CH_4	C_2H_4	C_2H_6	C_3H_6	C_3H_8	i-C_4	n-C_4
Water	89.29	7.03	0.48	0.66	0.20	0.20	0.68	0.19	0.85	0.41
Water–isopropanol	75.56	9.86	1.00	0.51	0.20	0.25	11.79	0.83	-	-

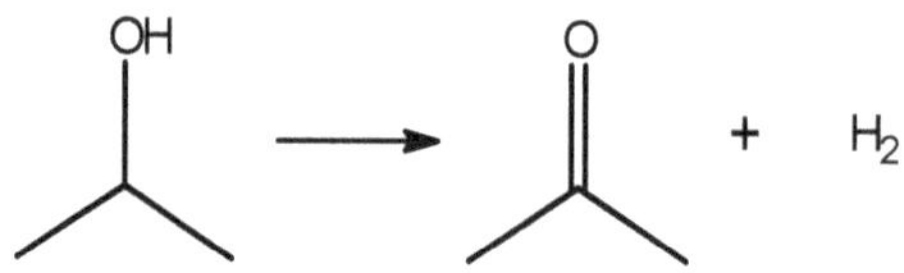

(2)

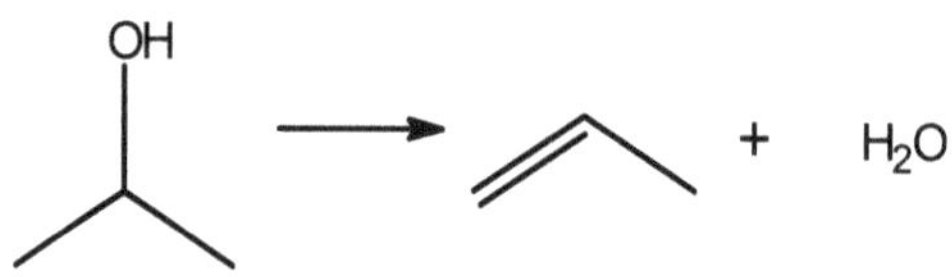

(3)

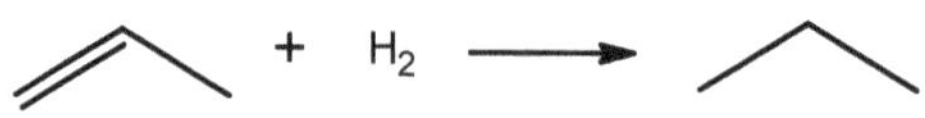

(4)

Thus, it is worth noting that the use of isopropyl alcohol positively impacts not only the noticeable increase in bio-oil yield but also entails a higher share of combustible hydrocarbons in the gaseous phase. This enhances the higher heating value and expands the potential applications of the gas phase. For instance, a CO_2-rich stream can be used as a gasifying agent for hydrochar, as previously proposed [28], while flammable compounds can constitute an internal source of energy.

4. Conclusions

Coffee in its numerous varieties is the most popular hot drink type around the world. Its popularity results in the generation of more than 9 million tons of coffee waste each year. This type of residual matter exhibits a high moisture content and a sufficiently homogeneous particle size, so it seems to be an excellent candidate for further conversion via hydrothermal processing routes. Thus, new routes for the effective conversion of these molecules into value-added bioproducts are being sought. In the present contribution, we investigate the efficiency of the conversion of SCGs through processing under subcritical water conditions. According to this, the effect of various processing conditions (temperature, residence time) and the addition of isopropanol to water as a hydrothermal liquefaction reaction medium was studied. We found that the introduction of monohydroxy alcohol could be beneficial in terms of the deep decomposition of the raw material, with a conversion ratio of up to 85%, which consequently resulted in a noticeable increase in bio-oil yield compared to processing the same feedstock in pure water. Thus, one can conjecture that the addition of oxygen compounds as liquefaction co-solvents creates favorable conditions for the efficient conversion of spent coffee grounds by altering the chemical nature of the reaction medium. Consequently, this leads to a supportive effect on the hydrolysis reaction and the stabilization of the formed intermediate products. As a result, a noticeable increase in the content of valuable esters in the resultant bio-oil was noted. The change in reaction mechanism in the case of the addition of co-solvent was also proven by the vast differences in the composition of the gas phase, which, for isopropyl, alcohol leads to the formation of combustible species like hydrogen, propene, and propane. This creates an excellent opportunity for the gas phase to be used for the generation of processing heat. Nonetheless, for all tested conditions, the obtained bioproducts such as bio-oil and hydrochars were of similar quality, exhibiting energy densification compared to the raw material. Bio-oils appeared as liquids with a high carbon content (71–74 wt.%) featuring a substantial energy value (approximately 35 MJ/kg). Also, hydrochars showed noticeable energy densification compared to the raw material, where the C content and HHV reached up to 73.8% by weight and 30 MJ/kg, respectively.

Supplementary Materials: The following supporting information can be downloaded at: https://www.mdpi.com/article/10.3390/en17246446/s1, The temperature and pressure profiles for liquefaction tests as a function of time for various runs can be found in the Supplementary Materials section.

Author Contributions: Conceptualization, M.W.; Methodology, M.W.; Formal analysis, M.W. and J.K.; Investigation, M.W., J.K. and M.P.; Writing—original draft preparation, M.W.; Writing—review and editing, J.K., R.J., M.L., M.P. and Ł.K.; Visualization: M.W.; Supervision, M.W.; Funding acquisition, M.W. All authors have read and agreed to the published version of the manuscript.

Funding: This research was funded by the National Centre for Research and Development in Poland, grant number LIDER/58/0241/L-11/19/NCBR/2020. Additionally, analysis of the bioproducts was possible thanks to the support of the program 'Excellence initiative—research university' (contract No. 5023) and research subsidy (No. 16.16.210.476) for the AGH University of Kraków.

Data Availability Statement: Data is contained within the article and Supplementary Materials. Raw dataset are available on request from the authors.

Conflicts of Interest: The authors declare no conflicts of interest.

References

1. Statista. Market Insights. In *Industries & Markets. Coffee: Market Data & Analysis*; Statista: Hamburg, Germany, 2024.
2. International Coffee Organization. *Coffee Report and Outlook*; International Coffee Organization: London, UK, 2023.
3. Kim, D.; Lee, K.; Bae, D.; Park, K.Y. Characterizations of Biochar from Hydrothermal Carbonization of Exhausted Coffee Residue. *J. Mater. Cycles Waste Manag.* **2017**, *19*, 1036–1043. [CrossRef]
4. Blinová, L.; Sirotiak, M.; Bartošová, A.; Soldán, M. Review: Utilization of Waste From Coffee Production. *Res. Pap.* **2017**, *25*, 91–102. [CrossRef]
5. Zhang, L.; Li, H.; Yang, L.; Du, X.; Zhou, Y.; Sun, G.; Liu, J. Carbon Footprints of Centralized and Decentralized Food Waste Utilization Pathways. *Renew. Sustain. Energy Rev.* **2025**, *208*, 115040. [CrossRef]

6. Annal, U.N.; R, V.; Natarajan, A.; Rajadurai, V.; Kumar, P.S.M.; Li, Y.Y. Electrolytic Biodiesel Production from Spent Coffee Grounds: Optimization through Response Surface Methodology and Artificial Neural Network. *J. Taiwan. Inst. Chem. Eng.* **2024**, *165*, 105697. [CrossRef]
7. Getachew, A.T.; Chun, B.S. Influence of Pretreatment and Modifiers on Subcritical Water Liquefaction of Spent Coffee Grounds: A Green Waste Valorization Approach. *J. Clean. Prod.* **2017**, *142*, 3719–3727. [CrossRef]
8. Atabani, A.E.; Al-Muhtaseb, A.H.; Kumar, G.; Saratale, G.D.; Aslam, M.; Khan, H.A.; Said, Z.; Mahmoud, E. Valorization of Spent Coffee Grounds into Biofuels and Value-Added Products: Pathway towards Integrated Bio-Refinery. *Fuel* **2019**, *254*, 115640. [CrossRef]
9. Dattatraya Saratale, G.; Bhosale, R.; Shobana, S.; Banu, J.R.; Pugazhendhi, A.; Mahmoud, E.; Sirohi, R.; Kant Bhatia, S.; Atabani, A.E.; Mulone, V.; et al. A Review on Valorization of Spent Coffee Grounds (SCG) towards Biopolymers and Biocatalysts Production. *Bioresour. Technol.* **2020**, *314*, 123800. [CrossRef]
10. Afolabi, O.O.D.; Sohail, M.; Cheng, Y.L. Optimisation and Characterisation of Hydrochar Production from Spent Coffee Grounds by Hydrothermal Carbonisation. *Renew. Energy* **2020**, *147*, 1380–1391. [CrossRef]
11. Maree, Z.; Strydom, C.A.; Bunt, J.R. Chemical and Physical Characterization of Spent Coffee Ground Biochar Treated by a Wet Oxidation Method for the Production of a Coke Substitute. *Waste Manag.* **2020**, *113*, 422–429. [CrossRef] [PubMed]
12. Espuelas, S.; Marcelino, S.; Echeverría, A.M.; del Castillo, J.M.; Seco, A. Low Energy Spent Coffee Grounds Briquetting with Organic Binders for Biomass Fuel Manufacturing. *Fuel* **2020**, *278*, 118310. [CrossRef]
13. Limousy, L.; Jeguirim, M.; Dutournié, P.; Kraiem, N.; Lajili, M.; Said, R. Gaseous Products and Particulate Matter Emissions of Biomass Residential Boiler Fired with Spent Coffee Grounds Pellets. *Fuel* **2013**, *107*, 323–329. [CrossRef]
14. Lee, X.J.; Ong, H.C.; Gao, W.; Ok, Y.S.; Chen, W.H.; Goh, B.H.H.; Chong, C.T. Solid Biofuel Production from Spent Coffee Ground Wastes: Process Optimisation, Characterisation and Kinetic Studies. *Fuel* **2021**, *292*, 120309. [CrossRef]
15. Codignole Luz, F.; Cordiner, S.; Manni, A.; Mulone, V.; Rocco, V. Biomass Fast Pyrolysis in Screw Reactors: Prediction of Spent Coffee Grounds Bio-Oil Production through a Monodimensional Model. *Energy Convers. Manag.* **2018**, *168*, 98–106. [CrossRef]
16. Barbanera, M.; Muguerza, I.F. Effect of the Temperature on the Spent Coffee Grounds Torrefaction Process in a Continuous Pilot-Scale Reactor. *Fuel* **2020**, *262*, 116493. [CrossRef]
17. Liu, Y.; Tu, Q.; Knothe, G.; Lu, M. Direct Transesterification of Spent Coffee Grounds for Biodiesel Production. *Fuel* **2017**, *199*, 157–161. [CrossRef]
18. Lee, S.; Choi, Y.; Kang, K. Application of Blended Fuel Containing Coffee Ground Pyrolysis Oil in a Diesel Generator. *Fuel* **2019**, *256*, 115998. [CrossRef]
19. Van Nguyen, Q.; Choi, Y.S.; Choi, S.K.; Jeong, Y.W.; Han, S.Y. Co-Pyrolysis of Coffee-Grounds and Waste Polystyrene Foam: Synergistic Effect and Product Characteristics Analysis. *Fuel* **2021**, *292*, 120375. [CrossRef]
20. Cervera-Mata, A.; Lara, L.; Fernández-Arteaga, A.; Ángel Rufián-Henares, J.; Delgado, G. Washed Hydrochar from Spent Coffee Grounds: A Second Generation of Coffee Residues. Evaluation as Organic Amendment. *Waste Manag.* **2021**, *120*, 322–329. [CrossRef] [PubMed]
21. Adan-Mas, A.; Alcaraz, L.; Arévalo-Cid, P.; López-Gómez, F.A.; Montemor, F. Coffee-Derived Activated Carbon from Second Biowaste for Supercapacitor Applications. *Waste Manag.* **2021**, *120*, 280–289. [CrossRef]
22. Kaya, M.; Atelge, M.R.; Bekirogullari, M.; Eskicioglu, C.; Atabani, A.E.; Kumar, G.; Yildiz, Y.S.; Unalan, S. Carbon Molecular Sieve Production from Defatted Spent Coffee Ground Using $ZnCl_2$ and Benzene for Gas Purification. *Fuel* **2020**, *277*, 118183. [CrossRef]
23. Dias, M.E.; Oliveira, G.H.D.; Couto, P.T.; Dussán, K.J.; Zaiat, M.; Ribeiro, R.; Stablein, M.J.; Watson, J.T.; Zhang, Y.; Tommaso, G. Anaerobic Digestion of Hydrothermal Liquefaction Wastewater from Spent Coffee Grounds. *Biomass Bioenergy* **2021**, *148*, 106030. [CrossRef]
24. Arauzo, P.J.; Lucian, M.; Du, L.; Olszewski, M.P.; Fiori, L.; Kruse, A. Improving the Recovery of Phenolic Compounds from Spent Coffee Grounds by Using Hydrothermal Delignification Coupled with Ultrasound Assisted Extraction. *Biomass Bioenergy* **2020**, *139*, 105616. [CrossRef]
25. Shakya, R.; Adhikari, S.; Mahadevan, R.; Shanmugam, S.R.; Nam, H.; Barbary, E.; Dempster, T.A. Influence of Biochemical Composition during Hydrothermal Liquefaction of Algae on Product Yields and Fuel Properties. *Bioresour. Technol.* **2017**, *243*, 1112–1120. [CrossRef]
26. Zhu, W.W.; Zong, Z.M.; Yan, H.L.; Zhao, Y.P.; Lu, Y.; Wei, X.Y.; Zhang, D. Cornstalk Liquefaction in Methanol/Water Mixed Solvents. *Fuel Process. Technol.* **2014**, *117*, 1–7. [CrossRef]
27. Martins-Vieira, J.C.; Lachos-Perez, D.; Draszewski, C.P.; Celante, D.; Castilhos, F. Sugar, Hydrochar and Bio-Oil Production by Sequential Hydrothermal Processing of Corn Cob. *J. Supercrit. Fluids* **2023**, *194*, 105838. [CrossRef]
28. Wądrzyk, M.; Grzywacz, P.; Janus, R.; Michalik, M. A Two-Stage Processing of Cherry Pomace via Hydrothermal Treatment Followed by Biochar Gasification. *Renew. Energy* **2021**, *179*, 248–261. [CrossRef]
29. Wadrzyk, M.; Berdel, M.; Janus, R.; Brilman, D.W.F. Hydrothermal Processing of Pine Wood: Effect of Process Variables on Bio-Oil Quality and Yield. *E3S Web Conf.* **2019**, *108*, 02004. [CrossRef]
30. Peng, N.; Liu, Z.; Liu, T.; Gai, C. Emissions of Polycyclic Aromatic Hydrocarbons (PAHs) during Hydrothermally Treated Municipal Solid Waste Combustion for Energy Generation. *Appl. Energy* **2016**, *184*, 396–403. [CrossRef]
31. Olszewski, M.P.; Arauzo, P.J.; Wądrzyk, M.; Kruse, A. Py-GC-MS of Hydrochars Produced from Brewer's Spent Grains. *J. Anal. Appl. Pyrolysis* **2019**, *140*, 255–263. [CrossRef]

32. Tekin, K.; Karagöz, S.; Bektaş, S. A Review of Hydrothermal Biomass Processing. *Renew. Sustain. Energy Rev.* **2014**, *40*, 673–687. [CrossRef]
33. Gollakota, A.R.K.; Kishore, N.; Gu, S. A Review on Hydrothermal Liquefaction of Biomass. *Renew. Sustain. Energy Rev.* **2018**, *81*, 1378–1392. [CrossRef]
34. Yang, L.; Nazari, L.; Yuan, Z.; Corscadden, K.; Xu, C.C.; He, Q.S. Hydrothermal Liquefaction of Spent Coffee Grounds in Water Medium for Bio-Oil Production. *Biomass Bioenergy* **2016**, *86*, 191–198. [CrossRef]
35. Liakos, D.; Chrysikou, L.P.; Triantafyllidis, K.; Bezergianni, S. Hydrothermal Liquefaction of Catering Wastes towards Biofuel Intermediates. *Biomass Convers. Biorefinery* **2024**. [CrossRef]
36. Yang, J.; Sophia, Q.; Corscadden, K.; Niu, H. The Impact of Downstream Processing Methods on the Yield and Physiochemical Properties of Hydrothermal Liquefaction Bio-Oil. *Fuel Process. Technol.* **2018**, *178*, 353–361. [CrossRef]
37. Yang, L.; He, Q.; Havard, P.; Corscadden, K.; Xu, C.; Wang, X. Co-Liquefaction of Spent Coffee Grounds and Lignocellulosic Feedstocks. *Bioresour. Technol.* **2017**, *237*, 108–121. [CrossRef]
38. Lozano, E.M.; Pedersen, T.H.; Rosendahl, L.A. Integration of Hydrothermal Liquefaction and Carbon Capture and Storage for the Production of Advanced Liquid Biofuels with Negative CO_2 Emissions. *Appl. Energy* **2020**, *279*, 115753. [CrossRef]
39. Dhalsamant, K.; Tirumareddy, P.; Borugadda, V.B.; Dalai, A.K. Utilization of Food Waste for Biocrude Production: A Review. *Bioresour. Technol. Rep.* **2023**, 24. [CrossRef]
40. Han, Y.; Hoekman, K.; Jena, U.; Das, P. Use of Co-Solvents in Hydrothermal Liquefaction (HTL) of Microalgae. *Energy* **2019**, *13*, 124. [CrossRef]
41. Wadrzyk, M.; Janus, R.; Korzeniowski, L.; Plata, M. Thermochemical Co-Liquefaction of Fruit Pomace's Blends in a Binary Solvent System towards Value-Added Bioproducts. *J. Energy Resour. Technol. Trans. ASME* **2023**, *146*, 030906. [CrossRef]
42. Wądrzyk, M.; Korzeniowski, Ł.; Plata, M.; Janus, R.; Lewandowski, M.; Borówka, G.; Maziarka, P. Solvothermal Liquefaction of Blackcurrant Pomace in the Water-Monohydroxy Alcohol Solvent System. *Energy* **2023**, *16*, 1127. [CrossRef]
43. Wądrzyk, M.; Janus, R.; Vos, M.P.; Brilman, D.W.F. Effect of Process Conditions on Bio-Oil Obtained through Continuous Hydrothermal Liquefaction of *Scenedesmus* sp. Microalgae. *J. Anal. Appl. Pyrolysis* **2018**, *134*, 415–426. [CrossRef]
44. Wądrzyk, M.; Janus, R.; Rządzik, B.; Lewandowski, M.; Budzyń, S. Pyrolysis Oil from Scrap Tires as a Source of Fuel Components: Manufacturing, Fractionation, and Characterization. *Energy Fuels* **2020**, *34*, 5917–5928. [CrossRef]
45. Toor, S.S.; Rosendahl, L.; Rudolf, A. Hydrothermal Liquefaction of Biomass: A Review of Subcritical Water Technologies. *Energy* **2011**, *36*, 2328–2342. [CrossRef]
46. Al-Gharibeh, E.; Leathers, R.; Kumar, K.; Sung, C.J. Ignition and Combustion Characteristics of Decanoic Acid Derived Alkyl Esters in a Fuel Ignition Tester. *Fuel* **2020**, *276*, 117982. [CrossRef]
47. Díaz de Leon, J.N.; Cruz-Taboada, A.; Esqueda-Barron, Y.; Alonso-Nuñez, G.; Loera-Serna, S.; Venezia, A.M.; Poisot, M.E.; Fuentes-Moyado, S. Catalytic Dehydration of 2 Propanol over Al_2O_3-Ga_2O_3 and Pd/Al_2O_3-Ga_2O_3 Catalysts. *Catal. Today* **2020**, *356*, 339–348. [CrossRef]

energies

MDPI

Article

Co-Pyrolysis of Plastic Waste and Lignin: A Pathway for Enhanced Hydrocarbon Recovery

Vilmantė Kudelytė, Justas Eimontas *, Rolandas Paulauskas and Nerijus Striūgas

Laboratory of Combustion Processes, Lithuanian Energy Institute, LT-44403 Kaunas, Lithuania; vilmante.kudelyte@lei.lt (V.K.)
* Correspondence: justas.eimontas@lei.lt

Abstract: Various plastics and biomass wastes, such as polypropylene (PP), low- or high-density polyethylene (LDPE/HDPE), and lignin, have become some of the most concerning wastes nowadays. In this context, this study aimed to investigate the possibility of applying thermochemical processes for the valorization of these materials. The experiments were carried out using a thermogravimetric analyzer on individual plastic and lignin samples and their mixtures at different mass ratios of 1:1, 1:2, 1:3, and 1:4. The gaseous products evolved during the pyrolysis process were analyzed by combined thermogravimetric and Fourier-transform infrared spectroscopy (TG-FTIR) and chromatography-mass spectrometry (Py-GC/MS) to analyze the functional groups and chemical composition of the obtained pyrolysis products. The results showed that the main functional groups of lignin monitored by TG-FTIR were aromatic and aliphatic hydrocarbons, while all plastics showed the same results for hydrocarbons. The investigation confirmed that mixing these types of plastics with lignin at different mass ratios led to increased recovery of higher-value-added products. Py-GC/MS analysis showed that the greatest results of compound recovery were achieved with lignin and LDPE/HDPE mixtures at 600 °C. At this temperature and with a mass ratio of 1:3, the plastic's radicals enhanced the depolymerization of lignin, encouraging its wider decomposition to hydrocarbons that can be applied for the production of value-added chemicals and bio-based energy.

Keywords: plastic; waste; lignin; conversion; hydrocarbons; co-pyrolysis

1. Introduction

For the last decades, various plastics have made massive contributions in nearly every human activity, such as medicine, agriculture, transportation, and even in households [1]. Therefore, increased plastic demand and improper utilization of its waste has become one of the most environmentally concerning problems all around the world. Plastics, mainly polypropylene (PP) and low- and high-density polyethylene (LDPE and HDPE), are widely used in manufacturing and the service industry for different kinds of applications [2]. Despite the rapidly growing distribution, only a few percent of plastic waste are being recycled properly, while the greater part is landfilled or incinerated with municipal waste [3]. Unfortunately, plastic waste is not the only worldwide environmental problem that causes pollution. Lignin is one of the main components of lignocellulosic biomass, which during the past couple of years has become another rapidly increasing industrial waste. In many manufacturing processes, such as papermaking, lignin is separated from other lignocellulose components as it is a very stable polymer that cannot be easily processed and becomes non-degradable [4].

According to statistics, 57% of the significant polymer waste constitutes LDPE, PP, and HDPE as they also have the largest usage because of their light weight, durability, and cheaper processing cost [5]. Based on Yueqing Ren et al. [6], these types of plastics are classified as a thermoplastics that have low-temperature impact resistance as they are linear macromolecules, in which atoms and molecules are connected end-to-end to form a sequence of long, single carbon chains. Therefore, they cannot change their chemical structure when heated or molded multiple times. For this unique quality, these plastics are commonly used in main industries, such as in agriculture to cover greenhouses, in manufacturing and packaging, or even in medicine. Unfortunately, the enormous usage of plastic has led to almost a worldwide catastrophe of pollution. It is estimated that nearly 30% of total waste is disposed and landfilled in Europe [7]. As these types of plastics mainly consist of petroleum products, which are chemically composed of hydrocarbons and usually include additives like antioxidants, stabilizers, or even colorants, their incineration might release toxicants that incorporate cadmium or chloride. Additionally, plastics have the feature of long-lasting decomposition, which causes these toxic substances to contaminate the biosphere for an inadequate amount of time [8]. Studies on plastic pollution reveal that when microplastics are devoured, they start to amass in the tissues of organisms, and this phenomenon mostly affects fish; therefore, there is an everyday risk that through polluted water or aquatic animals the human body can be damaged, as the released toxins can cause cancer, immune and hormonal system problems, or even birth abnormalities [9].

Along with plastic, there is less detrimental, however rapidly growing, lignin waste. It is an industrial waste usually formed during the course of wood processing or can be collected from agricultural residues. Lignin is one of the components of lignocellulose; however, it is not that adaptable compared to other components, such as cellulose or hemicellulose, due to its complex structural characteristics [10]. It has been estimated that approximately 150 billion tons of lignin are being produced worldwide and almost 95% of the generated lignin waste after manufacturing processes is directly distributed into rivers or simply combusted [11]. Therefore, management and utilization of lignin waste involve serious problems.

Energy recovery from plastic and lignin wastes might be a promising solution for utilizing waste and reclaiming higher-value-added products as all of these materials have some significant properties that might be used to recover alternative energy and chemical products [12]. Based on numerous papers, the co-pyrolysis of plastics and lignin is wildly applied due to it being a simple, environment-friendly, and low-cost efficient method [13]. Pyrolysis is a commonly used thermochemical conversion technology that produces three main products, such as the solid fraction (biochar), a liquid fraction (bio-oil), and a mixture of gases, depending on the process conditions. According to the literature on the pyrolysis behaviors of different plastics and lignin mixtures using micro-scale thermogravimetry [14], lignin via thermochemical treatment can decompose into aromatic hydrocarbons as it is structurally enriched with phenylpropane units. These abundant hydrocarbon groups can be used as a precursor for fuel production and may also be adapted in the pharmaceutical or chemical sphere [15]. Unfortunately, the yield of lignin pyrolysis products can be quite low as they are prone to repolymerize; therefore, it was estimated that the addition of microplastics such as LDPE could modify the distribution of hydrocarbons and enhance the production yield to its highest potential [16]. Moreover, according to Wei Jin et al. [16], the distribution of hydrocarbons can be defined by the temperature. It was assumed that, by increasing the temperature, the range and amount of these compounds would expand during the co-pyrolysis of lignin and LDPE. Scientists indicate that the most significant temperature to produce hydrocarbons during micro-thermal analysis is 600 °C, at which the largest range of these compounds' distribution was detected. Authors

Liangliang Tao et al. [4] investigated that lignin alone might not fully degrade even at a high temperature. The reason for this phenomenon could be the remaining material of carbon deposition; therefore, it was investigated that the addition of LDPE, as a hydrogen donor, could decrease carbon deposition and increase lignin's degradation rate [17].

Based on the literature, co-pyrolysis of the corresponding materials accelerates pyrolysis reactions and might change the composition of compounds. However, not all investigations of plastic, especially PP or HDPE, and lignin co-pyrolysis have considered the analysis of various recovered products from their mixtures. The scientific novelty of this study is to expand the existing knowledge and address the gap in the literature regarding product formation during the co-pyrolysis of specific plastics and lignin. Therefore, in this study, the main aim was to explore the interaction between lignin and plastics such as PP, LDPE, and HDPE for pyrolysis tests and chemical composition analysis by applying thermogravimetry to analyze the mass and temperature range changes. Moreover, evaluating the composition of the formed volatile products using the combined system of thermogravimetry and FTIR was also incorporated. Additionally, to investigate the physical products, pyrolysis tests were also performed using Py-GC/MS, and to determine the possibilities of recovering higher-value-added products from waste and their facilities to be applied in different types of bio-based industrial fields. The aim of applying the PY-GC/MS technique was to analyze the chemical composition of the products and assess the potential for recovering high-value compounds. Additionally, different plastic substrates were compared to gain deeper insights. These data fill the gap in knowledge of different plastics selected as co-pyrolysis substrates for lignin, expanding the knowledge of plastic waste utilization for higher-value-added product recovery.

2. Methods and Materials

2.1. Characterization and Preparation of Feedstock

For this analysis, plastic wastes such as PP, LDPE, and HDPE were supplied by the waste recycling company "Ekobazė" from Vilnius, Lithuania. These plastic samples were different types of packaging, such as canisters, containers, or plastic rolls, pressed into small, approximately 1 cm diameter pellets (Figure 1). Before applying any experimental treatment, all plastics were milled to particle size using a cryogenic ball mill (Retzsch GmbH, Hann, Germany) (Fritsch pulverisette 0). Each sample was ground for 10 min, keeping the 3 mm amplitude to achieve powder consistency. Lignin was collected from a paper manufacturing company in Lithuania. As it has almost 60% moisture in its structure, a 2V–151 dryer ("Büchi", Flawil, Switzerland) (was used for the first feedstock processing to eliminate the greater part of the moisture before applying the thermal treatment.

Figure 1. Visualization of the plastic samples before milling.

Moreover, as a co-pyrolysis component, lignin was collected from a paper manufacturing company in Kedainiai, Lithuania (Figure 2). As it has almost 60% moisture in its structure, a 2V–151 dryer was used for the first feedstock processing to eliminate the greater part of the moisture before applying the thermal treatment.

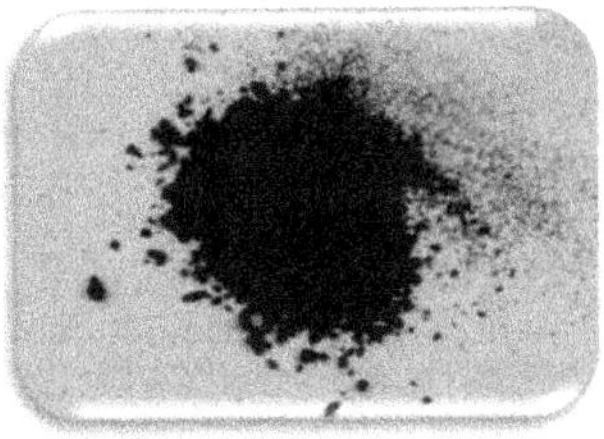

Figure 2. Visualization of the lignin sample.

The ultimate analysis of individual plastics was also incorporated and is presented in Table 1. Elemental analysis was performed to assess the nitrogen (N), sulfur (S), carbon (C), and hydrogen (H) contents in the samples using a Flash 2000 CHNS analyzer ("Thermofisher scientific", Milan, Italy) with all results reported on an as-received basis, which means that the feedstock for the samples was not affected by any process before the measurements. Furthermore, the calorific value was determined using an IKA C5000 calorimeter ("IKA", Staufen, Germany), following the LST EN 14918 standard [18]. The results showed that carbon was the highest compositional element in all samples—about 80 wt.% for all of the plastics. Another high compositional element was hydrogen, containing almost 14 wt.% for LDPE and PP and 14.37 wt.% for HDPE. However, there was not too much nitrogen content in all of the samples, except PP, which had 0.22 wt.%. The oxygen content was calculated based on Equation (1). The oxygen content for all samples was low—3.75% and 3.74% for LDPE and PP samples, respectively, and only 1.9% for HDPE. As all samples do not contain a lot of sulfur, they can be used for the production of gaseous products and, also, they can be adapted for fuel production as they mainly consist of carbon.

$$O_2 = 100\% - C - H - N - S - Cl - Ash \quad (1)$$

Table 1. Ultimate analysis of plastic waste.

Sample	N (wt.%)	C (wt.%)	H (wt.%)	S (wt.%)	O (wt.%)
HDPE	-	83.67 ± 0.24	14.37 ± 0.28	-	1.9 ± 0.26
LDPE	-	82.00 ± 0.34	13.91 ± 0.01	0.12 ± 0.04	3.75 ± 0.36
PP	0.22 ± 0.01	80.44 ± 0.07	13.40 ± 0.03	0.07 ± 0.01	3.74 ± 0.04

Proximate analysis of individual feedstocks was examined, and the results are presented in Table 2. The proximate analysis was conducted according to LST EN 15148 [19] for volatile content, LST EN 14775 [20] for ash content, and LST EN 14774–1 [21] for moisture content. The feedstocks for this analysis were not treated by any form of processes before the measurements—the results are outlined on an as-received basis. The performed experiments with lignin showed that it contains 1.59% moisture and only 41.77% volatile matter. During the pyrolysis process, formed fixed carbon reached 30.44% of the content, and after combustion, 26.15% was left. Therefore, it can be claimed that, in lignin's case, during its degradation, a lot of residue is formed. The analysis of the HDPE sample showed that no moisture or fixed carbon was detected. Additionally, almost 100% volatile matter was detected, with the absence of ash due to its degradation, with minimal formation of residue. Very similar results were obtained with the other plastics. No moisture or fixed carbon was detected in the LDPE sample, only 99.78% volatile matter with a slightly higher percent of ash—0.22. The PP sample, on the other hand, showed slightly higher contents of ash and fixed carbon, reaching 2.13% and 1.21%, respectively, with 96.66% volatile matter.

Table 2. Proximate analysis of lignin and plastic.

Sample	Moisture (%)	Volatile Matter (%)	Fixed Carbon (%)	Ash (%)
Lignin	1.59	41.77	30.44	26.20
LDPE	-	99.78	-	0.22
HDPE	-	99.94	-	0.05
PP	-	96.66	1.21	2.13

2.2. Thermal Analysis by TG-FTIR System

The selected and prepared feedstocks and their mixtures were used as samples for further experiments and the overview of the applied analysis methods in this research is presented in Figure 3.

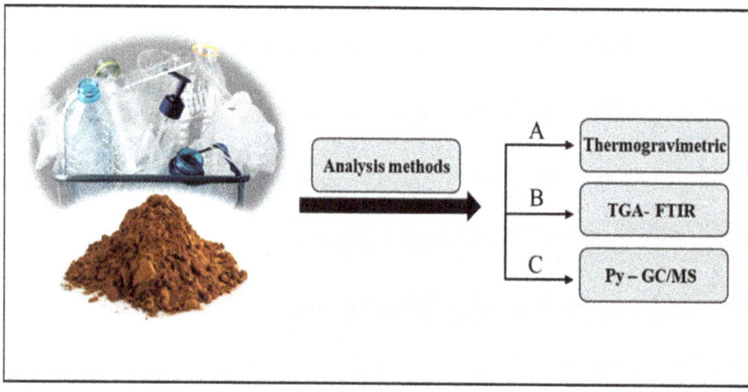

Figure 3. Flowchart of the current research methodologies.

Thermal analysis at the micro scale (A) of selected feedstocks was performed using a Netzsch Jupiter STA 449 F3 thermogravimetric analyzer produced in Selb, Germany, which is designed for precise measurements of mass changes as a function of temperature and time. The individual lignin and plastic materials and their mixtures were analyzed by detecting their thermal behaviors and decomposition points. PP, LDPE, HDPE, and lignin were analyzed using 10–15 mg of each sample, keeping the proportions of 1:1, 1:2, 1:3, and 1:4, and heated in an Al_2O_3 crucible with an inert ambient atmosphere supported by supplying nitrogen gas into the furnace at the flow rate of 60 mL/min. The temperature was gradually increased from 40 °C to 900 °C at a heating rate of 30 °C/min. The resulting thermogravimetric curves were analyzed using NETZSCH Proteus 8.0.3 software, which allowed for the extraction of data on temperature changes and weight loss during the thermal treatment. Maximum decomposition points were identified using the DTG curves derived from the first derivative of the TGA data. To study the evolved volatile functional groups during thermal treatment, the thermogravimetric analyzer was coupled with the Bruker Tensor 27 ("Bruker", Bremen, Germany) Fourier-transform infrared spectrometer system (TG-FTIR) (B). The FTIR system, capable of high sensitivity and resolution, was set to a spectral range of 4500–650 cm^{-1}, with a resolution factor of 4 cm^{-1}. During the experiment, volatile compounds released from the samples were swept into the heated FTIR gas cell, specifically at the moment when the maximum rate of gas evolution occurred during the pyrolysis process. This setup enabled real-time monitoring of the evolved functional groups.

2.3. Co-Pyrolysis Investigation by Py-GC/Ms

The Py-GC/MS system (C) was used to characterize the fast pyrolysis products of lignin, plastics, and their mixtures. The process was performed in a double-shot pyrolyzer (Frontier Laboratories EGA/PY -3030D, Koriyama, Japan) connected to a gas chromatography/mass spectrometry unit (Agilent 7890A GC/MS ("Agilent", Santa Clara, CA, USA) system). During the experimental process, the direct pyrolysis method was applied. Direct pyrolysis was carried out at 550, 600, and 650 °C. These temperatures were chosen based on the TGA analysis, as all collected samples decomposed at 500 °C, and according to the literature [3], as the highest amounts of the targeted products, such as aromatic compounds and hydrocarbons, are recovered at these temperatures. For each pyrolysis run, a precise amount of the sample was weighed into a quartz cup: 0.4 $\pm$ 0.1 mg for the mixtures and 0.3 $\pm$ 0.1 mg for the individual feedstock materials. High-purity helium was used as the carrier gas, with the split ratio set to 1:20, ensuring optimal transfer of volatiles during the experiment. The carrier gas flow was maintained at 1.5 mL/min, which efficiently transported the evolved volatiles from the pyrolyzer to the GC/MS system. The temperature of the gas transmission pipeline between the pyrolyzer and the GC injector was carefully controlled at 230 °C to prevent condensation of volatile compounds. The chromatographic separation was performed using an HP-5MS column ("Agilent", Santa Clara, CA, USA), with the column temperature initially set to 40 °C and then ramped to 270 °C at a heating rate of 10 °C/min, which is suitable for the separation of complex pyrolysis products. The mass spectra of the volatiles were recorded within the mass-to-charge (m/z) range of 30–600. Identification of the main compounds was achieved by comparing the obtained spectra to the National Institute of Standards and Technology (NIST) mass spectral library, providing reliable compound identification. The pyrolysis process using the Py-GC/MS system (Figure 4) begins when the sample is placed in the cup attached to the sample holder. At the start of the pyrolysis process, the cup is released and falls into the preheated pyrolyzer furnace, which has been set to the chosen temperature. The gases inside the furnace are swept by the helium carrier gas, and as the sample vapors are released, a portion of these vapors is transferred through the split in the system to the gas chromatography column. The compounds are separated by an ITF needle via the GC injection port. After separation, the individual compounds are sent to the mass spectrometer, where they are identified based on the NIST mass spectra library.

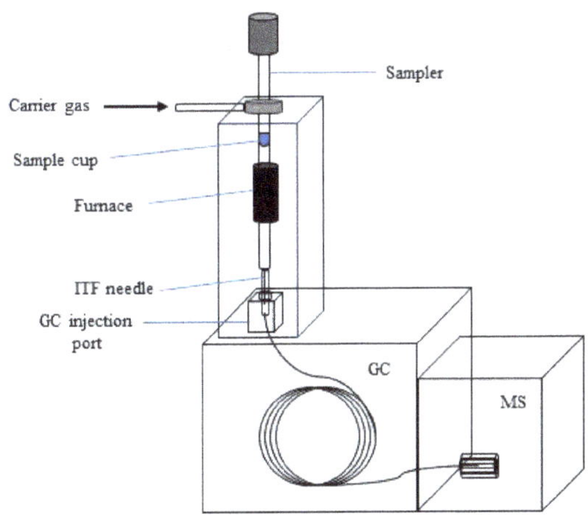

Figure 4. Instrumentation scheme of Py-GC/MS system.

3. Results and Discussion

3.1. Micro-Thermal Analysis of Plastic and Lignin Co-Pyrolysis

In the examined temperature interval, HDPE degraded in a single mass-loss zone with one decomposition peak at 500 °C, with a mass loss of about 99 wt.% and a residual mass of 0.22 wt.% (Figure 5A). This indicated that some amount of char was formed during the decomposition process. The evaluation of LDPE showed that there were two decomposition peaks described by the DTG (Figure 5B) curve at 390 °C and 510 °C, respectively. This temperature range belongs to the decomposition of polyethylene material. The two decomposition points observed are a result of the sample comprising a mixture of various types of low-density polyethylene (LDPE), each with distinct decomposition temperatures. The mass loss in total reached almost 98 wt.%, with a residual mass of 0.69 wt.%. Analyzing the PP curve, the decomposition peak was reduced to 480 °C and the residual mass increased comparing to the other plastics, up to 2.57 wt.%. According to a previous study [22], the examined PP decomposition temperature is 300–400 °C. The degradation temperature might have been increased by additives or fillers, making the material more thermally stable and leaving a greater residual mass due the non-volatile components. In addition, the obtained TGA curve of lignin (Figure 5A) showed a slight weight loss range from 65 °C to 190 °C, with the mass loss of almost 4 wt.%, which represented the evaporation of moisture. The curve itself was divided into one more mass loss zone corresponding to temperature and a mass loss of about 40 wt.% between 190 °C and 760 °C, with the residual mass of char of 53 wt.% at 900 °C. As claimed by Manara et al. [23], the mass loss that occurs after moisture evaporation is related to the purity of the material and the proportions of β-O-4 bonds. These bonds in lignin's structure are responsible for its thermal stability. Therefore, when lignin is not pure enough to decompose into carbohydrates, the higher the proportion of β-O-4 bonds in lignin's structure, the more it is condensed and the less thermally stable it is [24]. Moreover, according to the derivative of the DTG curve (Figure 5B), the feedstock degraded with a single peak of volatile evaporation at approximately 380 °C, indicating the decomposition of lignin. Lignin is more thermally stable than other lignocellulose components; therefore, it has a wider decomposition temperature interval (from 150 °C to 900 °C), related to the cleavage of the bonds [25]. If lignin decomposes at a lower temperature than 300 °C, it has a low activation energy in the bonds; however, degradation at a temperature higher than 300 °C correlates with the stronger chemical bonds in its structure, which need more energy to break.

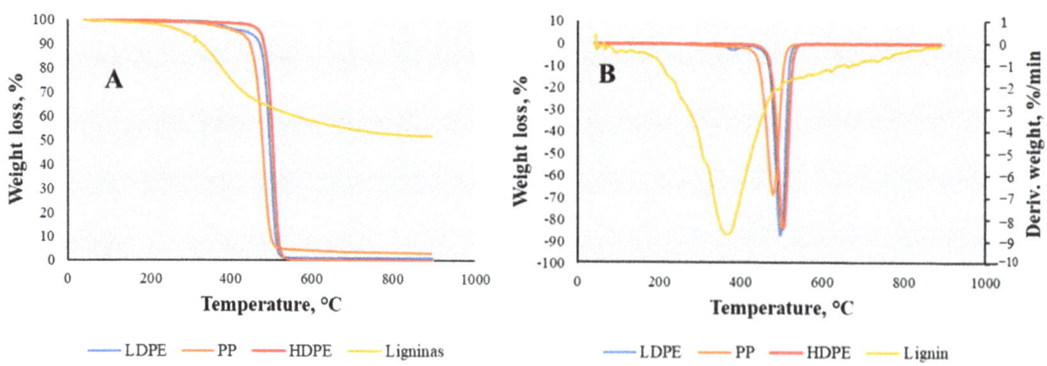

Figure 5. TGA (**A**) and DTG (**B**) curves of lignin, LDPE, HDPE, and PP samples.

The obtained TGA and DTG curves of individual lignin, HDPE, LDPE, and PP mixtures with mass ratios of 1:1, 1:2, 1:3, and 1:4 under an inert ambient atmosphere at a 30 °C/min heating rate are presented in Figure 6. As determined by Özge Çepelioğullar et al. [26], the lignocellulosic biomass tends to have a longer thermal degradation than polymers as it has more degradation stages; also, to break down the strong chemical structures, a higher temperature is needed. On the other hand, plastic materials lose almost 99% of their weight within the temperature range of 400–500 °C, unlike lignin. Comparison of the lignin and plastic mixtures with the experiments of individual feedstocks revealed the correlations between the thermal decomposition characteristics of both lignin and plastic materials. The lignin and LDPE mixture with a mass ratio of 1:1 showed a minor decomposition range in the temperature interval of 280–770 °C, with one volatile evaporation peak at 500 °C (Figure 6A). This temperature interval corresponded to the decomposition of both feedstocks as lignin has a wide decomposition temperature range and a mixture of different types of LDPE was used. The residual mass after co-pyrolysis was 29.42 wt.% of the initial weight. This could be acknowledged as the phenomenon of the plastic's impact on lignin's decomposition and the chemical structure's disintegration for greater targeted product recovery. As claimed by Aleksandr Ketov et al. [27], the pyrolysis of plastics such as LDPE or HDPE generates free radicals that can interact with lignin's macromolecular structure and promote depolymerization by enhancing the decomposition of lignin. In the mixture of lignin and HDPE, no significant changes were observed: the DTG peak was noted at 510 °C, which was slightly higher than that of the LDPE mixture. The mass loss during the decomposition temperature interval was approximately 66.99 wt.%, with a residual mass of 28.59 wt.%, which was marginally lower than that seen in the LDPE mixture. Therefore, the correlation of this mixture showed a little better results in the case of decomposing lignin. However, the least applicable results were obtained with mixtures with PP. The mass change in the decomposition range of 290–760 °C was only 59.19 wt.%, with the biggest residual mass of 36.02 wt.%. On the other hand, the decomposition peak temperature of the mixture was slightly lower—490 °C. Further experiments were conducted with a mass ratio of 1:2 (Figure 6B). With this mass ratio, the case of the mixtures of lignin with HDPE and PP showed very similar results as in Figure 6A. The mass change in the decomposition interval was a lot higher than when applying the mass ratio of 1:1—it was almost 75 wt.%—with a residual mass of 21 wt.% in both cases. The mixture of lignin and LDPE with this mass ratio showed slightly different results by reaching 79.68 wt.% of the mass change in the decomposition interval with a residual mass of 15.94 wt.%.

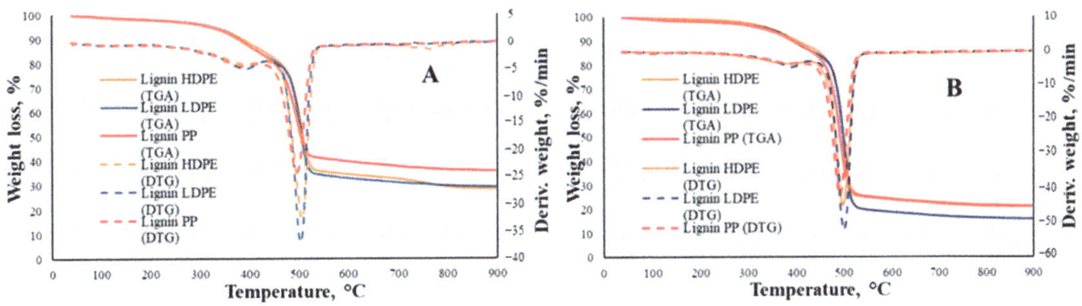

Figure 6. TGA-DTG curves of lignin/LDPE, lignin/HDPE, and lignin/PP mixtures of 1:1 (**A**) and 1:2 (**B**) mass ratios.

The analysis of the mixtures with mass ratios of 1:3 (Figure 7A) and 1:4 (Figure 7B) shown in Figure 7 showed quite correlating results. However, for the best results of the lowest residual mass and highest mass change in the decomposition interval, both (1:3

and 1:4) mass ratios showed that the mixtures of lignin and HDPE left just 10.44 wt.% and 10.87 wt.% of the initial weight, respectively. Therefore, according to the analyzed results, in order to activate the disintegration of lignin's chemical structure and the decomposition to targeted products such as hydrocarbons, the best applied polymer was HDPE with a mixture mass ratio of 1:4.

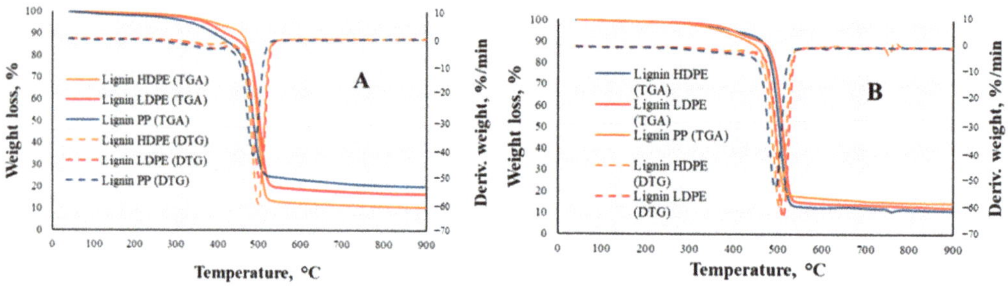

Figure 7. TGA -DTG curves of lignin/LDPE, lignin/HDPE and lignin/PP mixtures of 1:3 (**A**), 1:4 (**B**) mass ratios.

3.2. TGA-FTIR Analysis of Lignin and Plastic Co-Pyrolysis

The emitted gaseous products were analyzed using a combined TGA and Fourier-transform infrared spectroscopy (FTIR) system in order to analyze the main functional groups of the individual feedstocks. The results are depicted in Figure 8. Hydrocarbons were the main volatile products that evolved during the thermal decomposition of polyethylene and polypropylene samples mainly at about 500 °C. The temperature was decided based on the DTG curves, which showed the highest emission of volatile compounds. This could be proved by the IR absorption bands at 3050–2600 cm^{-1}, with blending vibrations in the range of 1450–1325 cm^{-1} indicating the formation of hydrocarbons and corresponding to aromatic and aliphatic compounds. According to Jude A. Onwudili et al. [28], in addition to the volatilization process, reactions such as side chain cracking from aromatic rings, isomerization, and polycondensation take place. As a result, all volatile components of polyethylene are decomposed into volatile hydrocarbons. Moreover, analyzing lignin's functional groups, the typical aromatic vibrations of lignin molecules were observed in the 1600–1400 cm^{-1} wavenumber range. Additionally, a prominent peak was also observed in the 2300 cm^{-1} wavenumber region, which corresponded to the vibration of CO_2. Furthermore, the absorption peaks at 1750–1700 cm^{-1} indicated the stretching of carbonyls (C=O), indicating the presence of aldehydes and ketones. According to the obtained TG-FTIR results, the main compounds that could be recovered from individual plastics were hydrocarbons, while lignin was enriched with aromatic compounds that can be applied in different types of bio-based industrial fields.

TGA-FTIR analysis of lignin and plastic mixtures was performed, and the results are presented in Figure 9. As can be seen, they show the lignin and LDPE mixtures keeping the mass ratios of 1:1, 1:2, 1:3, and 1:4. A wider absorption peak was found in the 1400 cm^{-1} wavenumber region that corresponded to aromatic vibrations, which was not very notable in the individual FTIR spectra of lignin and LDPE. Moreover, the absorption peak of aromatic hydrocarbons in the 3050 cm^{-1} wavenumber region was also promoted by the co-pyrolysis process. This confirmed that the addition of LDPE significantly promoted the production of aromatic compounds during lignin pyrolysis. Comparing the results of different mass ratios, the tendency was obvious that the more plastic (more specifically LDPE) was added, the wider the absorption peak of the aromatic compounds appeared.

Therefore, the optimal mass ratio of the mixture to produce the greatest amount of aromatic compounds was 1:4 lignin and LDPE, respectively.

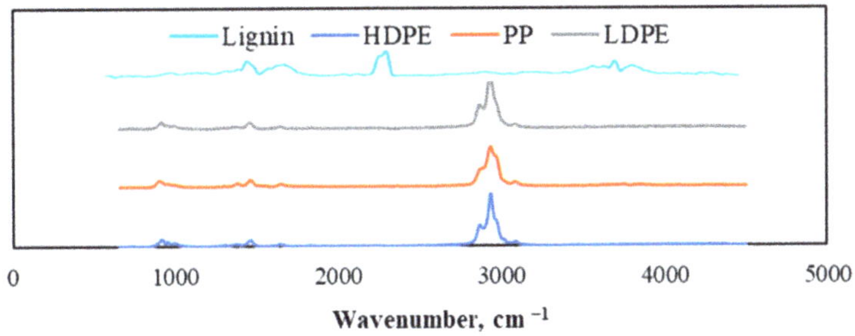

Figure 8. TGA-FTIR spectra of generated gaseous products from pyrolyzed lignin, LDPE, HDPE, and PP individual samples.

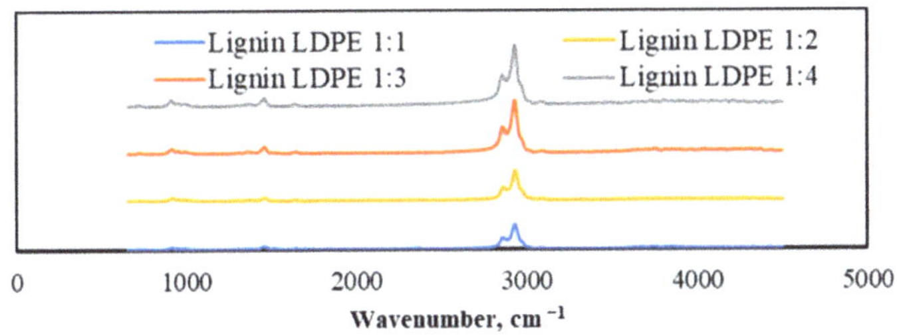

Figure 9. TGA-FTIR spectra of generated gaseous products from pyrolyzed lignin and LDPE mixtures with different mass ratios.

A similar tendency occurred in the lignin and HDPE mixture case presented in Figure 10. The co-pyrolysis of lignin and HDPE often also showed enhanced hydrocarbon production. Lignin, which is rich in oxygen, created an environment that facilitated the decomposition of HDPE, directing the process toward the formation of more valuable volatile products, such as aromatic hydrocarbons. The absorption peak observed at 3050 cm^{-1} corresponded to the presence of aliphatic and aromatic hydrocarbons. Additionally, minor oscillations around 1400 cm^{-1} provided further confirmation of the presence of aromatic hydrocarbons. Comparing the results of different mass ratios, the disposition was very similar to the lignin and LDPE mixture at the optimal mass ratio of 1:4, when the biggest amount of aromatic compounds was produced; however, the whole results showed lower efficiency of aromatic compound production compared with the lignin and LDPE mixtures.

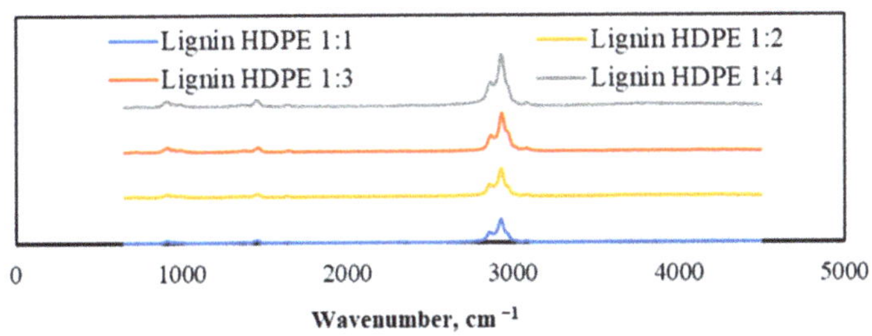

Figure 10. TGA-FTIR spectra of generated gaseous products from pyrolyzed lignin and HDPE mixtures with different mass ratios.

Furthermore, analyzing the occurring tendency of lignin and polypropylene, the results in Figure 11 showed that the co-pyrolysis process did not promote the emission of aromatic compounds and hydrocarbons as much as in the LDPE or HDPE cases, especially with the lower content of plastic in the mixture. The absorption range at 1400 cm^{-1} that corresponded to aromatic vibrations was almost non-existent, and the absorption peak of hydrocarbons at 3050 cm^{-1} was significantly lower.

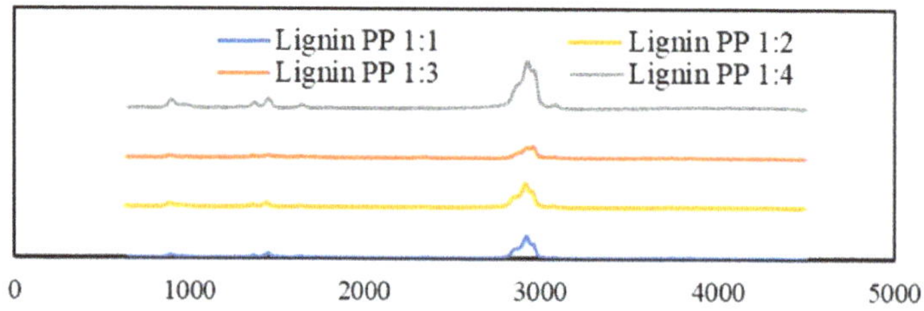

Figure 11. TGA-FTIR spectra of generated gaseous products from pyrolyzed lignin and PP mixtures with different mass ratios.

The TGA-FTIR analysis confirmed that the addition of plastics such as LDPE or HDPE as a hydrogen source changed the distribution of products in the co-pyrolysis process. The co-feeding of plastics increased and promoted the formation of the targeted products—aromatic hydrocarbons. Unfortunately, the experiments with the lignin and PP mixtures might suggest that the existence of lignin inhibits the decomposition of PP; therefore, the probability of recovering the targeted products of aromatic compounds in this case was very low compared with the lignin and LDPE or HDPE mixtures.

3.3. Py-GC/MS Analysis of Lignin and Individual Plastics

The composition of the feedstocks for accurate determination of chemical derivatives was analyzed using Py-GC/MS analysis, where all samples were introduced into three starting temperature ranges based on the TGA-DTG analysis of the materials' decomposition temperatures: beginning from 550 °C, up to 600 °C, and 650 °C. According to the results shown in (Figure 12A) lignin was composed of phenolic units, mainly consisting of phenol type (H-type), guaical (G-type), and hydroxyphenyl types [29]. The detected hydroxyphenyl and guaiacol structures were 2,4 methyl-phenol and 2 methoxy-phenol. The amounts of aromatic hydrocarbons and phenol-type compounds increased with increasing temperature. As shown in Figure 12A, the phenolic compounds reached 14.79% at 550 °C and, by increasing the pyrolysis temperature to 600 °C, the major aromatic compounds increased up to 28.93%. However, at the 650 °C temperature range, none of these phenolic units were detected, as the demethoxylation and alkylation reactions occurred at higher pyrolysis temperature, causing a decrease in aromatic compound formation [30]. Therefore, based on the analysis results, the optimal temperature for recovering aromatic compounds was 600 °C. The main abundance of phenol compounds caused by increasing the temperature was the guaiacol type, but the formation of phenol increased. Therefore, it can be proposed that, via the demethoxylation reaction, phenol, 2-methoxy type can produce phenol with increased temperature.

Analyzing the fast pyrolysis results of plastics showed that they all contained light hydrocarbons that consisted of 1 to 14 carbon atoms and heavy hydrocarbons having more than 20 carbon atoms in their chemical structures. Individual LDPE (Figure 12B) at 550 °C consisted mostly of heavy hydrocarbons ranging from light gases to long chain waxy compounds, such as 1-heptadecene and 1-octadecane, reaching 10.27% and 13.33% respectively. These high-molecular-weight hydrocarbons reflected the incomplete breakdown of the polymer chains. Moreover, another highly elevated compound was behenic alcohol, containing almost 32.78%. Behenic alcohol, as one of the structural components, is usually used to enhance the texture and appearance of the products; also, it might be used as thickener or stabilizer to regulate the viscosity [31]. By increasing the temperature range up to 600 °C, some changes were detected. The amount hydrocarbons increased more than twice—up to 33.69% of octadecanes; however, 1-heptadecanes were not detected at all. The main reason for this tendency was the decomposition process of high-molecular-weight compounds into low-molecular-weight compounds, which was initiated by the increased temperature and heat flux. At this temperature range, more light hydrocarbons also formed, such as 1-hexene, 1-nonene, and 1-heptane, making up 22.34% of all formed compounds. These hydrocarbons resulted from the random cleavage of C-C bonds in the polymer backbone [20]. In addition, the amount of behenic alcohol at this temperature also decreased to 25.39%. At a temperature of 650 °C, the quantity of octadecanes slightly decreased to 30.56%, 1-hepatdecanes and behenic alcohol were also not detected, but the amount of light hydrocarbons slightly increased to 22.55%. However, cyclohexane 1,1′-(2-methyl-1,3-propanediyl)bis-compound appeared and reached 25.62%. This component also works as a plasticizer and improves the flexibility of polymers. Overall, the analysis indicated that the optimal temperature for the compound recovery process was 600 °C, as this resulted in the highest yield of targeted products.

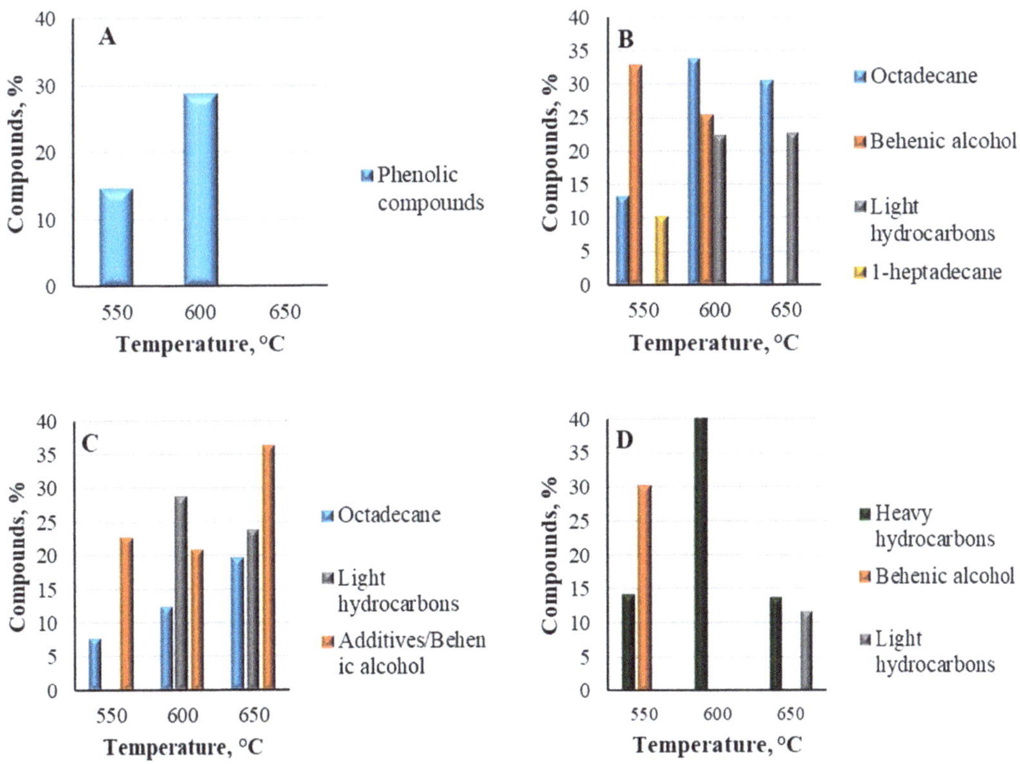

Figure 12. Py-GC/MS compounds generated from pyrolyzed lignin (**A**), LDPE (**B**), HDPE (**C**) and PP (**D**) samples.

Furthermore, the experiments were performed across the same temperature ranges with another polymer, HDPE (Figure 12C). At a starting temperature of 550 °C, the main compounds were identified as hydrocarbons, such as 13-octadecenal, (Z), reaching 7.72%, and were the same as for LDPE, including cyclohexane compounds and oxyrane, hexadecyl-, which are used for enhancing the structural properties of polymers, reaching 10.84% and 11.78%, respectively. At this temperature, no light hydrocarbons were detected. With the increased temperature of 600 °C, the quantity of light hydrocarbons increased to 28.68%. Moreover, hydrocarbons such as 1-octadecane reached 12.26%, while the additive compounds such as behenic alcohol reached 20.86%. With the temperature increased to 650 °C, the quantity of heavy hydrocarbons increased to 19.53%. Furthermore, the formation of light hydrocarbons slightly decreased to 23.69%. The main additive components of behenic alcohol also increased to 36.26%. According to the obtained results, the optimal temperature for the highest recovery of product quantities for HDPE was 650 °C, unlike the previously analyzed LDPE case.

Moreover, the analyzed PP sample (Figure 12D) showed quite similar results. The feedstock mainly consisted of various hydrocarbons and additives for structural improvement. At 550 °C, the main formed hydrocarbons were 1-pentadecane, 1-heptadecane, and 1-nonadecane, reaching 14.3%. As in the previous analysis of different plastics, the main formed compound was additive behenic alcohol, comprising 30.23% of all formed compounds. By increasing the temperature up to 600 °C, the formation of heavy hydrocarbons increased to 40.78%. No light hydrocarbons were detected at any lower temperature. Finally, analyzing the samples at 650 °C, the quantity of hydrocarbons heavily decreased to only 13.8%. However, a slight formation of light hydrocarbons, such as heptane, propene, and 2-butene, occurred with 11.66%. Therefore, the main products of hydrocarbons formed at their highest quantities at 600 °C. According to the results obtained from the Py-GC-MS analysis, the optimal temperature for all samples was 600 °C as it produced the greatest amounts of targeted products, such as aromatic phenolic compounds in the case of lignin and various hydrocarbons in the case of all plastic wastes.

3.4. Py-GC/Ms Analysis of Lignin-Plastic Co-Pyrolysis

The experiments were performed with the lignin and different plastic mixtures with the same mass ratios of 1:1, 1:2, 1:3, and 1:4 at different temperatures starting from 550 °C to 650 °C. The results are depicted in Figure 13. The main compounds identified at each mass ratio were similar and correlated. The combination of heavy hydrocarbons, mainly octadecane and 1-heptadecane, and light hydrocarbons, such as heptane, octene, butene, or pentene, were the main targeted products of all mixtures at different temperatures and mass ratios. At a starting temperature of 550 °C with a mass ratio of 1:1, the recovery of hydrocarbons from the lignin and LDPE mixture (Figure 13A) was 20.72% of all identified compounds. However, the results obtained with the mixture of lignin and HDPE (Figure 13B) under the same conditions were quite different. The same compounds were identified; however, the total quantity was only 13.32%. Nevertheless, the mixture of lignin and PP (Figure 13C) showed quite similar results as the lignin/LDPE mixture. The quantity of hydrocarbons decreased down to 13.98% with a mass ratio of 1:1. By adding more plastic to the mixture and increasing the mass ratio up to 1:2 and 1:3, the quantities of hydrocarbons increased up to 15.99% and 19.75%, respectively, and with a mass ratio of 1:4, the quantity of hydrocarbons increased to 32.92%. The mixtures of lignin/LDPE and lignin/HDPE with the same mass ratios also showed quite correlating results, increasing to 24.71% and 14.69%, respectively, with a mass ratio of 1:2. Increasing the mass ratio to 1:3, the quantities of hydrocarbons also increased up to 33.27% and 20.89%, respectively. With the same mixtures applying the mass ratio of 1:4, the quantities of hydrocarbons increased up to 33.42% and 35.33%, respectively. Lower hydrocarbon yields in mixtures with lower plastic contents aligned with the fact that lignin, in this case, still dominated the feedstock, leading to insufficient hydrogen availability from the plastics. As a result, lignin-derived radicals were less likely to stabilize and more prone to forming char or undergoing polymerization into larger, less volatile compounds [32]. Meanwhile, the increased plastic ratio reflected the optimal synergy between lignin's aromatic precursors and the plastic's hydrogen-rich radicals. The balance allowed efficient stabilization of lignin-derived fragments into lighter hydrocarbons.

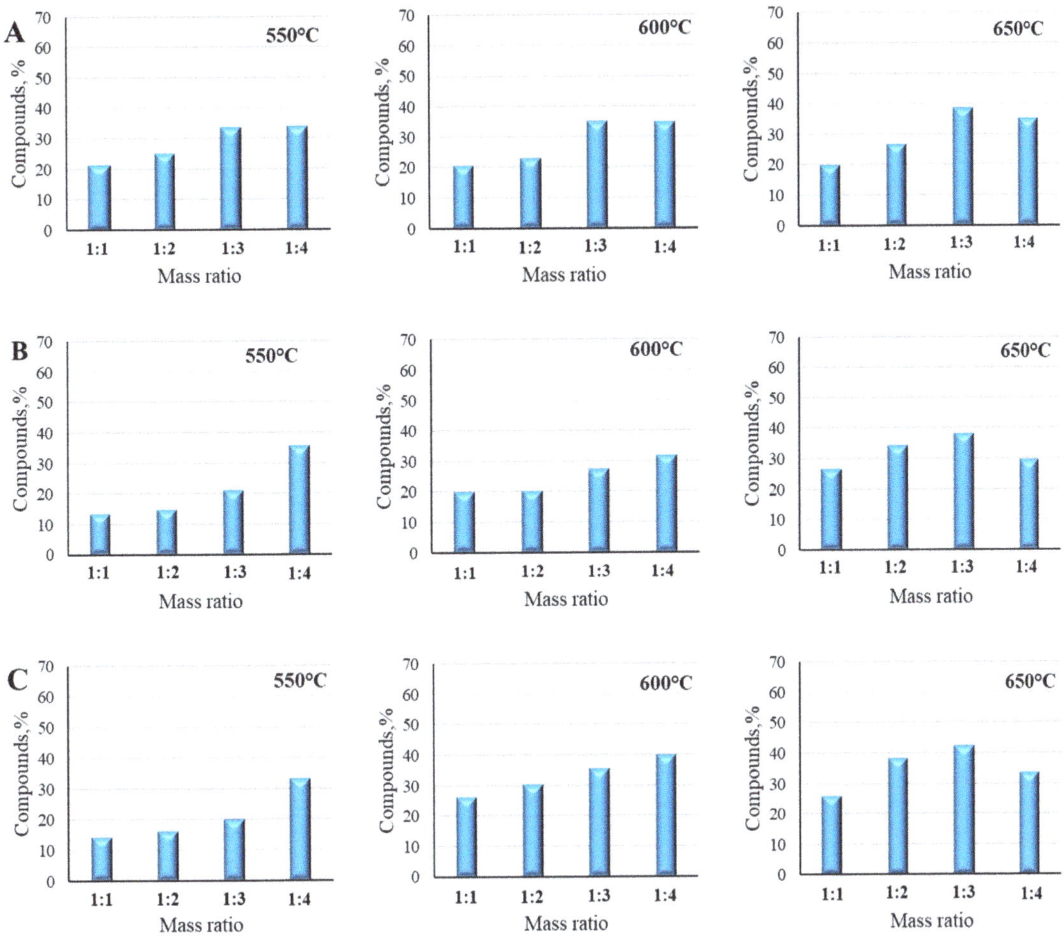

Figure 13. Py-GC/MS compounds generated from pyrolyzed lignin/LDPE (**A**), lignin/HDPE (**B**), and lignin/PP (**C**) mixtures at 550 °C, 600 °C, and 650 °C temperatures.

By rising the temperature to 600 °C, the quantity of targeted products also significantly increased. The mixture of lignin and LDPE polymer with a mass ratio of 1:1 reached 20.29% of hydrocarbons. Almost identical results were found with the lignin-HDPE mixture: the discrepancies between these two sample values were negligible and reached 2–3%. The most significant results were achieved with the mixture of lignin and PP. The quantity of hydrocarbons reached 26.17% applying a mass ratio of 1:1. With mass ratios of 1:2 and 1:3, the tendency of decreased quantity of hydrocarbons was the same as at the 550 °C temperature. The quantity of hydrocarbons from the mixture of lignin and LDPE was 20.11%, and slightly more, 27.17%, with the mixture of lignin and HDPE, with mass ratios of

1:2 and 1:3, respectively. The mixture of lignin and LDPE showed quite correlating results, reaching 22.72% and 34.94% with mass ratios of 1:2 and 1:3, respectively. However, in this temperature case, the highest quantities of hydrocarbons were obtained with the lignin and PP mixtures with 1:2 and 1:3 mass ratios, reaching 36.28% and 39.31%, respectively. Moreover, when comparing these mass ratios at the 600 °C temperature with the results obtained at 550 °C, the lignin and PP mixtures showed quite different results. With mass ratios of 1:2, 1:3, and 1:4, the evaporation of hydrocarbons increased up to 30.16%, 35.91%, and 39.78%, respectively. By increasing the mass ratio up to 1:4, the quantities of hydrocarbons with the mixtures of lignin/HDPE and lignin/LDPE significantly increased up to 31.46% and 34.58%, respectively. The temperature of 600 °C optimally drove the decomposition of both lignin and plastics, producing a high number of radicals and enabling efficient interactions. Lignin-derived aromatic fragments were more effectively stabilized by plastic-derived hydrogen radicals, resulting in a peak in hydrocarbon production [33].

At a final temperature of 650 °C, all mixtures showed very correlating results, elevating from 30–40% of hydrocarbons in each case. The greatest amount of hydrocarbons was obtained from the mixture of lignin and PP with a mass ratio of 1:3, reaching 42.41%. Very similar results were captured with the mixture of lignin and LDPE with the same mass ratio of 1:3. The mass ratios of lower plastic content, such as 1:2, in all mixture cases revealed the least adaptable results, reaching only 20.29%, 20.11%, and 26.17% with the mixtures of lignin and HDPE/LDPE/PP, respectively. With a mass ratio of 1:4, the greatest amount of hydrocarbons at this temperature was obtained with the mixture of lignin and LDPE, reaching 35.20% of all identified compounds; however, even with a higher plastic content, the quantity of elevated hydrocarbons decreased. This phenomenon could be explained by the impact of higher temperature, when the secondary reactions begins and over-cracking of hydrocarbons appears, and larger hydrocarbons break down into smaller volatile gases [34].

The tendencies of all experiments were very similar and the synergy of co-pyrolysis products with all mass ratios at different temperatures was also correlated. The production of hydrocarbons during the co-pyrolysis of lignin and plastic mixtures is governed by complex interactions between the decomposition products of both materials, particularly through the balance of hydrogen availability from plastics and the aromatic precursors provided by lignin. Plastics, especially polyolefins, such as polyethylene (PE) and polypropylene (PP), are rich in hydrogen and break down through chain scission during pyrolysis to produce alkyl and allyl radicals. These radicals act as hydrogen donors, which are crucial for stabilizing lignin-derived radicals that would otherwise recombine into larger, less volatile structures, such as char or polyaromatic hydrocarbons [35]. As the plastic content increases, more hydrogen radicals are available to participate in stabilization, which enhances the formation of smaller hydrocarbons while simultaneously reducing the tendency of lignin to form char. Lignin, being a complex polymer composed of phenolic units, primarily decomposes into aromatic compounds such as phenols, guaiacols, and syringols during pyrolysis. These lignin-derived intermediates interact with plastic-derived radicals, leading to reactions that produce hydrocarbons. However, at lower plastic–lignin ratios, the hydrogen supply is insufficient to fully stabilize lignin-derived radicals, resulting in higher char yields and fewer hydrocarbons [36]. Furthermore, without adequate plastic content, lignin's decomposition intermediates are more likely to undergo polymerization into large aromatic structures instead of being converted into simpler hydrocarbons. Temperature significantly influences these processes, as it affects both the extent of decomposition and the nature of radical interactions. At lower pyrolysis temperatures, such as 550 °C, lignin decomposition is incomplete, favouring the formation of char and larger aromatic fragments over smaller hydrocarbons. Plastics, at these temperatures, also only partially decompose, leading to reduced availability of hydrogen radicals. At moderate temperatures of 600 °C, both lignin and plastics decompose more fully, allowing optimal interaction between lignin-derived aromatics and plastic-derived radicals, maximizing hydrocarbon production. By increasing the temperature to 650 °C, secondary cracking reactions break down larger hydrocarbons into smaller volatile compounds, which can increase gaseous product yields at the expense of liquid hydrocarbon production [37]. Excessive temperatures may also lead to over-cracking, reducing the selectivity for desired hydrocarbon products.

Based on the results, the co-pyrolysis of lignin and LDPE/PP, compared with all mixtures, most dynamically promoted the decomposition of their chemical structures to produce the highest amounts of hydrocarbons, such as octadecane, that can be applied as precursors for bio-fuel production [38] or used as phase-change materials for thermal energy storage [39], as they have relatively high melting points. Also, it was estimated that the optimal temperature and mass ratio for greater compound recovery in mixture cases was the same as for individual materials—600 °C with a mass ratio of 1:3. According to Fujin Mo et al. [29], a temperature around 600 °C favors the cracking of polymer chains and lignin's aromatic structures, maximizing the recovery of valuable products, as lower temperatures may not fully break down the complex lignin structure while higher temperatures could limit production yields. The interaction between lignin and plastic radicals also highlights the importance of achieving the right mass ratio to maximize hydrocarbon production. Lower plastic ratios, such as 1:1 or 1:2, are limited by insufficient hydrogen availability, which reduces the stabilization of lignin-derived intermediates and promotes the formation of solid char. At a very high plastic content, like 1:4, the dilution effect can reduce lignin's contribution to aromatic hydrocarbons, as the reactions are dominated by plastic decomposition. The optimal synergy is typically observed at a 1:3 lignin-to-plastic ratio, where the balance between lignin's aromatic precursors and the plastic's hydrogen radicals results in the highest hydrocarbon yields [40].

4. Conclusions

This study investigated the correlations between two industrial wastes, PP, LDPE, and HDPE plastics and lignin, to determine the possibility of their utilization by applying thermochemical processes and recovering higher-value-added products. After performing all fundamental analyses of the individual materials and their mixtures using TGA-DTG, TGA-FTIR, and Py-GC/MS analyzers, it was determined that lignin has a wide range of decomposition temperatures, starting from 190 °C to 760 °C; however, even at a high temperature, lignin's structure is not fully disintegrated, leaving a residual mass of 53 wt.%. In analyzing the plastic thermal decomposition, it was clear that all plastics decomposed at about 500 °C and almost fully cracked into volatiles. Therefore, further analysis of material mixtures performed with different mass ratios of up to 1:4 wt./wt. revealed that mixing both materials promoted the decomposition of aromatic and aliphatic compounds by radical interactions from polymers, which enhanced the depolymerization of lignin's structure. The most significant results were achieved with the lignin and LDPE/HDPE mixtures, for which the residual mass decreased almost down to 10% with a mass ratio of 1:4.

The experiments and analysis of the Py-GC/MS experiments also confirmed that the mixtures of both materials stimulated the formation of aromatic and aliphatic hydrocarbons from lignin, most commonly heavy hydrocarbons such as octadecane. The greatest results of recovering higher-value-added products were reached at 600 °C with the lignin and LDPE/PP mixtures, applying a mass ratio of 1:3, recovering almost 40% of targeted products.

According to the obtained results, mixtures of lignin and plastics such as LDPE and PP increase the recovery of valuable products. Therefore, it can be confirmed that the pyrolysis process is a promising option for the utilization of two large industrial wastes and recovery of higher-value-added products, such as aromatic and aliphatic hydrocarbons. Lignin and plastic co-pyrolysis presents a comprehensive strategy for waste management, renewable energy generation, the synthesis of high-value chemicals, and the mitigation of environmental impacts. The integration of lignin and plastic waste feedstocks in this process facilitates the recovery of valuable resources, enhances economic viability, and contributes to environmental sustainability.

Author Contributions: V.K.: Conceptualization, Data curation, Formal analysis, Investigation, Methodology, Resources, Software, Supervision, Writing—original draft, Writing—review & editing. J.E.: Conceptualization, Data curation, Formal analysis, Investigation, Methodology, Writing—review & editing. N.S.: Conceptualization, Data curation, Formal analysis, Investigation. R.P.: Data curation, Investigation, Methodology, Formal analysis. All authors have read and agreed to the published version of the manuscript.

Funding: This work was supported by Nordic Energy Research and the Ministry of Energy of the Republic of Lithuania through the funding of Plastic Waste Conversion to Methanol Through Water Vapor Plasma Gasification and Advanced Catalytic Synthesis, 189738.

Data Availability Statement: The raw data supporting the conclusions of this article will be made available by the authors on request.

Conflicts of Interest: The authors declare no conflicts of interest.

References

1. Jankowska, E.; Gorman, M.R.; Frischmann, C.J. Transforming the Plastic Production System Presents Opportunities to Tackle the Climate Crisis. *Sustainability* **2022**, *14*, 6539. [CrossRef]
2. Achilias, D.S.; Roupakias, C.; Megalokonomos, P.; Lappas, A.A.; Antonakou, V. Chemical Recycling of Plastic Wastes Made from Polyethylene (LDPE and HDPE) and Polypropylene (PP). *J. Hazard. Mater.* **2007**, *149*, 536–542. [CrossRef] [PubMed]
3. Lau, W.W.Y.; Shiran, Y.; Bailey, R.M.; Cook, E.; Stuchtey, M.R.; Koskella, J.; Velis, C.A.; Godfrey, L.; Boucher, J.; Murphy, M.B.; et al. Evaluating Scenarios toward Zero Plastic Pollution. *Science* **2020**, *369*, 1455–1461. [CrossRef] [PubMed]
4. Tao, L.; Ma, X.; Ye, L.; Jia, J.; Wang, L.; Ma, P.; Liu, J. Interactions of Lignin and LDPE during Catalytic Co-Pyrolysis: Thermal Behavior and Kinetics Study by TG-FTIR. *J. Anal. Appl. Pyrolysis* **2021**, *158*, 105267. [CrossRef]
5. Sogancioglu, M.; Yel, E.; Ahmetli, G. Pyrolysis of Waste High Density Polyethylene (HDPE) and Low Density Polyethylene (LDPE) Plastics and Production of Epoxy Composites with Their Pyrolysis Chars. *J. Clean. Prod.* **2017**, *165*, 369–381. [CrossRef]
6. Ren, Y.; Sun, X.; Chen, L.; Li, Y.; Sun, M.; Duan, X.; Liang, W. Structures and Impact Strength Variation of Chemically Crosslinked High-Density Polyethylene: Effect of Crosslinking Density. *RSC Adv.* **2021**, *11*, 6791–6797. [CrossRef]
7. Patel, D.; Mamtora, D.; Kamath, A.; Shukla, A. Rogue One: A Plastic Story. *Mar. Pollut. Bull.* **2022**, *177*, 113509. [CrossRef] [PubMed]
8. Pham, N.T.H. Characterization of Low-Density Polyethylene and Ldpe-Based/Ethylene-Vinyl Acetate with Medium Content of Vinyl Acetate. *Polymers* **2021**, *13*, 2352. [CrossRef]
9. Subaramaniyam, U.; Allimuthu, R.S.; Vappu, S.; Ramalingam, D.; Balan, R.; Paital, B.; Panda, N.; Rath, P.K.; Ramalingam, N.; Sahoo, D.K. Effects of Microplastics, Pesticides and Nano-Materials on Fish Health, Oxidative Stress and Antioxidant Defense Mechanism. *Front. Physiol.* **2023**, *14*, 1217666. [CrossRef] [PubMed]
10. Khan, R.J.; Lau, C.Y.; Guan, J.; Lam, C.H.; Zhao, J.; Ji, Y.; Wang, H.; Xu, J.; Lee, D.J.; Leu, S.Y. Recent Advances of Lignin Valorization Techniques toward Sustainable Aromatics and Potential Benchmarks to Fossil Refinery Products. *Bioresour. Technol.* **2022**, *346*, 126419. [CrossRef]
11. Zhao, C.; Chen, A.; Jiang, E.; Qin, L. Pyrolysis of Industrial Waste Lignin: Analysis of Product Yields and Character. *Energy Sources Part A Recovery Util. Environ. Eff.* **2017**, *39*, 458–464. [CrossRef]
12. Charusiri, W.; Phowan, N.; Vitidsant, T. Pyrolysis of Lignocellulosic Biomass with High-Density Polyethylene to Produce Chemicals and Bio-Oil with High Liquid Yields. *Sustain. Chem. Pharm.* **2022**, *25*, 100567. [CrossRef]
13. Eimontas, J.; Yousef, S.; Striūgas, N.; Abdelnaby, M.A. Catalytic Pyrolysis Kinetic Behaviour and TG-FTIR-GC–MS Analysis of Waste Fishing Nets over ZSM-5 Zeolite Catalyst for Caprolactam Recovery. *Renew. Energy* **2021**, *179*, 1385–1403. [CrossRef]

14. Zhao, J.; Xiuwen, W.; Hu, J.; Liu, Q.; Shen, D.; Xiao, R. Thermal Degradation of Softwood Lignin and Hardwood Lignin by TG-FTIR and Py-GC/MS. *Polym. Degrad. Stab.* **2014**, *108*, 133–138. [CrossRef]
15. Wang, S.; Wan, Z.; Han, Y.; Jiao, Y.; Li, Z.; Fu, P.; Li, N.; Zhang, A.; Yi, W. A Review on Lignin Waste Valorization by Catalytic Pyrolysis: Catalyst, Reaction System, and Industrial Symbiosis Mode. *J. Environ. Chem. Eng.* **2023**, *11*, 109113. [CrossRef]
16. Jin, W.; Shen, D.; Liu, Q.; Xiao, R. Evaluation of the Co-Pyrolysis of Lignin with Plastic Polymers by TG-FTIR and Py-GC/MS. *Polym. Degrad. Stab.* **2016**, *133*, 65–74. [CrossRef]
17. Liu, H.; Shen, Y.; Cui, C.; Zhu, L.; Zhou, Z.; Qi, F. Catalytic Co-Pyrolysis of Low-Density Polyethylene and Lignin over Cu-Modified HZSM-5: Insight with Online Photoionization Mass Spectrometry. *Fuel Process. Technol.* **2023**, *251*, 107945. [CrossRef]
18. *EN 14918:2009*; Solid Biofuels—Determination of Calorific Value. Slovenian Institute for Standardization: Ljubljana, Slovenia, 2009.
19. *EN 15148:2009*; Solid Biofuels—Determination of the Content of Volatile Matter. Slovenian Institute for Standardization: Ljubljana, Slovenia, 2009.
20. *EN 14775:2009*; Solid Biofuels—Determination of Ash Content. Slovenian Institute for Standardization: Ljubljana, Slovenia, 2009.
21. *EN 14774-1:2009*; Solid Biofuels—Determination of Moisture Content—Oven Dry Method—Part 1: Total Moisture—Reference Method. Slovenian Institute for Standardization: Ljubljana, Slovenia, 2009.
22. Mourad, A.H.I. Thermo-Mechanical Characteristics of Thermally Aged Polyethylene/Polypropylene Blends. *Mater. Des.* **2010**, *31*, 918–929. [CrossRef]
23. El Moustaqim, M.; El Kaihal, A.; El Marouani, M.; Men-La-Yakhaf, S.; Taibi, M.; Sebbahi, S.; El Hajjaji, S.; Kifani-Sahban, F. Thermal and Thermomechanical Analyses of Lignin. *Sustain. Chem. Pharm.* **2018**, *9*, 63–68. [CrossRef]
24. Kim, J.Y.; Oh, S.; Hwang, H.; Kim, U.J.; Choi, J.W. Structural Features and Thermal Degradation Properties of Various Lignin Macromolecules Obtained from Poplar Wood (*Populus albaglandulosa*). *Polym. Degrad. Stab.* **2013**, *98*, 1671–1678. [CrossRef]
25. Uslu, A.; Faaij, A.P.C.; Bergman, P.C.A. Pre-Treatment Technologies, and Their Effect on International Bioenergy Supply Chain Logistics. Techno-Economic Evaluation of Torrefaction, Fast Pyrolysis and Pelletisation. *Energy* **2008**, *33*, 1206–1223. [CrossRef]
26. Çepelioğullar, Ö.; Pütün, A.E. Thermal and Kinetic Behaviors of Biomass and Plastic Wastes in Co-Pyrolysis. *Energy Convers. Manag.* **2013**, *75*, 263–270. [CrossRef]
27. Ketov, A.; Korotaev, V.; Sliusar, N.; Bosnic, V.; Krasnovskikh, M.; Gorbunov, A. Baseline Data of Low-Density Polyethylene Continuous Pyrolysis for Liquid Fuel Manufacture. *Recycling* **2022**, *7*, 2. [CrossRef]
28. Onwudili, J.A.; Insura, N.; Williams, P.T. Composition of Products from the Pyrolysis of Polyethylene and Polystyrene in a Closed Batch Reactor: Effects of Temperature and Residence Time. *J. Anal. Appl. Pyrolysis* **2009**, *86*, 293–303. [CrossRef]
29. Mo, F.; Ullah, H.; Zada, N.; Shahab, A. A Review on Catalytic Co-Pyrolysis of Biomass and Plastics Waste as a Thermochemical Conversion to Produce Valuable Products. *Energies* **2023**, *16*, 5403. [CrossRef]
30. Ohra-Aho, T.; Tenkanen, M.; Tamminen, T. Direct Analysis of Lignin and Lignin-like Components from Softwood Kraft Pulp by Py-GC/MS Techniques. *J. Anal. Appl. Pyrolysis* **2005**, *74*, 123–128. [CrossRef]
31. Jiang, G.; Nowakowski, D.J.; Bridgwater, A.V. Effect of the Temperature on the Composition of Lignin Pyrolysis Products. *Energy Fuels* **2010**, *24*, 4470–4475. [CrossRef]
32. Kanduri, P.K.; Seethamraju, S. In Situ and Ex Situ Catalytic Co-Pyrolysis of Lignocellulosic Biomass and Plastics (Low-Density and High-Density Polyethylene) Using Spent FCC Catalyst. *Waste Biomass Valorization* **2023**, *14*, 1737–1751. [CrossRef]
33. Baranitharan, P.; Elaiyarasan, U.; Sakthivel, R.; Sriariyanun, M.; Tamilarasan, N. Energy out of Waste: Kinetics and Thermolysis of Co-Pyrolysis of Biomass and Municipal Plastic Waste. *Biomass Convers. Biorefinery* **2024**. [CrossRef]
34. Serrano, D.P.; Aguado, J.; Escola, J.M.; Rodríguez, J.M.; San Miguel, G. An investigation into the catalytic cracking of LDPE using Py–GC/MS. *J. Anal. Appl. Pyrolysis* **2004**, *74*, 370–378. [CrossRef]
35. Singh, M.; Salaudeen, S.A.; Gilroyed, B.H.; Al-Salem, S.M.; Dutta, A. A Review on Co-Pyrolysis of Biomass with Plastics and Tires: Recent Progress, Catalyst Development, and Scaling up Potential. *Biomass Convers. Biorefinery* **2023**, *13*, 8747–8771. [CrossRef]
36. Kawamoto, H. Lignin Pyrolysis Reactions. *J. Wood Sci.* **2017**, *63*, 117–132. [CrossRef]
37. Lu, X.; Gu, X. A Review on Lignin Pyrolysis: Pyrolytic Behavior, Mechanism, and Relevant Upgrading for Improving Process Efficiency. *Biotechnol. Biofuels Bioprod.* **2022**, *15*, 106. [CrossRef] [PubMed]
38. Meller, E.; Green, U.; Aizenshtat, Z.; Sasson, Y. Catalytic Deoxygenation of Castor Oil over Pd/C for the Production of Cost Effective Biofuel. *Fuel* **2014**, *133*, 89–95. [CrossRef]
39. Liu, L.; Niu, J.; Wu, J.Y. Formulation of Highly Stable PCM Nano-Emulsions with Reduced Supercooling for Thermal Energy Storage Using Surfactant Mixtures. *Sol. Energy Mater. Sol. Cells* **2021**, *223*, 110983. [CrossRef]
40. Akubo, K.; Nahil, M.A.; Williams, P.T. Co-Pyrolysis–Catalytic Steam Reforming of Cellulose/Lignin with Polyethylene/Polystyrene for the Production of Hydrogen. *Waste Dispos. Sustain. Energy* **2020**, *2*, 177–191. [CrossRef]

energies

MDPI

Article

Characteristics of High-Temperature Torrefied Wood Pellets for Use in a Blast Furnace Injection System

Richard Deutsch [1,2], Norbert Kienzl [1,*], Hugo Stocker [3], Christoph Strasser [1] and Gernot Krammer [2]

1 BEST—Bioenergy and Sustainable Technologies GmbH, 8010 Graz, Austria; richard.deutsch@best-research.eu (R.D.); christoph.strasser@best-research.eu (C.S.)
2 Institute of Process and Particle Engineering, Graz University of Technology, 8010 Graz, Austria; krammer@tugraz.at
3 voestalpine Stahl Donawitz GmbH, 8700 Leoben, Austria; hugo.stocker@voestalpine.com
* Correspondence: norbert.kienzl@best-research.eu; Tel.: +43-5-02378-9233

Abstract: As the iron and steel industry needs to cut its CO_2 emissions drastically, much effort has been put into establishing new—less greenhouse-gas-intensive—production lines fueled by hydrogen and electricity. Blast furnaces, as a central element of hot iron production, are expected to lose importance, at least in European production strategies. Yet, blast furnaces could play a significant role in the transitional phase, as they allow for the implementation of another CO_2-reducing fuel, carbonized wood reducing agents, as a substitute for coal in auxiliary injection systems, which are currently widely used. Wood carbonization yields vastly differing fuel types depending on the severity of the treatment process, mainly its peak temperature. The goal of this study is to define the lowest treatment temperature, i.e., torrefaction temperature, which results in a biogenic reducing agent readily employable in existing coal injection systems, focusing on their conveying properties. Samples of different treatment temperatures ranging from 285 to 340 °C were produced and compared to injection coal regarding their chemical and mechanical properties. The critical conveyability in a standard dense-phase pneumatic conveying system was demonstrated with a sample of pilot-scale high-temperature torrefaction.

Keywords: torrefied wood pellets; iron ore reduction; pulverized coal injection; coal substitution; fuel suitability

1. Introduction

Currently, it is well recognized that reducing CO_2 emissions without negative effects on global development is one of the imminent challenges to science, industry, and nations. The iron and steel sector is understood to account for roughly 8% of global anthropogenic CO_2 emissions [1]. At the same time, the steel industry is a major force of social development [2], being predicted to grow by about 30% until 2050 [3]. Thus, a reduction in specific emissions of 23% would be needed simply to avoid an increase in emissions from this sector on a global scale. In the EU, the goal was set to decrease industry-related emissions by about 85% in the very same year [4]. For the iron and steel sector, this goal is unachievable without a radical transformation of the production routes themselves. In response to this, the first direct reduced iron (DRI) European plants with renewable hydrogen and electric arc furnaces (EAFs) have been projected and built to substitute the inherently CO_2-intensive blast furnace route [5,6]. To date, no reliable timeframe can be predicted within which European hydrogen and electricity production can be ramped up to meet the demands of

Energies **2025**, *18*, 458 https://doi.org/10.3390/en18030458

European steel makers [7]. Large-scale imports of hydrogen [8] and even DRI [9], mainly as hot briquetted iron (HBI) [10], are expected.

While this envisaged production route is able to almost eliminate direct steel-related CO_2 emissions, and the blast furnace route can only be partially optimized, the latter is currently widely used. In 2020, roughly 60% of Europe's primary steel was produced with blast furnaces [11]; if a fuel substitution proves technically feasible, this will allow instant CO_2 emissions reductions instead of the lengthy and costly construction of new facilities beforehand. In this respect, the presented approach of substituting injection coal with a pulverized torrefied biomass, i.e., bioreductant, is an often-discussed option [12,13]. Modern blast furnaces use coke as a primary fuel, which is fed in at the top and burned with a hot blast in front of the tuyeres in the lower-blast-furnace region. Additionally, auxiliary fuels are primarily injected into this region to produce reducing gas. Most commonly, pulverized coal injection (PCI) is used. Almost 50% of the world's blast furnaces are believed to employ coal injection technology [14]. There are particular strengths to substituting injection coal with upgraded biomass. It is a short-to-mid-term transitional solution [15] with a possibility to reduce blast furnace CO_2 emissions by about 18–40% [16] without significantly changing up-to-date infrastructure, thereby enabling this fast implementation which is essential in adhering to the 1.5 °C goal [17]. Amidst the establishment of a hydrogen market and a general electrification of processes, the use of an alternative fuel can help overcome problems of acquisition and, also, relieve pressure on named markets.

The use of thermally treated wood—in this case, charcoal—has a long tradition in iron production, causing severe deforestation in many regions, until the introduction of coke as the main blast furnace fuel [18]. Today, the specific fuel requirement is far lower than the 2.6 tons of charcoal per ton of iron that was common at the time [19] due to the envisaged usage in injection systems with supposed average injection rates of 0.15 tons per ton of iron. Nevertheless, the overall wood demand would be undoubtedly high as annual European iron production has increased from 200 kt in 1750, based on rough estimates [20], to over 200 Mt in EU27 and CIS countries [21]. The high efficiency of wood carbonization, i.e., solid yields higher than 20 m-$\%_{db}$, as in traditional kilns [22], is mandatory, as is the realization that the sustainable production of carbonized wood is minimal or has yet to be established. Wood production in short-rotation plantations might be a viable solution; however, it is unlikely to provide sufficient amounts of wood for the use of bioreductants to be more than a niche strategy. Thus, supply-sided volatilities are to be expected, and maintaining the ability to temporarily use conventional fuel—PCI coal—is important. Both aspects of fast implementation and resilience motivate tailoring the bioreductant substitute to be suitable for coal conveying systems.

The essential characteristic of this envisaged bioreductant is the ability to provide a flow of energy and reduction potential to an extent similar to coal, with the same conveying and injection equipment. This general requirement can be divided into specific attributes concerning chemical characteristics and particle and bulk characteristics, especially where the development of volatile content, particle sphericity, and fluidizability [23] with increased torrefaction temperatures is the main differentiator. The specific chemical composition, in terms of carbon content in relation to oxygen content, is of top priority for its suitability as a reducing agent. In addition, the volatile content affects the bioreductant's burnout in the raceway [24,25] and the efficiency of its usage [26]. Forty mass percent is generally understood to be the upper limit in this regard [27]. Generally, the ash and sulfur content are of importance as well; however, biomass and charcoal are superior to coal in this respect. Only the higher alkaline content must be addressed in the operation of the blast furnace [25,28]. Equally important are the particle and bulk characteristics. PCI systems usually employ a dense phase conveying system, which requires the feedstock to be fluidized before feeding it to the pipe system. This fluidizability is fundamental and can be achieved or improved with a proper particle size distribution, high enough bulk density, and particle shape [29,30]. Not every torrefaction temperature, however, allows for the tailoring of all these aspects [31]. Fluidization and conveying are performed with a preferentially low conveying gas stream, as excess injection velocities impair the combustion in the raceway [32]. Therefore, it is crucial that the achieved mass proportion of the conveyed solid—the gas loading ratio—is high. These aspects of proper particle size distributions and particle sphericities eventually determine the proper conveyability and affect the achievable specific mass flow of bioreductant and, together with its chemical composition, the specific input of reduction potential via the PCI system.

The aim of this study is to determine the lowest torrefaction temperature that provides all necessary bioreductant characteristics compared to a PCI coal. From over 100 material tests, the results of four different lab-scale torrefaction samples, which were produced, comminuted, analyzed, and tested for their bulk characteristics in comparison with PCI coal, are discussed. The lowest-temperature sample, which exhibited the required characteristics, was reproduced at pilot scale for a conveying test to validate its conveyability and assess its suitability as a PCI substitute.

2. Materials and Methods

2.1. Sample Preparation

A sample of the actual PCI coal of the studied case, labeled "PC" and already comminuted in an industrial vertical roller mill, was acquired from voestalpine Stahl Donawitz. This pulverized coal is perceived as exhibiting all necessary properties regarding combustion performance and conveyability to be used in a dense-phase pneumatic conveying system for blast furnace injection.

For the production of the bioreductant samples, eight tons of poplar pellets were produced and analyzed: poplar wood chips, including natural bark, were acquired from the company pan forst (Donaudorf, Austria), milled to 4 mm sieve size, and dried at Cycleenergy Gaishorn (Gaishorn, Austria). Eventually, pellet production was performed by Holzforschung Austria (Stetten, Austria) on an Amandus Kahl flat die press mill (35 mm press channel length, 6 mm diameter) without additional binders.

For the laboratory tests, a small portion of the pellets was taken to perform the torrefaction tests and material characterization. Four different samples—MF285, MF310, MF325, and MF340, where each sample's peak temperature (sample surface temperature) is signified by the name's number—were produced in a lab-scale pyrolysis reactor in a muffle furnace (Metronik Forno MK-FM35AT) with roughly 1 kg of educt each, loosely packed in steel mesh baskets. The retort used for torrefaction was flushed with N_2 (5–10 times the retort's empty volume per hour) to ensure an oxygen-depleted atmosphere until the samples were cooled below 100 °C. With a heating rate below 10 K/min, the samples were heated to the peak temperature, held at this temperature for twenty minutes, and finally cooled very slowly due to the high heat capacity of the furnace's lining. To generate bioreductant samples with grain sizes that are suitable for usage in a PCI system, all carbonization products were, subsequently, pulverized with a rotor mill (Fritsch Pulverisette 14 classic line; 20k rpm and a sieve size of 0.2 mm) for further testing. The chosen milling setup and parameters were specified by a prior test campaign.

The pilot-scale-sized sample "NF" was torrefied in a semi-industrial-scale rotary drum kiln at NextFuel (Frohnleiten, Austria) [33] at approximately 330 °C and—in an attempt to resemble the laboratory milling equipment to the largest possible extent—comminuted with a bench-scale impact classifier mill at Neuman & Esser Process Technology (Übach-Palenberg, Germany). The laboratory mill used for these tests was an impact mill type with a horizontal rotor orientation, central feed, and material outlet through a sieve ring around the perimeter. In contrast to this, the bench-scale mill is fed from the perimeter and only the rejects of the classifier are fed from the center. This choice should ensure that the milling is performed with impact forces, no imposed particle orientation towards these forces, and little particle residence time within the mill. The integrated classifier, not included in the lab mill, is typical for PCI mills [25].

2.2. Chemical Analysis

Proximate analysis was performed with all fuel types. The water content was measured according to ISO 18134-3 [34] by placing two samples per fuel type in a ventilated drying oven overnight and analyzing the weight difference of the samples after drying. For the measurement of the ash content, the already dried samples were heated to 550 °C and weighed again according to ISO 18122 [35]. The volatile matter was measured according to ISO 18123 [36]. The fixed carbon $w_{cfix,db}$ was calculated by difference according to Equation (1). The parameter fuel ratio is often used to evaluate the expected burnout of the reducing agent in the raceway [37,38], and it can be calculated with the results of the proximate analysis.

$$w_{cfix,db} = 1 - w_{ash,db} - w_{volatiles,db} \tag{1}$$

The chemical main components (C, H, N) of all fuel types were analyzed with a vario MACRO cube (Elementar Analysensysteme GmbH, Langenselbold, Germany). Oxygen contents were calculated by difference according to Equation (2).

$$w_{O,db} = 1 - w_{ash,db} - w_{C,db} - w_{H,db} - w_{N,db} \tag{2}$$

Alternative reducing agents for the blast furnace process can be assessed in terms of their reduction potential (ReP) [39], which can be used for calculating the mass ratio of coke that can be replaced by the injected reductant. According to Equation (3), the ReP is calculated by the weight proportions of carbon, hydrogen, and oxygen (with w_i being the weight fraction of component i), either on a *db* (dry basis) or *daf* (dry and ash-free) basis. Whereas the ReP calculation is based on a stoichiometric calculation of the blast furnace's reduction process, coke replacement ratios (CRRs) are commonly estimated by the reductants' chemical composition based on empirical data. All samples were evaluated by CRR formulae found in [25,40,41]. Empirical correlations might have the advantage of overcoming the strict limitation to stoichiometry purported by ReP but often refer to experiences with coal only. To avoid too narrow a focus, both approaches are discussed.

$$\text{ReP} = 250 \cdot w_C + 500 \cdot w_H - 125 \cdot w_O \tag{3}$$

All portions torrefied in the muffle furnace were weighed prior to and after torrefaction to measure the solid yield, and together with the mentioned analyses, the specific yields according to Equations (4) and (5) were calculated.

$$\text{yield}_{db} = \frac{m_{out}}{m_{in} \cdot \left(1 - w_{moisture,pellets}\right)} \tag{4}$$

$$\text{yield}_{daf} = \text{yield}_{db} \cdot \frac{1 - w_{ash,sample}}{1 - w_{ash,pellets}} \tag{5}$$

In [42], a mathematical correlation between ReP_{daf} and the solid yield_{daf} was found, which is noted in Equation (6). As both product parameters are mainly governed by carbonization intensity, a mutual dependency was inferred as a direct correlation. The torrefaction products considered were compared to this correlation in order to assess their validity.

$$\mathrm{ReP_{daf}} = 267.76 \cdot e^{-1.0913 \cdot \mathrm{yield_{daf}}} \quad (6)$$

2.3. Particle Analysis

Particle size and sphericity distributions were measured using QICPIC/L02 equipped with the dry dispersion system RODOS/L-Q (Sympatec). For the particle size, the QICPIC parameter EQPC was used. This parameter is based on the measurement of a particle's projection area and expresses the diameter of a spherical particle of equal projection area. The sphericity(s), being an indicator of the particle shape, relates the circumference of the projection area of this calculated spherical particle with the circumference of the real particle. The values are, thus, between 0 and 1; the more spherical the particles are, the closer this value is to 1.

Additionally, all samples were tested for bulk density and angle of repose in the style of the respective guidelines. In the style of ISO 60 [43], the bulk density was measured by means of stirring the specimen through a funnel into a cylindric vessel. The vessel had an inner diameter of 59 mm and an inner height of 36.4 mm, which amounted to a capacity of roughly 100 mL. The funnel was placed above the vessel so that its 11 mm outlet was 60 mm above the vessel's upper edge. Before weighing the specimen, all excess powder was removed by scraping it off with a straight metal blade without densifying the specimen. This procedure was repeated three times per sample and the average density was rounded to 10 $kg \cdot m^{-3}$. The angle of repose was measured in the style of ISO 4324 [44] on a Dr. Pfrengle type of apparatus. The mentioned vessel was turned upside down and served as an elevated circular base with a radius of 33 mm on which the specimen was stirred again through a funnel. The funnel outlet's distance to the base was set to 75 mm. The specimen was stirred through the funnel until a proper powder cone fully covering the base formed. By measuring the distance between the funnel's outlet to the tip of the cone, its height was determined, and, consequently, the outer angle of the powder cone could be calculated by the arctan of cone height to the base radius.

The NF and PC samples were tested in a ring shear cell (Schulze Ringschergerät RST Mk II) to compare their flow characteristics in a non-fluidized state. Both tests were performed with a shear cell with a surface of 94 cm^2, with 3500 Pa as the normal load for pre-shear and 200, 900, 1600, and 3000 Pa for the shear points.

Lastly, to estimate the grindability of the semi-industrial bioreductant NF, an equally produced sample was tested according to ISO/TS 21596 [45] by DMT GmbH (Essen, Germany) for its "thermally treated biomass grindability index" (TTBGI), which is understood to be correlated with the Hardgrove Index (HGI) and, thus, provides equal grindability values but is evaluated with different sieve sizes which are thought to better suit torrefied biomass.

2.4. Fluidization and Conveying Properties

A PCI dense-phase conveying system essentially consists of fuel fluidization in a pressurized vessel, generating a mixture of solids and part of the conveying gas prior to feeding the fuel into the conveying pipe system, which conveys the fuel to the blast furnace injection. Therefore, fluidizability is an indispensable prerequisite of each bioreductant; in general, the fluidization behavior is the first aspect to be tested with an unknown material considered for dense-phase pneumatic conveying [30]. A series of measurements were performed by means of a custom-built fluidization device [46], which allows fluidization tests to be performed as described in the literature (e.g., [30] pp. 95–97). This device consists of a vertical acrylic glass tube with an inner diameter of 110 mm which is equipped with a permeable steel disk (GKN sinter metals, 1.440 steel, 10 mm thickness, type: -R 5 AX) at the bottom, a closable sample outlet at the center of this disk with an inner diameter of 9.5 mm, and a measurement of differential pressure from the upper edge of the disk to the top of the tube (one value per second). A sketch of the device is shown in Figure 1. By placing specimens on the permeable disk and applying a carefully controlled gas stream through the disk and, thus, also through the specimen, the force exerted by the air on the sample and vice versa in dependence on the superficial air velocity was determined by measuring the pressure difference above and below the specimen. Respective signals were recorded with 1 Hz frequency. Equally, by visually observing the change in bed height, the corresponding expansion of the powder bed was evaluated. A measurement was taken by incrementally increasing the gas velocity in small steps, with each step waiting for a steady pressure drop across the bed after initial and sometimes spontaneous fluctuations and denoting the steady-state bed height, until full fluidization was established. Then, the gas velocity was lowered stepwise to zero again. The superficial gas velocity, where full fluidization was achieved, was noted as the sample's minimum fluidization velocity ($U_{m,f}$) [30]. The very limited amount of time the bioreductant has to combust in the raceway [47] requires it to be comminuted thoroughly; in fact, the Sauter mean diameter (SMD) of all samples was very close to the Geldart A/C border differentiating between good fluidization behavior and excess cohesiveness. Fluidization tests are difficult to interpret for this kind of fuel dust. While interpretations of a so-called pseudo-$U_{m,f}$, where an intercept of the linearized pressure signals of the transitional phase and of the fluidized state is inferred, would be a possibility [48], another strategy is presented in this paper. Comparisons based on a pseudo-$U_{m,f}$ are potentially misleading in our respect as the main interests are—first—the fluidizability below a certain superficial gas velocity, 2 cm/s here, and—secondly—the general reliability of achieving fluidization. Apart from standard fluidization analysis, additional information from the pressure drop fluctuations was included. Certain presumptions were made:

(1) While both the state at rest and the fluidized state are characterized by homogeneity regarding the bed porosity and resistance to a gas stream, the phase until $U_{m,f}$ is reached is considered inhomogeneous and transitionary.
(2) Gas velocities approaching $U_{m,f}$ should result in increasing homogeneity; otherwise, the risk of randomly encountering troubles in fluidization cannot be ruled out.
(3) Lower pressure-drop fluctuations signify a higher degree of homogeneity within the powder and gas mixture.
(4) The fluidization graph should either signify the pressure drop value encountered most often per gas velocity or averaged values; however, a certain level of congruence should be observable. Both procedures are influenced by the operator, but the former is less influenced by spontaneous semi-fluidization.
(5) During fluidization, additional adhesive forces must be overcome, which are not relevant in de-fluidization [48]. Thus, the pressure drop signal at decreasing gas velocities from $U_{m,f}$ should not be higher than at increasing gas velocities. Contrary observations would signify a faulty fluidization behavior—such as blow-throughs—up to $U_{m,f}$.

$U_{m,f}$ was understood to be reached in the following instances:

- The pressure drop stopped increasing with increases in superficial gas velocity;
- The bed height stopped increasing with increases in superficial gas velocity;
- The pressure drop did not fluctuate anymore according to (1) and (3).

Reliability in fluidization was defined by comparing the spread of the pressure drop signal from one gas velocity to the next: samples leading to decreasing spread when drawing near to $U_{m,f}$ (according to (2)) exhibited a very good repeatability in fluidization results. The comparison of pressure signals from increasing and decreasing gas velocities (according to (5)) and $m_{fluidized}$ (calculated as the ratio of the measured pressure difference at $U_{m,f}$ to the counteracting area-specific weight force of the powder) were used to evaluate the correctness of the postulated $U_{m,f}$.

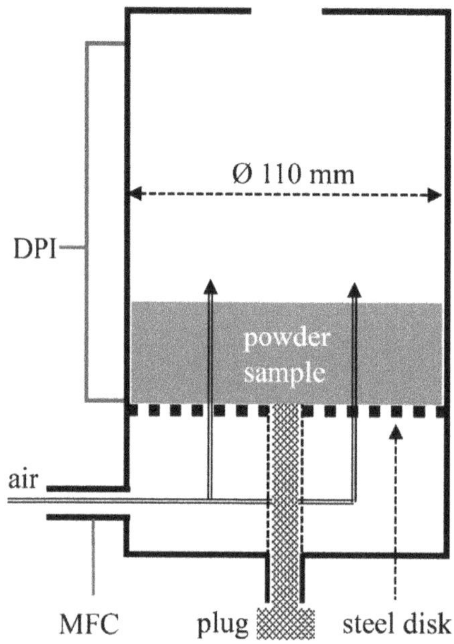

Figure 1. Sketch of the fluidization device (DPI and MFC signify differential pressure indication and mass flow control, respectively).

The ratio of the bed height (rel_{rho}) at $U_{m,f}$ to the initial bed height was also denoted. Each test run comprised a first sequence of fluidization and a subsequent stepwise decrease in gas velocity, immediately followed by a repeated fluidization and deaeration without exchanging the material. During this deaerating stage, the gas retention capacity (*grc*) was measured by suddenly stopping the gas flow and measuring the time until the powder fully de-aerated. Finally, a third fluidization step was performed and the central sample discharge plug was removed at $U_{m,f}$. The extent to which the material was able to exit the device under these conditions was observed.

Fluidization and dense-phase conveying have very similar requirements for the material to be processed; however, requirements from the latter are slightly more restrictive [30]. Thus, although fluidizability is an indispensable prerequisite for dense-phase conveying, conveyability itself must be validated by actual conveying tests. The NF sample produced by semi-industrial equipment was tested for its ability to be conveyed in a dense-phase pneumatic conveying system by E.S.C.H. (Unterwellenborn, Germany). The test installation comprised two pressure vessels connected by a pipe with a length of 51 m and an inner diameter of 25.7 mm. This pipe comprised mainly horizontal but also vertical passages and seven 90° angles. A vertical cylindrical vessel with a conical lower end was filled with 380 kg of the sample, pressurized with N_2, and used for partially fluidizing the sample. Only within the conical lower end are gas velocities sufficiently high to achieve this—therefore—partial fluidization; whilst conveying, the sample is discharged at the bottom so that the remaining sample slowly flows into the fluidization zone and is also discharged and conveyed via the pipe to the receiving vessel. In several repetitions of the conveying procedure, parameters, such as gas flows and pressures, were altered to maximize the specific mass stream conveyed.

3. Results

3.1. Chemical Compositions

The results of the proximate analyses for all seven fuels of interest are shown in Table 1. Additionally, the torrefaction temperatures and solid yields are indicated for the samples produced in the muffle furnace. The characteristics of the pilot-scale sample NF which was produced according to the preliminary laboratory test results are shown as well to compare it to the lab samples. Proximate analysis of PC serves as a baseline of a PCI coal for further comparison. These data are also shown and are markedly different from the bioreductant types, as expected.

Table 1. Proximate analyses and torrefaction solid yields (* as temperature measurement is not comparable to MF samples, only an estimation based on product analyses is indicated; ** solid yields could not be measured, due to high losses in the warm-up phase of the reactor).

	Pellets	MF285	MF310	MF325	MF340	NF	PC
T peak/°C	-	285	310	325	340	320–340 *	-
Moisture/$\%_{mass}$	7	2	2	2	3	2	1
Ash/$\%_{mass,db}$	2	3	4	5	5	5	7
Volatiles/$\%_{mass,db}$	82	64	53	42	41	40	20
Fixed Carbon/$\%_{mass,db}$	16	33	43	53	53	55	73
Fuel Ratio/-	0.20	0.52	0.83	1.28	1.3	1.4	3.65
$yield_{solid}$/$\%_{mass}$	-	65	50	42	38	- **	-
$yield_{solid,db}$/$\%_{mass}$	-	70	54	45	41	- **	-
$yield_{solid,daf}$/$\%_{mass}$	-	70	53	44	40	- **	-

Considering the MF samples, a couple of observations are worth pointing out. As expected, the volatile content and the solid yield decrease with increased torrefaction temperature. Consequently, the fixed carbon content increases. It has to be stressed that the bioreductants' organic volatile content is two to three times higher than that of the PC, which is potentially unfavorable. The $yield_{solid,daf}$ is calculated by relating the proportion of organic matter of each sample to that from the input pellets; by doing so, the yield calculations are unaffected by the educts' ash content and thus better comparable. It has to be emphasized that the analyzed torrefaction temperature range causes the steepest specific decrease in $yield_{solid,daf}$: towards 285 °C, a mass loss of 30% on the *daf* basis is observed, and the decomposition of a further 30% to a $yield_{solid,daf}$ of 40% takes place by only increasing the peak temperature by 55 °C to 340 °C. With the further increase in the treatment temperature to pyrolysis conditions, the decrease in yield again significantly levels off [49].

The semi-industrial NF sample was meant to replicate the MF340 sample, and the resemblance regarding the volatile and fixed carbon content was satisfying. In contrast to the PC sample, all torrefied samples have a significantly higher volatile content. However, MF325, MF340, and NF are close to the mentioned 40 m-% of volatiles, which are understood to be acceptable for blast furnace injection.

The results for the chemical analysis are shown in Table 2. All MF samples exhibit a continuous increase in carbon content with temperature. Even from MF325 to MF340, this increase is observable, although their proximate analyses are almost identical. The NF sample comprises a composition very similar to MF340, as intended. Notably, PC has a higher C/O ratio and thus also a higher ReP. When maintaining equal specific injection rates on the *daf* basis, the ReP injected per mass of iron would drop to at most 81% when replacing PC with one of these bioreductants. Therefore, an evaluation of the reasonability of this coal substitution approach must consider potentially increased coke demand, which certainly minimizes this approach's ecological and economic impact. One approach of economically evaluating this implication on bioreductant design was presented elsewhere. There, the resulting difference in fuel costs and efficiency was shown to be outweighed by presumably lower emission certificate costs for prices expected at the end of this decade [42]. Nevertheless, the emission savings do not fully correspond to the amount of coal substituted but have to take increased coke rates into account. Moreover, a reasonable usage of the non-solid wood carbonization products—i.e., liquids and gasses—is another relevant factor for an ecological assessment.

Table 2. Elemental composition (C/H/N/O) and reduction potential (ReP).

	Pellets	MF285	MF310	MF325	MF340	NF	PC
$C/\%_{mass,db}$	48.7	59.5	64.3	68.2	69.7	68.6	78.8
$H/\%_{mass,db}$	6.0	5.2	4.8	3.9	3.6	4.4	3.8
$N/\%_{mass,db}$	0.2	0.2	0.3	0.3	0.3	0.3	2.2
$O/\%_{mass,db}$	42.9	32.2	26.8	23.0	21.2	21.3	8.2
ReP_{db}/t_{db}^{-1}	98.2	134.5	151.0	161.3	165.5	166.3	205.6
ReP_{daf}/t_{daf}^{-1}	100.4	138.4	157.2	169.1	174.8	175.9	221.1

The reduction potential ReP also increases with torrefaction temperature, as the carbon content increases and oxygen content decreases. The hydrogen content, which positively contributes to the ReP, also decreases by a similar factor, yet is less relevant due to its lower mass fraction. In [42], a regression analysis with a multitude of published torrefaction and pyrolysis tests resulted in an estimated ReP_{daf} as a function of the $yield_{daf}$. The current samples are plotted together with this function in Figure 2. Apart from MF285, all samples are pretty close to the regression result, with MF325 and MF340 even being within the 95% confidence interval. This comparison affirms the results' appropriateness to published torrefaction and pyrolysis data.

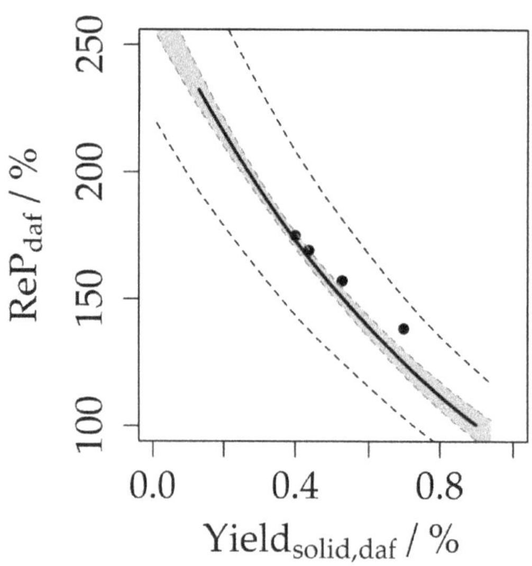

Figure 2. Reduction potential and yield of MF samples (MF285 to MF340 from right to left; gray area: 95% confidence interval, dashed lines: 95% prediction interval).

Several suggested formulae were used to calculate the CRR for PC and all bioreductant samples. Figure 3 shows the CRR of the bioreductants relative to that of the PC. CCR 'a'–'c' were calculated with formulae from [40], while CRR 'a' needed further input from a blast furnace mass and energy model [25]. CRR 'd' used a formula solely based on ash and moisture content [41], CRR 'e' relates the respective RePs already calculated, and CRR 'f' and 'g' are based on empirical formulae [25]. As the CRRs are only tentative in nature, the vast differences are to be expected; however, it shall be highlighted that the mean value of all calculated relative CRRs is almost equivalent to CRR 'e', which is based on the ReP calculation. This observation again emphasizes that the bioreductants are expected to contribute 20% less to the reduction process compared to the PC sample on an equal mass basis. Considering samples MF325 and MF340, the overlap of the error indicators shows only a slight improvement with higher torrefaction temperatures in this regard.

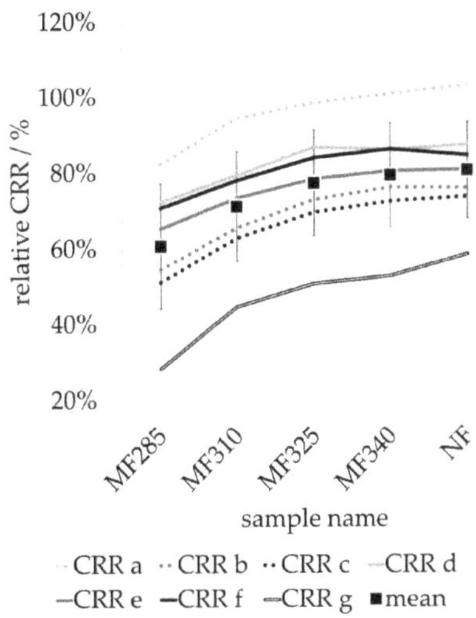

Figure 3. Relative coke replacement ratios.

In the Van Krevelen representation, as shown in Figure 4, it is apparent that the chemical composition of the bioreductant types becomes more similar to PC with higher treatment temperatures. Comparing this diagram with other, more general diagrams of that kind, such as [50], shows that the pellets already comprise an unusual low O/C ratio, probably explaining the higher-than-expected ReP_{daf} for lower-temperature samples in Figure 2. Interestingly, this appears to affect the products' RePs at lower treatment temperatures but not at elevated ones. An aspect complementing the ReP is the H/C ratio: the lower the H/C ratio is, the less steam is generated by oxidation of the bioreductant in the raceway, and the energy demand of the blast furnace is decreased [12]. The samples MF325 and MF340 are almost at the same level as PC and more favorable in this respect. The O/C ratio of all bioreductant samples is higher than that of PC, which is already reflected in the inferior ReP values.

3.2. Bulk and Particle Characteristics

In contrast to the comparative specimen PC, a median particle diameter of 70 μm was targeted for the bioreductant samples due to their tendency to produce a large fraction of ultra-fines and thereby resulting in SMDs below Geldart's "A" class [29]. This diameter is significantly larger than that of PC. Generally, the largest particle diameter suitable for sufficient combustion under raceway conditions is chosen to decrease the comminution expense and conveying difficulties. It has already been shown that increased particle size is acceptable for bioreductants concerning their raceway burnout, as their higher porosity outweighs the size's influence to some extent [51]. Table 3 lists all measured mechanical characteristics of the pulverized samples. As can be seen, MF285 drastically fell short of the targeted $x_{50,3}$. Considering the particle size range of Geldart's class "A" particles and the transition range to the cohesive class "C", the particle size distributions of all other MF samples were practically identical; most importantly, the proportions of fines are equally similar [23] as are the mean diameters $x_{50,3}$. Particle size distributions are shown in Figure 5a. The NF sample's particle size distribution is slightly lower than intended due to restrictions at the semi-industrial mill. With an increase in torrefaction temperature, the angle of repose and bulk density could also be reliably de- and increased, respectively. Measuring these aspects, however, involved stirring the samples through a funnel with a spoon, as all samples exhibited a tendency to bridging. A low angle of repose benefits the bulks' flow- and fluidization behaviors [52,53], and high bulk density facilitates achieving high gas loading ratios. Interestingly, the NF sample's bulk density fell short of that of the MF340 sample. Smaller particle sizes and slightly narrower size distributions serve as explanations for this observation [54], but probably not to that extent. It can be speculated that due to the rotary drum unit, the NF sample might be carbonized less uniformly than the MF samples, and less carbonized fractions exert greater influence on the overall bulk density. In both respects, the coal sample has intermediary qualities compared to the MF samples.

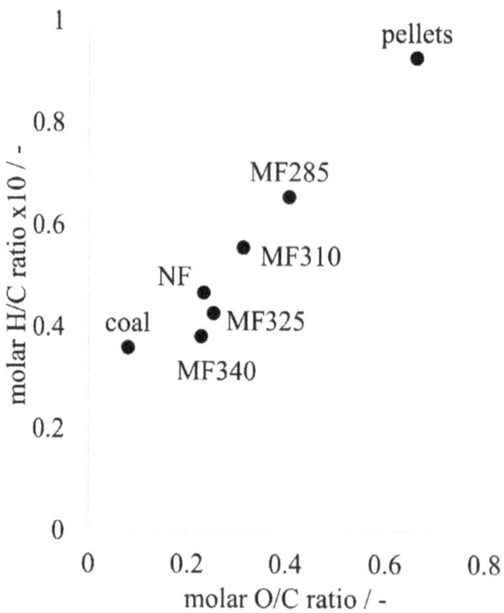

Figure 4. Van Krevelen diagram of all samples used.

The sphericity, also shown in Figure 5b, does also improve significantly with higher torrefaction temperature. For powders with low sphericity, bad flow- and fluidization characteristics can also be expected [55]. A significant improvement is achieved especially between the two lower-temperature samples MF285 and MF310. A further increase in temperature does result in even higher sphericities; however, the improvement is less pronounced from MF310 to MF325. There is no significant difference between MF325 and MF340; it appears as if a certain threshold in this respect is approached in this investigated region of torrefaction temperatures. Similar improvements in sphericity with treatment temperatures have already been reported [12]. In comparison to the coal sample, all produced bioreductant samples are inferior regarding the sphericity.

Table 3. Pulverized samples: bulk characteristics and QICPIC results.

	MF285	MF310	MF325	MF340	NF	PC
Angle of repose/°	48	46	43	39	41	44
bulk density/kg·m^{-3}	400	420	480	500	450	480
$x_{10,3}$/µm	14.5	11.55	10.88	10.68	11.67	8.01
$x_{50,3}$/µm	101	67.17	71.18	68.94	58.39	35.79
$s_{50,3}$/-	0.766	0.806	0.818	0.822	0.821	0.845
s < 0.7/m-%	29	16	11	11	10	6
SMD/µm	39.31	30.45	29.67	29.22	29.73	19.65

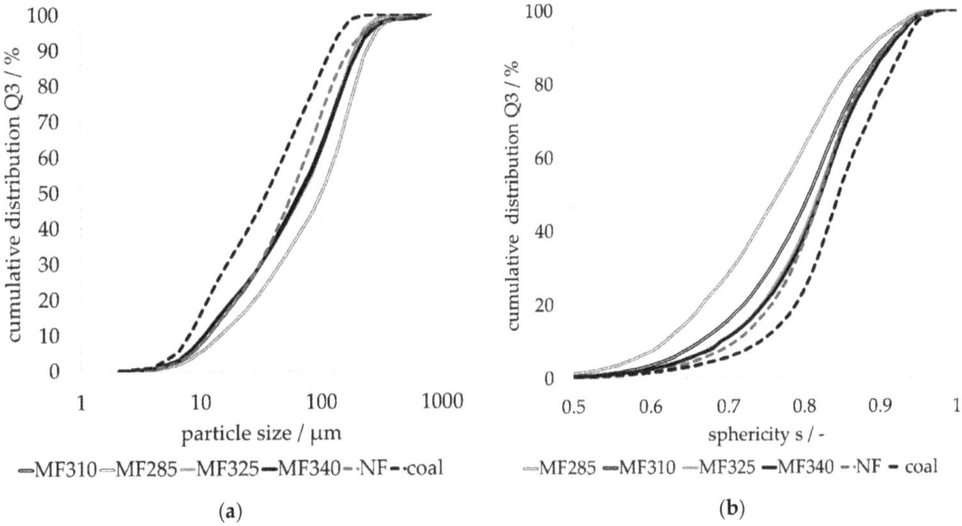

Figure 5. QICPIC results for all samples: (**a**) particle size distributions; (**b**) particle sphericity distribution.

Grindability and shear tests have only been performed with NF and PC. The grindability of the NF sample provided a TTBGI of 112, which is significantly higher than the usual HGI of <70 of a comparable PCI coal. Similar results were obtained elsewhere [24]. This means that the NF sample takes very little effort to grind. On one hand, this positively affects the energy demand of the milling equipment, but on the other hand, the high grindability might cause problems in material handling and transport due to particle breakage and dust generation [15]. It has been reported that lower treatment temperatures result in a lower TTBGI and less grindability [56], and might help to tackle possible problems. The co-grinding and co-injection of the bioreductant and coal might be of interest for industrial implementation but are not within the scope of this study. Studies on the co-grinding of bioreductants and coal questioned the equivalence of HGI and TTBGI values, and advised performing such assessments on an HGI basis alone [57]. Since information from the literature is not fully conclusive on this matter, specific tests need to be performed to see whether and, if so, how and to what extent the bioreductant's high TTBGI (as a precursor for HGI) influences pneumatic transport properties in potential co-grinding applications.

Shear cell tests showed a significant difference in the bulk flow characteristics between the NF and PC regarding the flow function (compressed) (ff_c) value. The NF sample had an ff_c of 8.6 and PC had a value of only 4.8. Both values fall within the range of 'easy-flowing' powders from 4 to 10; however, PC's 4.8 is close to the 'cohesive' class [58]. This might be due to the significantly smaller particle size. Elsewhere, even higher flow values for thermally treated biomass were retrieved, and potential problems such as dust leakages in valves were pointed out [12].

3.3. Fluidization and Conveying Results

The lab samples were tested for their fluidization behavior in order to determine a minimal torrefaction temperature needed for a large-scale conveying test. Table 4 shows all corresponding results as well as the results from the NF and coal sample. The fluidization diagrams are shown in Figure 6.

Table 4. Fluidization results.

	MF285	MF310	MF325	MF340	NF	PC
$U_{m,f}/cm{\cdot}s^{-1}$	>3	1.3	1	1	1.3	1.3
$m_{fluidized}/\%$	81–90	92	91.5	90	90	88.3
$rel_{rho}/\%$	73–78	59	54	58	55	51
$grc/s{\cdot}cm^{-1}$	0.45	0.8	1	0.89	1.39	1.33
discharge/%	87	95	95	94	93	91

The most important parameter in fluidization testing is the minimum superficial gas velocity $U_{m,f}$ needed for fluidizing a powder bed. However, this parameter is not precisely determinable for samples which have a low and broad particle size distribution and a tendency for bridging. For samples with perfect fluidization behavior, this $U_{m,f}$ is easily detectable as a peak in the differential pressure measurement at the moment of transition from the fixed to fluidized bed [30]. The presented measurements did not result in such peaks, but the transition to the fluidized bed appears to occur gradually. Therefore, $U_{m,f}$ was estimated as described in Section 2.4. The MF285 sample did not fluidize properly, most probably because of its decidedly lower particle sphericities. With superficial gas velocities of up to 3 $cm{\cdot}s^{-1}$, the bed height did not stabilize nor did the differential pressure. Its $U_{m,f}$ was indicated with >3 cm s^{-1}, although a partial fluidization was observed and material discharge was possible; but with Δp main dec being higher than Δp main, it must be concluded that no prior sufficient fluidization was achieved. These results also explain why the interpretation cannot rely on only one indicator: The value of Δp main and the corresponding $m_{fluidized}$ are sufficiently high; however, all other observations indicate that friction between the sample and the cylinder probably affected this measurement. The $U_{m,f}$ of all other samples could be estimated more precisely and were in the same range as that of PC. The $m_{fluidized}$ of the PC sample is slightly different: although the pressure signal along with the visual observation of constant bed height signifies a fully fluidized bed, $m_{fluidized}$ is slightly lower. This could indicate that a larger portion of particles rests on the steel disk, neither participating in fluidization nor interfering with it. Indeed, this was visually observable in some portions of the powder bed's perimeter. However, in coherence with the MF285 interpretation, this lower $m_{fluidized}$ can also signify less friction between PC and the device's cylinder.

In the fluidized state, the bed height of all samples increases significantly. The relative fluidization density is calculated as a fraction of the initial bed height to the fluidized bed height. It is less than 60% for all fluidized samples. The measured values for gas retention capacity *grc* are related to the initial bed height. Only the NF sample comprises a *grc* comparable to PC's.

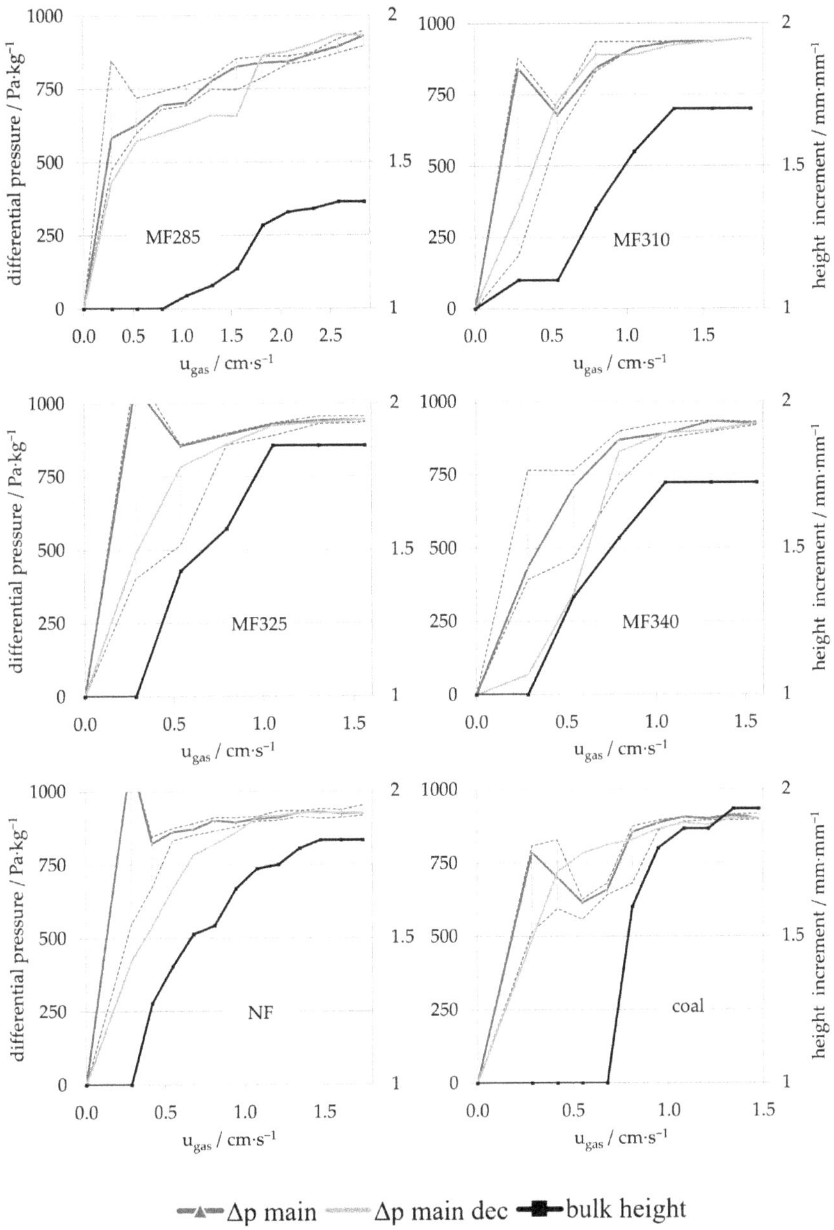

Figure 6. Fluidization diagrams (Δp main is the most often measured value; respective maximal and minimal pressure differences are also plotted as dashed lines).

The peak in pressure drop in Figure 6 at the start in MF310, MF325, NF, and PC is due to the initial levitation of the whole powder bed. These levitations continue occurring with most samples, but mostly collapse again quickly, and are most probably caused by too small an inner diameter of the testing device for this kind of powder. Fortunately, the behavior with initial gas velocities does not affect the assessment of the general fluidizability and the minimum fluidization velocity $U_{m,f}$. The latter can also be roughly estimated by decreasing the gas velocities after reaching fluidization [30], as shown in Figure 6 as Δp main dec. Except MF285, all samples exhibited similar behavior in this regard, largely resembling Δp main well below 1 $cm \cdot s^{-1}$ u_{gas}. Data for PC are peculiar as Δp main dec is above Δp main at lower u_{gas}; this relates to PC's little resistance to lower gas streams after collapsing from a levitating bulk. As stated in Section 2.4, such instances might signify a tendency for unreliable fluidization behavior; however, as the respective gas velocities are low and bulk levitations are unlikely in partial fluidization, as in PCI, this observation appears irrelevant.

The lowest recorded pressure drops often do not relate to the resistance of the powder bed as a whole, but rather signify the channeling of the applied gas stream through one or more channels across the bed. All graphs have in common that the measured maxima and minima, shown as dashed lines in Figure 6, draw closer with increasing gas velocities. This phenomenon is interpreted to signify growing homogeneity regarding the fluidization state within the powder bed. The facts that this is reliably observable, the good congruence of Δp main and Δp main dec, and that the measured specific $U_{m,f}$ is sufficiently repeatable from one test to another foster confidence in the sufficient fluidization behavior of all bioreductant samples excluding MF285. Possibly, this failure in fluidizing at gas velocities below 3 $cm \cdot s^{-1}$ is due to the larger particle sizes after comminution, and a repetition with MF285 with $x_{50,3}$ = 70 μm might have provided positive fluidization results with already lower gas velocities. However, as other characteristics, especially the volatile content, were also inferior, a repetition of testing MF285 with lower particle size was deemed expendable in the course of this study.

Finally, the testing of the NF sample's conveying characteristics was successful in terms of verifying its general ability to be conveyed pneumatically in the dense phase. Moreover, it was possible to achieve a solid loading ratio of more than 150 g_{solids}/g_{gas} by attuning pressure levels and conveying gas amounts. This is higher than the usually employed loading ratios in the PCI system, and the achieved conveying rates on the test apparatus were also sufficiently high with acceptable gas rates. In general, the conveying behavior of the NF sample largely resembled the coal conveying test results with the same equipment.

4. Discussion of the Minimal Torrefaction Temperature

The discussion of the results is separated into two sections, namely a discussion of the aspects relating to the injection system—that is, grinding, fluidization, and conveying—and aspects relating to the blast furnace itself. In both sections, a synopsis of the presented results and additional observations reported in the literature is offered. Where applicable, the implications of the bioreductant types on full or partial replacement are differentiated.

4.1. Implications of Torrefaction Temperature on Usage in PCI System

The comminution of the bioreductant-type NF is expected to demand significantly less specific energy than coal, as the TTBGI test shows. This is in accordance with many other results [59–61]. For bioreductant types of lower torrefaction temperature, somewhat lower grindability values are reported [56]. However, the mill type has proven to affect the particle shape and size distribution of thermally treated biomass [15,46]. As these studies were performed with wood chips, their applicability on already pelletized material, as in this study, is unclear. Nevertheless, it is important to emphasize that changing milling equipment is not only undesirable but will also most probably result in a decreased energy efficiency. Roller mills are very common in PCI systems [25] and they are understood to perform very efficiently [62]. It has been reported [63,64] and also observed in previous tests that it is more difficult to avoid oversized particles after the classifier with bioreductants, perhaps due to irregular particle shapes and very low density. Again, it is not yet clear to what extent this also applies to torrefied pellets, as the density is most influential on the classifier's selectivity. The grindability TTBGI is beyond the usual values for PCI systems [25]. Implications of this were already discussed in Section 3.2. High-HGI coals tend to exhibit stronger caking behavior and are therefore usually not considered for PCI applications [65]. No such correlation between the grindability and caking behavior of bioreductants is known to the authors; thus, a high TTBGI should not cause such reservations. For possible co-grinding applications in a partial substitution approach, a vastly differing grindability might be problematic nevertheless. In this case, opting for lower treatment temperatures in an attempt to decrease grindability might prove reasonable in this respect [66].

The fluidization of most samples worked properly. Only MF285 had no proper fluidization behavior. Using this bioreductant type in a dense-phase conveying system is not recommended without further fluidization aids. Considering a partial substitution, however, might be reasonable for very low proportions in blends. MF310 exhibited similar fluidization behavior to the higher-temperature samples, and also the particle shape was only slightly inferior. Apparently, major changes take place within the bioreductant in the torrefaction temperature range between 285 and 310 °C. The chemical analyses back this assumption. Other parameters, such as the angle of repose and bulk density, do not encourage considering MF310 for usage for the targeted purpose; clearly, some drawbacks regarding flowability in contrast to the higher-temperature samples and PC are evident, although the measured $U_{m,f}$ was equal to PC's. Again, while MF310 might be an option for the partial replacement of coal, the minimum advisable torrefaction temperature for full substitution in this respect should be higher than 310 °C up to 325 °C. From MF325 to MF340, no notable improvement in all tests regarding the fluidizability were encountered, where the results of both samples were not at least equivalent to the PC's.

The conveying of the bioreductant was tested only for the NF sample, which largely resembles the MF340 sample. MF285 and MF310 were excluded because of the aspects mentioned in the previous paragraphs. MF325 was also excluded as MF340's bulk density was notably higher, although, unfortunately, NF's bulk density was lower again. What caused this unexpected drop in bulk density is unclear, as the influence of the different mill type could be ruled out by small NF educts pulverized on the lab-scale mill as well: here, the bulk density was even lower than that of the final semi-industrial product. However, the conveying test was successful inasmuch as the fluidization step worked reliably and conveying rates and solid loading ratios were satisfying. No observations regarding negative implications on the conveying behavior compared to coal were made; only the slightly inferior bulk density to the MF340 sample should be addressed in further test runs.

4.2. Implications of Torrefaction Temperature on Usage in Blast Furnace

A general implication on the whole PCI system's contribution to the blast furnace process is the lower energy density and reduction potential in the bioreductant compared to PC. Potential improvements in, e.g., the energy demand in grinding might be offset by higher overall expenditures in relation to the reductants' avail in the blast furnace's fuel supply. The most important parameters discussed for these implications are the ReP and the CRR, with both being an estimation of an alternative reducing agent's contribution to the reduction of iron ore and the thermal needs of the blast furnace. Both, and by definition, especially the ReP, can only estimate the theoretical maximum contribution based on the chemical composition and stoichiometric requirements. Concerning this maximum, the samples fall short from the PC results, with MF325 and MF340 exhibiting a CRR of 0.8 relative to PC. Along with the reduced bulk- and conveying density, this will result in a drop in injection performance of roughly 30% in this regard when maintaining the same volumetric injection rate. For the samples of lower treatment temperature, this drop is even more severe.

However, there are more factors affecting the effective replacement ratio. Leaving it at that would oversimplify the processes in the raceway which comprise pyrolysis [67], devolatilization of the alternative reducing agent (ARA), cracking and oxidation of the released volatiles and the remaining char, and solution loss reaction (i.e., Boudouard reaction) of the char with the generated carbon dioxide [26,68]. While ReP and CRR valuably estimate a theoretical optimum, the ability of ARA to achieve this optimum within given conditions cannot be assessed easily, let alone generically. Cameron et al. [25] insist that only a prolonged period of in-process testing can provide reliable conclusions. Certain basic requirements can be postulated, nevertheless. One such requirement is the general reactivity, describing the readiness of ARA to oxidize with O_2 and CO_2 [65]. High reactivity can provide gains in effective CRR as losses of unused chars in the bosh gas or slag can be minimized [68], but it also contributes positively to CO generation at lower temperatures than coke [69], enabling an increase in overall efficiency [70]. There appears to be a consensus that thermally treated biomass and their chars have a higher reactivity than coal [71,72]. Its positive effect on burnout behavior has already been demonstrated [73,74], as well as for torrefied materials with especially higher torrefaction temperatures [75]. Another parameter on which to base the bioreductant assessment is the volatile content. This volatile matter is related to a reduced temperature in the raceway, due to devolatilization and endothermic cracking [26], which was related especially to the injectant's tar content in an early study [76]. Arising additional thermal demand reduces the ARA's aggregated contribution to the process. Neither aspect allows for a precise definition of a recommended minimal torrefaction temperature. However, both imply that the lower the treatment temperature is below 325 °C, the more caution in process configuration is advisable.

Aspects relating to ash composition, such as potentially improved blast furnace efficiency due to the basicity of ash [77], or the detrimental effects of the biomass' alkali content [24], are not considered, as the ash composition should not be affected by torrefaction temperature and, thus, does not contribute to defining the minimum suitable torrefaction temperature.

5. Conclusions

To define a minimum torrefaction temperature for bioreductant production, four different samples of bioreductants were produced with intermediate and high torrefaction temperatures, comprising great variations in chemical composition and characteristics in bulk handling. Testing the mechanical characteristics and fluidization behavior provided satisfying results for the samples with torrefaction temperatures of 325 °C and above. The results do not provide evidence that, with these types of bioreductants, a substitution of coal in an ordinary PCI system should not be feasible. On the contrary, pilot-scale conveying tests of a 340 °C semi-industrial sample emphasize the similarity to coal in this respect. Because of the lower reduction potential of the samples of all tested treatment temperatures and their lower bulk density, the need for higher-than-usual injection rates might arise to maintain the PCI system's avail. To address this topic without increasing injection rates, bioreductant treatment temperatures must be above the presented temperature range; assessing the reasonability of doing so, however, is an economic consideration beyond the scope of this report.

Although the analyzed range of torrefaction temperature is small, a pronounced difference in volatile matter content was observed. The lowest measured value for the 340 °C sample was emphasized to be merely within the usual range considered for injection coals.

The resulting possible implications on the blast furnace efficiency were delineated, but they have not been fully evaluated yet. Future research is suggested to focus on the impact different bioreductants—produced at the presented minimal carbonization temperature and above—have on the in-raceway conversion, the respective flame temperatures, and blast furnace thermal control, in order to fully assess a potential fuel substitution.

Author Contributions: R.D.: Conceptualization, methodology, investigation, writing—original draft preparation, visualization; N.K.: conceptualization, resources, project administration, funding acquisition; G.K.: formal analysis, methodology, writing—review and editing, supervision; H.S.: data curation, validation, writing—review and editing, funding acquisition; C.S.: writing—review and editing. All authors have read and agreed to the published version of the manuscript.

Funding: The research activities are financed by the "Industrienahe Dissertation" program of FFG (Austrian research Promotion Agency) within the project "PyroReductant" (FFG project number F0999899743).

Data Availability Statement: Data will be made available upon request.

Conflicts of Interest: Authors Richard Deutsch, Norbert Kienzl and Christoph Strasser are employed by the company BEST—Bioenergy and Sustainable Technologies GmbH. Author Hugo Stocker is employed by the company voestalpine Stahl Donawitz GmbH. The remaining author declares that the research was conducted in the absence of any commercial or financial relationships that could be construed as potential conflicts of interest.

References

1. Sun, Y.; Tian, S.; Ciais, P.; Zeng, Z.; Meng, J.; Zhang, Z. Decarbonising the Iron and Steel Sector for a 2 °C Target Using Inherent Waste Streams. *Nat. Commun.* **2022**, *13*, 297. [CrossRef] [PubMed]
2. Fan, Z.; Friedmann, S.J. Low-Carbon Production of Iron and Steel: Technology Options, Economic Assessment, and Policy. *Joule* **2021**, *5*, 829–862. [CrossRef]
3. Holappa, L. A General Vision for Reduction of Energy Consumption and CO_2 Emissions from the Steel Industry. *Metals* **2020**, *10*, 1117. [CrossRef]
4. EUROFER Low Carbon Roadmap—Pathways to a CO_2-Neutral European Steel Industry. Available online: https://www.eurofer.eu/assets/Uploads/EUROFER-Low-Carbon-Roadmap-Pathways-to-a-CO2-neutral-European-Steel-Industry.pdf (accessed on 20 August 2024).
5. Hartbrich, I. H2 Green Steel Baut Das Elektrisierte Stahlwerk. VDI Nachrichten 2023. Available online: https://www.vdi-nachrichten.com/technik/werkstoffe/h2-green-steel-baut-das-elektrisierte-stahlwerk/ (accessed on 20 December 2024).
6. ArcelorMittal Hamburg H2—Working Towards the Production of Zero-Carbon Emissions Steel with Hydrogen. Available online: https://corporate.arcelormittal.com/climate-action/decarbonisation-technologies/hamburg-h2-working-towards-the-production-of-zero-carbon-emissions-steel-with-hydrogen (accessed on 25 October 2024).
7. Parkes, R. Green Hydrogen Is Too Expensive to Use in Our EU Steel Mills, Even Though We've Secured Billions in Subsidies; Hydrogen Insight. Available online: https://www.hydrogeninsight.com/industrial/green-hydrogen-is-too-expensive-to-use-in-our-eu-steel-mills-even-though-weve-secured-billions-in-subsidies/2-1-1601199?zephr_sso_ott=8f3mAP (accessed on 10 October 2024).
8. Wachsmuth, J.; Aydemir, A.; Döscher, H.; Eckstein, J. The Potential of Hydrogen for Decarbonising EU Industry; Scientific Foresight Unit (STOA). 2021. Available online: https://www.europarl.europa.eu/RegData/etudes/STUD/2021/697199/EPRS_STU(2021)697199_EN.pdf (accessed on 20 December 2024).
9. Rystad-Energy European Steel Mills Waver on Decarbonization Despite Promise of Public Funding 2024. Available online: https://www.rystadenergy.com/news/european-steel-mills-waver-on-decarbonization-despite-public-funding (accessed on 20 December 2024).
10. Lopez, G.; Galimova, T.; Fasihi, M.; Bogdanov, D.; Breyer, C. Towards Defossilised Steel: Supply Chain Options for a Green European Steel Industry. *Energy* **2023**, *273*, 127236. [CrossRef]
11. Ito, A.; Langefeld, B.; Goetz, N. The Future of Steelmaking—How the European Steel Industry Can Achieve Carbon Neutrality. *FOCUS Roland Berg.* 2020. Available online: https://www.rolandberger.com/en/Insights/Publications/Europe's-steel-industry-at-a-crossroads.html (accessed on 20 December 2024).
12. Farrokh, N.T.; Suopajaervi, H.; Mattila, O.; Sulasalmi, P.; Fabritius, T. Characteristics of Wood-Based Biochars for Pulverized Coal Injection. *Fuel* **2020**, *265*, 117017. [CrossRef]

13. Safarian, S. To What Extent Could Biochar Replace Coal and Coke in Steel Industries? *Fuel* **2023**, *339*, 127401. [CrossRef]
14. Babich, A. Blast Furnace Injection for Minimizing the Coke Rate and CO_2 Emissions. *Ironmak. Steelmak.* **2021**, *48*, 728–741. [CrossRef]
15. Liu, Y.; Shen, Y. Modelling and Optimisation of Biomass Injection in Ironmaking Blast Furnaces. *Prog. Energy Combust. Sci.* **2021**, *87*, 100952. [CrossRef]
16. Feliciano-Bruzual, C. Charcoal Injection in Blast Furnaces (Bio-PCI): CO_2 Reduction Potential and Economic Prospects. *J. Mater. Res. Technol.* **2014**, *3*, 233–243. [CrossRef]
17. Lei, T.; Wang, D.; Yu, X.; Ma, S.; Zhao, W.; Cui, C.; Meng, J.; Tao, S.; Guan, D. Global Iron and Steel Plant CO_2 Emissions and Carbon-Neutrality Pathways. *Nature* **2023**, *622*, 514–520. [CrossRef]
18. Straka, T.J. Charcoal as a Fuel in the Ironmaking and Smelting Industries. *Adv. Hist. Stud.* **2017**, *06*, 56–64. [CrossRef]
19. Wakelin, D.H.; Rickets, J.A. The Nature of Ironmaking. In *Materials Science, Engineering*; The AISE Steel Foundation: Pittsburgh, PA, USA, 1999.
20. Fremdling, R. European Iron Industry from the Late 17th to the Middle of the 19th Century—The Struggle between an 'advanced Organic Economy' (Based on Wood) and a 'mineral-Based Energy Economy' (Based on Coal). In *Internationale Studien zur Geschichte von Wirtschaft und Gesellschaft*; Hardach, K., Ed.; P.I.E.-Peter Lang: Frankfurt am Main, Germany, 2012; pp. 211–227.
21. EUROFER European Steel in Figures 2023. Available online: https://www.eurofer.eu/assets/publications/brochures-booklets-and-factsheets/european-steel-in-figures-2023/FINAL_EUROFER_Steel-in-Figures_2023.pdf (accessed on 20 August 2024).
22. Antal, M.J.; Croiset, E.; Dai, X.; deAlmeida, C.E.; Mok, W.S.-L.; Norberg, N.; Richard, J.-R.; Majthoub, M.A. High-Yield Biomass Charcoal. *Energy Fuels* **1996**, *110*, 652–658. [CrossRef]
23. Cocco, R.; Chew, J.W. 50 Years of Geldart Classification. *Powder Technol.* **2023**, *428*, 118861. [CrossRef]
24. Dang, H.; Wang, G.; Wang, C.; Ning, X.; Zhang, J.; Mao, X.; Zhang, N.; Wang, C. Comprehensive Study on the Feasibility of Pyrolysis Biomass Char Applied to Blast Furnace Injection and Tuyere Simulation Combustion. *ACS Omega* **2021**, *6*, 20166–20180. [CrossRef]
25. Cameron, I.; Sukhram, M.; Lefebvre, K.; Davenport, W. *Blast Furnace Ironmaking: Analysis, Control, and Optimization*; Elsevier: Amsterdam, The Netherlands, 2021; ISBN 978-0-12-814227-1.
26. Sundqvist Ökvist, L.; Lundgren, M. Experiences of Bio-Coal Applications in the Blast Furnace Process—Opportunities and Limitations. *Minerals* **2021**, *11*, 863. [CrossRef]
27. Anand, A.; Gautam, S.; Ram, L.C. Feedstock and Pyrolysis Conditions Affect Suitability of Biochar for Various Sustainable Energy and Environmental Applications. *J. Anal. Appl. Pyrolysis* **2023**, *170*, 105881. [CrossRef]
28. Wei, R.; Zhang, L.; Cang, D.; Li, J.; Li, X.; Xu, C.C. Current Status and Potential of Biomass Utilization in Ferrous Metallurgical Industry. *Renew. Sustain. Energy Rev.* **2017**, *68*, 511–524. [CrossRef]
29. Geldart, D. The Effect of Particle Size and Size Distribution on the Behaviour of Gas-Fluidised Beds. *Powder Technol.* **1972**, *6*, 201–215. [CrossRef]
30. Hilgraf, P. *Pneumatische Förderung: Grundlagen, Auslegung Und Betrieb von Anlagen*; Springer: Berlin/Heidelberg, Germany, 2019; ISBN 978-3-662-58407-1.
31. Pachón-Morales, J.; Colin, J.; Pierre, F.; Puel, F.; Perré, P. Effect of Torrefaction Intensity on the Flow Properties of Lignocellulosic Biomass Powders. *Biomass Bioenergy* **2019**, *120*, 301–312. [CrossRef]
32. Schott, R. Optimization Strategies for Pulverized Coal Injection into the Blast Furnace. *Stahl Eisen* **2016**, *136*, 39–47.
33. Trattner, K. ACB—A Brief Introduction. Proceedings of 19th European Biomass Conference & Exhibition, Berlin, Germany, 6–10 June 2011; pp. 1892–1893.
34. *ISO 18134-3:2023*; Solid Biofuels—Determination of Moisture Content—Part 3: Moisture in General Analysis Sample. ISO: Geneva, Switzerland, 2023.
35. *BS EN ISO 18122:2022*; Solid Biofuels. Determination of Ash Content. ISO: Geneva, Switzerland, 2022.
36. *ISO 18123:2023*; Solid Biofuels—Determination of Volatile Matter. ISO: Geneva, Switzerland, 2023.
37. Chen, W.-H.; Du, S.-W.; Tsai, C.-H.; Wang, Z.-Y. Torrefied Biomasses in a Drop Tube Furnace to Evaluate Their Utility in Blast Furnaces. *Bioresour. Technol.* **2012**, *111*, 433–438. [CrossRef] [PubMed]
38. Du, S.-W.; Chen, W.-H.; Lucas, J.A. Pretreatment of Biomass by Torrefaction and Carbonization for Coal Blend Used in Pulverized Coal Injection. *Bioresour. Technol.* **2014**, *161*, 333–339. [CrossRef] [PubMed]
39. Shinotake, A. Reduction Potential Evaluation Index of Various Reducing Agents in Blast Furnace. *ISIJ Int.* **2014**, *54*, 1546–1551. [CrossRef]
40. Campos, A.M.A.; Barbosa, J.P.E.; Assis, P.S. Blast Furnace's Replacement Rate Calculation for Biomasses Based on Chemical and Thermal Properties. *Int. J. Adv. Eng. Res. Sci.* **2021**, *8*, 125–133. [CrossRef]
41. COALTECH PTY LTY Calculation of Replacement Ratio 2005. Available online: https://www.coaltech.com.au/LinkedDocuments/Replacement%20Ratio.pdf (accessed on 20 December 2024).

42. Deutsch, R.; Kienzl, N.; Krammer, G.; Stocker, H.; Strasser, C. Carbonized Wood as a Blast Furnace Pulverized Coal Substitute: A Techno-Economic Assessment. *Energy* **2024**, *313*, 133955. [CrossRef]
43. *ISO 60:2023*; Plastics—Determination of Apparent Density of Material That Can Be Poured from a Specified Funnel. ISO: Geneva, Switzerland, 2023.
44. *ISO 4324:1977*; Surface Active Agents—Powders and Granules—Measurement of the Angle of Repose. ISO: Geneva, Switzerland, 1977.
45. *ISO/TS 21596:2021*; Solid Biofuels—Determination of Grindability — Hardgrove Type Method for Thermally Treated Biomass Fuels. ISO: Geneva, Switzerland, 2021.
46. Deutsch, R.; Krammer, G.; Strasser, C.; Kienzl, N. Effects of Varying Comminution Techniques and Pyrolysis Temperatures on the Particle Shape of Pyrolyzed Wood Powders and Its Implication on Fluidisation Behaviour. In Proceedings of the 29th European Biomass Conference and Exhibition, EUBCE, Marseille, France, 26–29 of April 2021; pp. 1053–1056.
47. Ng, K.W.; Giroux, L.; Todoschuk, T. Value-in-Use of Biocarbon Fuel for Direct Injection in Blast Furnace Ironmaking. *Ironmak. Steelmak.* **2018**, *45*, 406–411. [CrossRef]
48. Kamranian Marnani, A.; Bück, A.; Antonyuk, S.; van Wachem, B.; Thévenin, D.; Tomas, J. The Effect of the Presence of Very Cohesive Geldart C Ultra-Fine Particles on the Fluidization of Geldart A Fine Particle Beds. *Processes* **2019**, *7*, 35. [CrossRef]
49. Chen, D.; Li, Y.; Cen, K.; Luo, M.; Li, H.; Lu, B. Pyrolysis Polygeneration of Poplar Wood: Effect of Heating Rate and Pyrolysis Temperature. *Bioresour. Technol.* **2016**, *218*, 780–788. [CrossRef]
50. Carrasco, S.; Silva, J.; Pino-Cortés, E.; Gómez, J.; Vallejo, F.; Díaz-Robles, L.; Campos, V.; Cubillos, F.; Pelz, S.; Paczkowski, S.; et al. Experimental Study on Hydrothermal Carbonization of Lignocellulosic Biomass with Magnesium Chloride for Solid Fuel Production. *Processes* **2020**, *8*, 444. [CrossRef]
51. Babich, A.; Senk, D. Biomass Use in the Steel Industry: Back to the Future? *Stahl Eisen* **2013**, *133*, 57–67.
52. Geldart, D.; Abdullah, E.C.; Hassanpour, A.; Nwoke, L.C.; Wouters, I. Characterization of Powder Flowability Using Measurement of Angle of Repose. *China Particuology* **2006**, *4*, 104–107. [CrossRef]
53. Saluja, G.; Mallick, S.S.; Karmakar, S. Predicting Pneumatic Conveyability and Flowability of Fly Ash Using Bulk Property Characterization. *Part. Sci. Technol.* **2023**, *42*, 482–494. [CrossRef]
54. Stieß, M. *Mechanische Verfahrenstechnik 1*; Springer: Berlin/Heidelberg, Germany, 1995; ISBN 978-3-662-08600-1.
55. Liu, B.; Zhang, X.; Wang, L.; Hong, H. Fluidization of Non-Spherical Particles: Sphericity, Zingg Factor and Other Fluidization Parameters. *Particuology* **2008**, *6*, 125–129. [CrossRef]
56. Mościcki, K.J.; Niedźwiecki, Ł.; Owczarek, P.; Wnukowski, M. Commoditization of Biomass: Dry Torrefaction and Pelletization—A Review. *J. Power Technol.* **2014**, *94*, 233–249.
57. Huang, X.; Ng, K.W.; Giroux, L. Grindability of Biocarbon and Coal Blends in Rolling Mill. *Int. J. Coal Prep. Util.* **2020**, *42*, 1651–1663. [CrossRef]
58. Schulze, D. *Ring Shear Tester RST-01.Pc*; Dr. Dietmar Schulze Schüttgutmesstechnik: Wolfenbüttel, Germany, 2018.
59. Ng, K.W.; Huang, X.; Giroux, L.; Todoschuk, T. Grindability of Solid Biocarbon. In Proceedings of the AISTech2019 Proceedings of the Iron and Steel Technology Conference, Pittsburgh, PA, USA, 6–9 May 2019; AIST: Pittsburgh, PA, USA, 2019.
60. Manouchehrinejad, M.; van Giesen, I.; Mani, S. Grindability of Torrefied Wood Chips and Wood Pellets. *Fuel Process. Technol.* **2018**, *182*, 45–55. [CrossRef]
61. Bridgeman, T.G.; Jones, J.M.; Williams, A.; Waldron, D.J. An Investigation of the Grindability of Two Torrefied Energy Crops. *Fuel* **2010**, *89*, 3911–3918. [CrossRef]
62. Tontu, M. An Investigation of Performance Characteristics and Energetic Efficiency of Vertical Roller Coal Mill. *Int. J. Coal Prep. Util.* **2020**, *41*, 248–262. [CrossRef]
63. Mathieson, J.G.; Rogers, H.; Somerville, M.A.; Jahanshahi, S. Reducing Net CO_2 Emissions Using Charcoal as a Blast Furnace Tuyere Injectant. *ISIJ Int.* **2012**, *52*, 1489–1496. [CrossRef]
64. Suopajärvi, H.; Umeki, K.; Mousa, E.; Hedayati, A.; Romar, H.; Kemppainen, A.; Wang, C.; Phounglamcheik, A.; Tuomikoski, S.; Norberg, N.; et al. Use of Biomass in Integrated Steelmaking—Status Quo, Future Needs and Comparison to Other Low-CO_2 Steel Production Technologies. *Appl. Energy* **2018**, *213*, 384–407. [CrossRef]
65. Guo, H.; Su, B.; Zhang, J.; Zhu, M.; Chang, J. Extracting the Core Indicators of Pulverized Coal for Blast Furnace Injection Based on Principal Component Analysis. *Int. J. Miner. Metall. Mater.* **2013**, *20*, 246–252. [CrossRef]
66. Gil, M.V.; García, R.; Pevida, C.; Rubiera, F. Grindability and Combustion Behavior of Coal and Torrefied Biomass Blends. *Bioresour. Technol.* **2015**, *191*, 205–212. [CrossRef] [PubMed]
67. Wu, L.; Paterson, N.; Dugwell, D.R.; Kandiyoti, R. Simulation of Blast-Furnace Tuyere and Raceway Conditions in a Wire Mesh Reactor: Extents of Combustion and Gasification. *Energy Fuels* **2007**, *21*, 2325–2334. [CrossRef]
68. Agrawal, A.; Tiwari, R.K.; Kumar, S.; Chatterjee, R.; Singh, B.K.; Singh, R.S.; Tripathi, V.R.; Kundu, S.; Padmapal; Ramna, R.V.; et al. Technological Advancements in Evaluating the Performance of the Pulverized Coal Injection through Tuyeres in Blast Furnace. *Metall. Res. Technol.* **2020**, *117*, 611. [CrossRef]

69. Babich, A.; Senk, D.; Fernandez, M. Charcoal Behaviour by Its Injection into the Modern Blast Furnace. *ISIJ Int.* **2010**, *50*, 81–88. [CrossRef]
70. Mathieson, J.G.; Somerville, M.A.; Deev, A.; Jahanshahi, S. Utilization of Biomass as an Alternative Fuel in Ironmaking. In *Iron Ore*; Elsevier: Amsterdam, The Netherlands, 2015; pp. 581–613.
71. Babich, A.; Arnsfeld, S.; Kowitwarangkul, P.; Senk, D. Biomass Use in Ironmaking: Options and Limits. In Proceedings of the 6th International Congress on the Science and Technology of Ironmaking, Rio de Janiero, Brazil, 14–18 October 2012.
72. Machado, J.G.M.d.S.; Osório, E.; Vilela, A.C.F. Reactivity of Brazilian Coal, Charcoal, Imported Coal and Blends Aiming to Their Injection into Blast Furnaces. *Mater. Res.* **2010**, *13*, 287–292. [CrossRef]
73. Mathieson, J.G.; Somerville, H.R.M.; Ridgeway, P.; Jahanshahi, S. Use of Biomass in the Iron and Steel Industry—An Australian Perspective. In Proceedings of the 6th European Coke and Ironmaking Congress, Duesseldorf, Germany, 27 June–1 July 2011.
74. Pohlmann, J.G.; Borrego, A.G.; Osório, E.; Diez, M.A.; Vilela, A.C.F. Combustion of Eucalyptus Charcoals and Coals of Similar Volatile Yields Aiming at Blast Furnace Injection in a CO_2 Mitigation Environment. *J. Clean. Prod.* **2016**, *129*, 1–11. [CrossRef]
75. Arnsfeld, S.; Rodriguez Correa, C.; Choi, S.-M.; Babich, A.; Senk, D.; Gudenau, H.-W. Investigations of the Reaction Kinetics of Torrefied Biomass for Metallurgical Applications. In Proceedings of the 6th International Congress on the Science and Technology of Ironmaking, Rio de Janiero, Brazil, 14–18 October 2012.
76. Bortz, S. Coal Injection into the Blast Furnace. Technical Steel Research, Report EUR 8544 EN. Comm. Eur. Communities. 1983. Available online: https://aei.pitt.edu/44147/ (accessed on 20 December 2024).
77. Suopajärvi, H.; Kemppainen, A.; Haapakangas, J.; Fabritius, T. Extensive Review of the Opportunities to Use Biomass-Based Fuels in Iron and Steelmaking Processes. *J. Clean. Prod.* **2017**, *148*, 709–734. [CrossRef]

MDPI AG
Grosspeteranlage 5
4052 Basel
Switzerland
Tel.: +41 61 683 77 34

Energies Editorial Office
E-mail: energies@mdpi.com
www.mdpi.com/journal/energies

www.ingramcontent.com/pod-product-compliance
Lightning Source LLC
Chambersburg PA
CBHW061501210926
51850CB00008B/1786